Daylight Science and Daylighting Technology

Richard Kittler • Miroslav Kocifaj
Stanislav Darula

Daylight Science and Daylighting Technology

Springer

Richard Kittler
Institute of Construction and Architecture
Slovak Academy of Sciences
Bratislava, Slovakia
usarkit@savba.sk

Miroslav Kocifaj
Institute of Construction and Architecture
Slovak Academy of Sciences
Bratislava, Slovakia
kocifaj@savba.sk

Stanislav Darula
Institute of Construction and Architecture
Slovak Academy of Sciences
Bratislava, Slovakia
usarsdar@savba.sk

ISBN 978-1-4419-8815-7 e-ISBN 978-1-4419-8816-4
DOI 10.1007/978-1-4419-8816-4
Springer New York Dordrecht Heidelberg London

Library of Congress Control Number: 2011938016

Printed on acid-free paper

Springer is part of Springer Science+Business Media (www.springer.com)

Preface

Daylight is a critically important component of the immense flux of short wave photonic energy that flows continuously onto the surface of our Earth from our Sun. One current challenge for society is to discover how to make much more effective use of a globally free natural resource, the fluxes of energy from the sun. This book on the *Science of Daylight and Daylighting Technology* covers on a wide range of scientific topics relating to daylight and provides an outline of the climate data needs to be compiled to assess daylight availability in different parts of the world. There a strong emphasis on the daylighting of buildings. However the Science of Daylighting is important for many other fields of science.

The book enables daylighting climate data to be linked to the practical problems implicit in applied Daylighting Technology. This requires a focused structured application of Daylight Science. The mathematical emphasis is on the provision of sound scientific analytic methods for the assessment of daylight in occupied spaces in buildings. The issues of daylight assessment within cities are especially well discussed.

The text provides a detailed account of the various science based methodologies that can be applied to advance the sustainable use of daylight in conjunction with the optimisation of the conjoint use of artificial light sources. The complex scientific knowledge available on daylight climates is linked to available knowledge on how to achieve effective human vision within buildings using daylight. The book aims to help readers understand the goals of decision making in the field of visual environmental design. The technological design aim in building daylighting design is to provide indoor illumination at appropriate levels and to promote indoor visual fields of quality that enable the human eye to form satisfactory and pleasing images of the scene in the room on the retina of the eye for processing in the human brain.

The entire text is underlain by a systematic consistent analytic logic that interlinks the ground based solar geometry to the beam light coming from the sun, to the luminance distribution of the sky vault and to the light from reflected from the ground. Clear sky conditions, overcast condition and partially cloudy conditions are discussed in detail. The subsequent analysis of the consequences of light obstruction on daylighting availability in the built environment establishes the methods available for the numerical assessment of the amount of daylight transmitted through openings in buildings on obstructed sites. The concern for meeting both the quantity of daylight needed and the quality of the daylight within the visual field is addressed. Importantly the book includes descriptions of eye based visual comfort issues like the effects of sky glare on visual comfort. There is a detailed chapter on the response of the human eye and brain to the impacts of surrounding luminance environment. The references to the international literature are extensive. These include many references concerning the history of daylighting science which are of great interest.

As we live on the surface of an opaque sphere and not on the surface of a transparent globe, the consequence at most latitudes is that we can receive daylight for only part of each day at any specific place. Biologically this basic concept of day and night light cycles is fundamental to our organism and is embedded in our chemical metabolism through chemical messages created in our eyes. Our internal metabolic time systems have evolved biologically to respond photo-chemically to local day night illuminance variations. Our metabolism is regulated photo chemically to keep the diurnal body chemistry in step with the local day–night light diurnal cycle. The desire for sleep, an essential metabolic function, is linked to the retinal impacts of the natural cycles of light and darkness. These patterns of solar energy induced photosynthesis and subsequent synthetic decay at the back of the eye drive our own locally tuned body based internal timing system.

The length of the sunlit day depends on latitude and the time elapsed since the Northern hemisphere Vernal Equinox on March 21st. This is one of 2 days in the year when the daylength is the same at all latitudes in the world. In the two Polar regions it is possible to have days with no sunshine and also days with 24 h sunshine lasting a month or more. These situations of 24 h daylight and 0 h daylight create metabolic stress. The length of these two periods increases as one moves from the Polar Circles to the Poles, each of which have 6 months with sunlight and 6 months without sunlight in anti-phase to the other pole. In higher latitudes the length of the sunlit day in winter is usually substantially shorter than the length of the working day. Society then becomes much more dependent on artificial light during normal working hours.

One important function of daylighting climatology is to provide statistics to help designers become more aware of the periods of the day at different times of year when natural daylight can or cannot meet defined lighting needs at specific locations. Much design proceeds by selecting critical data for assisting decision making rather than attempting to access all the available data all at once. It is therefore helpful to prepare specific daylight studies for defined types of sky. The most important day types for practical daylight illuminance design are cloudless sky days and overcast sky days. Days of high temporal variance are important too as discussed in the main text. Cumulative statistics of the overcast day daylight availability provide a critical element in daylighting climatology. When considering glare from clouds, it is useful to assess sky luminance under partially clouded conditions when clouds have the highest luminance.

Important outputs for daylighting designers are values of the cloudless day beam illuminance normal to the beam as a function of time and day in different seasons, also the associated diffuse illuminance on horizontal planes and the zenith luminance. The geography of atmospheric clarity determines the balance between cloudless day beam illuminance and cloudless day diffuse horizontal illuminance. Turbid climates produce weak beam values and high diffuse values. The strength of the blueness of the cloudless sky always provides a visual clue about the actual local clarity. Basically the deeper the blue, the lower the visual turbidity.

This book explains in detail how the photonic energy reaching the outer edge of the atmosphere has first to pass through the mix of gases that compose our atmosphere to reach the surface. The transparency of a cloudless atmosphere depends on the path length of the solar beam through the atmosphere, the depth of the column of ozone in the ozone layer and the nature of the aerosols in the solar path below the ozone layer. Water vapour absorbs sunlight but the critical absorption bands lie in the infra region. On a horizontal surface there are two incoming components, the directional beam resolved on the horizontal plane, i.e. beam horizontal sunlight illuminance and the diffuse sky illuminance, i.e. skylight on the horizontal plane. On inclined surfaces there will be ground reflected daylight as well. The primary diffuse scattered energy in the horizontal skylight is reinforced by the ground reflected photonic energy which is first backscattered upwards from the surface to sky and then reflected back down again by multiple inter reflection. The reflectance of the ground determines the magnitude of this back scattered component. Clean snow cover, for example, enhances horizontal daylight levels through strong ground sky inter-reflections. The mapping of snow cover is therefore relevant for daylight design in cold climates as it alters the luminance distribution of the sky.

The next stage in climatic analysis of daylight involves introducing the impacts of cloud on daylight availability at differently located sites. Clouds vary immensely in their basic transmittance according cloud type. The proportion of the sky covered by clouds of different types also varies according to location. Desert climates are normally relatively cloud free. In Equatorial climates there may never be single cloudless day, for example Singapore. The variability of daylight from day to day as a consequence of the random nature of cloud cover indicates the need to evolve statistically based data outputs that embrace the nature of the scientific variance, presenting it to users in forms that will enlighten their daylighting design decision making.

One needs to know three things to study the impacts of daylight on the human eye. First of all one needs to understand how light is spectrally processed in the human eye to induce visual perceptual signals to brain. Knowledge of the wavelength dependent spectral response of the human eye is needed to integrate the physical spectral radiance per unit wavelength width into light signals of defined magnitude. This book thus includes an interesting account of the development of photometry and the critical role of the spectrally defined $V(\lambda)$ curve in defining objectively illuminance levels from different light sources. Secondly one needs to know about the spatial field of view and the distinctions between rod induced vision and cone induced vision. Finally one needs to know what conditions of light produce uncomfortable viewing to be avoided in design.

The book includes an interesting account of the history of daylight photometry and also a clear explanation of the scientific logic behind the historic evolution of the definition of lighting units. The book includes discussion of daylight climatology in different parts of world and the development of appropriate input daylight design data from climate records.

The various chapters assemble the lifelong scientific experience of the lead author Richard Kittler on the complex topics that need to be addressed within a single volume. Richard's role as the lead author has been ably supported by his two colleagues from Slovakia, Miroslav Kocifaj and Stanislav Darula, both of whom, though younger than Richard, have an impressive list of internationally refereed publications concerning the Science of Daylight, though their experience is not as long as Richard's. Additional support on the book has come from Professor David MacGowan. David started his academic career on lighting in my Department of Building Science in the University of Sheffield about 45 years ago.

I have had the very pleasant privilege of exchanging ideas with Richard Kittler for nearly 50 years now. When I first met Richard, his native country was in the Soviet block. I soon realised that lighting researchers from what was then Czechoslovakia and from many other Soviet block countries were very strong on analytic mathematics, especially complex integration. So, ever since first meeting Richard, I have used his published papers to cross check my own mathematical analyses. After nearly 50 years of cross checking I can confirm I nearly always found Richard's theoretical analysis was correct. I commend the quality of the mathematical analysis throughout the book. (I check many papers in journals like Solar Energy. Unfortunately a significant number contain fatal algorithmic errors).

It is easy to forget in the modern world of digital computing that exact mathematical analysis remains important. It is so easy to generate numbers using digital computers that are not reliable. So anyone who uses this book for analysis faces the simple challenge to programme accurately. The carefully checked reliable algorithms in this book provide a very significant tested resource for daylighting design analysis world wide.

The text throughout aims to enlighten the cultural understanding of daylight design by including well researched historical information concerning the development of daylight applications in buildings in different civilisations. The history presented is always based on the authors' wide historic reading so enriching the reader's understanding of why current status quo in this field of study is what it is. The text demonstrates how ideas have become what they are now. For example, the legal clarity of the Roman approach to urban Rights of Light, set down in detail in an early part of this book, continues to bear much relevance to present day challenges for regulating access to daylight in the contemporary overcrowded rapidly expanding cities of the world by legal measures based on geometric concepts.

Understanding the photonic impacts of daylight on human vision is needed to deliver effective daylight designs. The exact nature of the physiological impacts of incoming photonic radiation from our Sun on the eyes of different zoological species in different parts of world as a function of time of day and time of year is complex. The impacts of daylight discussed is this book centre concern on the impacts of daylight on the eyes of Homo Sapiens, a species which now typically lives in increasingly dense built environments. However, one must point out that much of the information in this book can be used in the study of the light response of other zoological species, though, as the eyes of different species have different spectral responses to light energy, this requires perceivable light to be treated as a species defined spectral concept, each having their own $V(\lambda)$ curves. The book does not discuss light and photosynthesis. So there is no detailed discussion of photo-actinic light in PAR units. Much of the book will nevertheless still be of great interest to scientists concerned with plant growth based on photosynthesis as it does provide significant guidance on the assessment of the impacts of light obstruction on daylight availability in shaded situations.

The sustenance of practically all forms of life on earth depends either directly or indirectly on the photonic energy fluxes received from the sun. The incoming solar photonic energy cannot be stored in the motion of photons. One of two things can happen when the sun generated photons impact at the surface of the earth. Let us start with consideration of the conversion of solar photonic energy into the energetic motions of molecules in the atmosphere, i.e. the Sun considered as the primary driver of general circulation of the atmosphere of the Earth. Thermodynamically we can identify the initial impacts of the photons from sun as inducing modifications in local air and surface temperatures. The temperature rise on any irradiated surface will depend on photon absorptance factor at the irradiated surface. We can choose black surfaces if we want to receive and store most of the incoming short wave thermal energy and white surfaces if we wish to reject a significant proportion of the incoming short wave photons. Photonic energy can be returned to space by choice of surface colour.

The energy absorbed from the incoming solar photons causes an increase in the mean molecular velocities of the irradiated mix of gases in the lower atmosphere. These increases in the local air temperature in turn cause reductions in the local air densities. The Equatorial zones are more highly irradiated by photons than the zones closer to the Poles so low latitudes experience higher mean air temperatures. The consequent global assembly of the air density differences drives convectively the general circulation of the Earth's atmosphere. The aerodynamic surface drag generated by the continuous East West rotation of the surface of the earth about its North South axis distorts the direction of fundamental sun generated North South convective flow patterns. The circulating bundles of air pick up water by evaporation especially from the oceans. Subsequent atmospheric cooling forces the condensation of water into cloud droplets spread by the winds. Under the right conditions, the water falls out of the clouds as precipitation.

The global patterns of wind flow are extremely complex. These winds are strongly influenced by the aerodynamic nature of the surfaces below. The surfaces they flow over consist of land of various heights of differing aerodynamic roughness surrounded by vast oceans. As a consequence, the basic Sciences of Meteorology and Oceanography are dominated by fluid mechanical analyses.

Nothing to do with Daylight Science? Most people concerned with Daylight Science fail to recognise with adequate clarity that the climatology of the global atmospheric circulation has a huge influence on the availability and temporal variability of daylight at different places. Except in desert areas, the clear path of the Sun's rays through the atmosphere to the surface is frequently obstructed by cloud cover. Weather has the capacity to generate localised cloud covers of many different types. Such clouds abstract more or less daylight from the beam according to cloud type and cloud thickness. Simultaneously some of the scattered beam light is returned to the surface as increases in the diffuse sky luminance. This light scattering effect is most obvious with high cirrus cloud covers. Where there is broken cloud, the surface illuminance is dynamically linked to local wind speed and cloud base height above ground level. Daylight designers therefore need to make themselves aware of the local cloud and sunshine climatologies when they are addressing daylight design for specific places. It is important to compile site specific climatologies of cloud cover and relative sunshine duration as crucial design formative topics in Daylight Science and applied Daylighting Design.

Now let us now follow the second strand of photonic impacts by discussing the photochemical impacts of sunlight. Long term storage of the photonic energy coming from the Sun is the critical process in the development of all plants. The Science of Life is thus closely linked to the Science of Photochemistry. Plant photosynthesis has come about through evolution of suitable chemical light receiving cells in plants to provide appropriate chemical photo-synthesising and chemical energy storage structures. The typical conversion efficiency is quite low, about 1%. Some crops like sugar cane are much more energy storage efficient, at around 5%. These critical photosynthetic processes in the world of Botany are described under the umbrella term Natural Photosynthesis. However our eyes also use photochemistry not to store energy but to send chemical messages to our central nervous system for immediate visual decoding. In this case, unlike plants, long term storage of the chemicals synthesised would have an adverse effect on vision, because the eye has evolved to provide almost instantaneous information about the surrounding visual environment. Importantly, however, the spectral response of the eye at different visible wavelengths is not the same as the photo-actinic spectral response of plants. We therefore always need to know which sort of daylight we are actually talking about. Botanists use the concept of Photosynthetically Active Radiation (PAR), the intensity of light at 400–700 nm, measuring as μmol quanta/m^2/s. This book however essentially addresses illuminance for human vision expressed in lux using the energy weighted V_λ eye light response function as defined by C.I.E. The other components of the incoming solar spectrum are Ultra Violet solar radiation, which can be divided into two parts UVA and UVB radiation and also Infra Red solar radiation. While it would be useful to have all these elements described fully within one book, this book concentrates on the daylighting component. Fortunately much of the basic science provided is applicable to the other wavebands mentioned. However, different transmittance functions have to be introduced for assessing transmittance within in these other wavebands.

As daylight is only available for part of each day, human society now makes extensive use of artificial daylight, normally shortened to artificial light. Such lighting requires suitable light sources and a source of electricity supply. Artificial lighting may only be in use during hours of darkness. However, more and more, artificial lighting is being used in conjunction with natural daylight during daytime. Now the daylighting and artificial lighting design professions having to face up to the new challenges of global climate stabilisation. This means addressing the challenge to promoting sustainable development by adopting more appropriate daylight and artificial light design strategies. This means aiming to reduce the impacts of emitted greenhouse gases generated in the powering of illumination systems. Helping secure future global climate has now become an essential daylight and artificial lighting design goal.

One critical practical design point in building interiors is that the satisfactory design of the indoor visual environment is strongly cross linked to the acceptable design of the indoor thermal environment. Designers need to assess the impacts of the linkages between solar radiation gains and achieved daylight illuminance levels. The eye and the body in occupied rooms share contact with one single external short wave radiation source common to both. While the building fabric retains after dark some thermal memory of the solar irradiation climate of the preceding day, no short wave photons from the sun remain available after nightfall. The use of artificial light then becomes unavoidable. All indoor artificial light sources contribute to the thermal warming of the indoor environment. In cold weather this lighting energy contributes positively to the thermal environment, however at a significant carbon emission cost. When over hot indoor conditions exist, the artificial lighting energy gains will make the indoor thermal environment even less comfortable. Energy efficient artificial lighting not only reduces the amount of energy required to achieve a given level of indoor illumination but also helps achieve cooler indoor environments during hot seasons. Unfortunately, in practical building design terms, the indoor environmental needs of the two human senses involved are not always compatible at all times and situations. As a consequence serious indoor environmental design conflicts may arise. These conflicts are most evident in hot sunny weather in buildings with large un-shaded windows. Tragically for progress on sustainability, at present the installation of air conditioning is too often adopted as the automatic response to the resolve these indoor comfort conflicts while at the same time designers often pay no significant attention to examining the energy efficiency of the lamps and luminaires being used. Such lack of interest in lighting energy performance usually implies accepting energy extravagant and therefore high carbon dioxide polluting lighting design responses in the attempt to resolve the identified thermal environmental control needs.

The scientific design of the windows, which includes appropriate choice of glazing transmission characteristics and any associated shading systems, requires design tools which interrelate the daylighting and solar radiation climatology of the site. Temperature data are needed too. It is important to resolve these basic window generated environmental conflicts at source by optimising window design to meet the human visual and indoor thermal climate needs without forcing the extensive use of air conditioning. Assessment of the impacts of sun shading devices on the indoor visual environment is an important task which has to include addressing the impacts of the changes in solar position on visual performance. Direct sunlight from a sun at low altitude shining in people's eyes will cause severe visual disability. The solutions depend on correctly understanding the implications of the effects of solar geometry at specific dates on the window radiative energy fluxes falling on glazed elements of different orientation at different latitudes. This book contains important information to help with this task.

At places where reliable electrical supplies are both available and affordable, artificial lighting is frequently used continuously during daylighting hours of work to supplement the daylighting supplied through the windows. Complex controls are sometimes installed to switch off artificial lighting automatically when enough daylight is available. In deeper buildings such artificial lighting may have to be kept on during all the hours of occupancy regardless of window design. Shading systems obviously particularly reduce daylight levels towards the back of rooms. Ceiling height exerts an important influence on the depth of effective daylight penetration.

The analysis of daylight design opportunities, importantly founded both on considerations of visual quality like avoidance of sky glare in addition to consideration of appropriate illuminance levels to match visual task requirements, has to be set in the context of addressing the simultaneous need to resolve the risks to human performance and health due to thermal overheating caused by penetrating broadband short wave radiation. Designers have to address simultaneously the provision of an adequate lighting environment and thermal environment, both of appropriate quality. Such a design risk analysis demands balancing design opportunities and threats through conjoint consideration of the radiative impacts on two separate human senses. It is necessary to decide whether the expected economic benefits of high daylighting standards are being adversely offset by the creation of unacceptably warm indoor thermal environments in hot weather and/or by the setting up excessively large heat losses in cold weather. Resolving these challenges requires making a suitable numerical analysis to assess whether the potential savings achievable by the reductions in the electricity demand for artificial lighting through larger windows are enough to justify the potential adverse thermal impacts of the larger daylighting apertures on indoor thermal comfort with the consequent impacts on any HEVAC systems installed.

Solar radiation shading systems that can be adjusted according weather conditions, can help reduce these conflicts. Effective shading systems are essential on solar exposed facades to help keep buildings cool enough in hot weather. Unfortunately, thermally effective sun shading systems tend to force the achieved indoor daylight levels adversely downwards so forcing increasing reliance on the artificial lighting systems during daylight hours. The shading systems also influence the quality of the views out from rooms. So one presently faces the common current paradox of un-shaded buildings with all their indoor artificial lighting running full on even though outside the luminance of the sky is very high. This practice is going on in a world that declares verbally that it has a strong interest in achieving global climate sustainability. Resolving such interactions successfully in the context of sustainability presents a complex design task. Clearly the impacts of window daylighting design on the energy economics of heating, ventilation and air cooling by refrigeration need better assessment.

These hot weather thermal environment problems are currently tending to become more acute in naturally ventilated buildings because building thermal insulation standards are being continuously upgraded to reduce demands for heat in winter. High thermal insulation standards, if not accompanied by appropriate hot weather control design features like substantially improved natural ventilation, will tend to make many buildings unacceptably hot in summer, because the entering radiative heat no longer can escape so easily as in the past, as a consequence of the improved high thermal insulation standards.

Obviously the quantitative assessment of artificial lighting energy performance in daylit buildings is complex. The European Committee for Standardisation has recently documented recommendations concerning the energy requirements for lighting as a component of their standards programme on the energy performance of buildings. Refer European Committee for Standardisation CEN EN 15193 (2009). The International Energy Agency (IEA) has also being studying the issues of energy efficient electric lighting for buildings within their programmes (Refer for example IEA ECBCS Annex 45 4th Expert meeting in Ottawa, IEA 2006). Downloading the individual papers in this reference allows an overview of facets of progress. The coordination of work on certain important new concepts within this annex has been led by Peter Dehoff (IEA 2006). The annual lighting energy performance is assessed using the Lighting Energy Numeric Indicator (LENI) expressed $kWh/m^2/a$. This index is designed to establish the annual lighting energy use per unit area not to be exceeded in order to meet the indicated maximum allowable unit area artificial illumination performance standards. Detailed daylighting data are obviously needed to do this. It has to be provided in forms suitable for simulation for example, the eQUEST approach embedded within the current US DOE2 energy design software with specific analytic documentation compiled by Jeff Hirsch in order to model quantitatively the extent to which the internal illumination is not required and so identify the periods which the artificial lighting can be switched off during daylight working hours using appropriate control strategies. Such simulation procedures however imply in the background the existence of an hour by hour test year illumination climatology to be incorporated into the simulation process. The short wave solar radiation data in existing Thermal Test Years can be converted onto Illumination Test Years using the luminous efficacy algorithms set down in this book. These data will then conveniently become linked in time with the corresponding basic data used for the parallel thermal performance simulations. Lighting designers are forced by regulation to face up to the proper consideration of energetics of future daylighting designs in a world that is having to face up to the need to achieve sustainable lighting systems that reduce the production of green house gases at affordable costs.

Drawing increasingly on mineral based fossil solar energy to energise lights through electric flows means drawing down the limited global resources of natural gas, coal and oil faster for electricity generation to supply the artificial lighting. Peak Oil supply is expected to be reached within a decade. As I write this Preface, this very day, at the end of May 2011, the International Energy Agency has drawn attention to the serious failure of current world efforts to reduce carbon dioxide emissions. Inappropriate lighting design worldwide is playing a significant role in the development of these patterns of failure.

As societies we aim, unfortunately presently mainly on paper rather than in practice, to counter these substantial shortfalls of achievement on sustainability, hoping to go more green, but *Going Green* in photo-metric terms is a strange concept. Plants are green light rejecters. They actually "go" for red blue. The fact is the world of plants can use only part of the incoming solar radiation for photosynthesis, rejecting some of the green light that they cannot use for photo synthesis. Plants also reject some of the near infra red solar energy which they also cannot use. These two botanic photonic energy rejections at the leaf surfaces improve the local heat balance so these surface based energy rejections help conserve water, often a short resource for plants. Some of this rejected solar energy from plants finds its way back into space. The rest of rejected energy is available to others. We humans can see some of it by using our human eyes which are very sensitive to the green light. Our eyes do not enable us to see infra red solar radiation. You have to fall back to view Nature programmes on TV, many of which use infra red light so as not to disturb animals during the night time viewing. You can see the animals walking on your screen in a "plant white" infra red environment. We can look at our successful plantings and boast we have gone green. We have essentially got to see just more green but for the most part we have yet to understand, in terms of natural energy balances, the fundamental energetic logic of the differential spectral energy responses of plant surfaces whose absorptances have evolved to take advantage of the differences between photo-chemically useful energy and photo-chemically non usable energy. The spectral surface absorption/transmission properties of our building materials certainly need more scientific attention.

There are other physiological impacts of radiation which might be loosely defined as daylight induced, ignoring the formal CIE definition of light. Many cold blooded creatures like crocodiles use the near infra red solar radiation to warm up in the morning sunrise to regain mobility through body warming. Near infrared radiation also penetrates the human body too, hence the immediate impact of the radiation from the open wood fire and the dawn warm up in the early sun's rays when camping. So one has to define carefully exactly what impact of daylight is under discussion and then what are the critical impacting wavelengths. The main agenda of this book rests firmly on photonic topics related to human vision in the three dimensional world of dense human habitats located in different parts of the globe, each with its own specific daylight climate.

I would like now to conclude this preface by adding some of my own comments about the great importance of improving globally our understanding of the Daylighting Climatology of Cities. We live in an age of rapidly expanding world population. The result across the world is people are moving from the countryside into ever expanding large cities. The land shortages are forcing the adoption adversely high land use densities. Daylighting goals have a significant bearing on quality of human life in cities. Geometrically current cities have unfortunately become random collections of sun and daylight obstructing solid angles assembled with totally inadequate analytic attention being paid to assessing what would be the preferred whole community based daylight outcomes. Many of the current developments are being pushed forward by people whose main aim is to increase their personal wealth without making any contribution to secure the future of human well being of all people living within the city. Town planners, world wide, mostly lack any scientific education and are in a poor position to offer sound scientific advice on the complexities of daylighting control in modern cities. So their capability to offer sensible local government led citizen protection from adverse developments linked with urban daylight is scientifically weak.

In my perception, we have how somehow become persuaded that sky scrapers improve the urban radiative environment and also save land while at the same time while comprehensively overlooking the costly creation of a long range transport needed to populate them daily with workers from afar. The resultant urban transport infrastructures required for staff travel almost always become highly subsidised systems. Presently these costs are being largely met by the general community rather than by the ever enriched developers.

However, in my design world which is dominated by solar radiation logic, I view sky scrapers as the daylight stealers of contemporary urbanism. They steal free solar radiation from the rest of the urban community, often on their way also creating huge air conditioning loads for their owners by their poorly detailed design. Such buildings leave the surrounding low buildings and streets highly light deprived. There is also a serious absence of logical approaches to the regulation of the outdoor urban thermal environment at street level and very poor consideration of 'extremely environmentally adverse' winds around the base of tall buildings. Sunlight and daylight for people at low level is simply not a priority. Nor is it good news for trees or plants in towns. *How do you go green with no daylight?*

Over the last few years I have come to see the need to construct an entirely new approach to urban climatic design using the premise that the microclimate of towns is primarily driven by the radiative climate. I see the achieved urban microclimate as a by-product of the urban radiative balances struck between incoming short wave radiation and out going long wave radiation (Fig. 1). My consequent conclusion is it would be a significant environmental advance to approach the climatic design of cities by starting with the view point that the impacting photonic environment is the critical climatic driving force. If urban heat island temperature effects are three times as great as global warming effects are we on the right track to prioritise adaptation to global warming?

As a volunteer advisor to the World Meteorological Organisation on Building Climatology over many years, I perceive there is an important current need to enhance substantially the international status of Daylight Climatology in cities. This topic to date has been addressed with the most vigour in northern Europe. However, the development of urban daylighting climatology has made much less progress in the most of the developing world. The daylight is free. We simply need

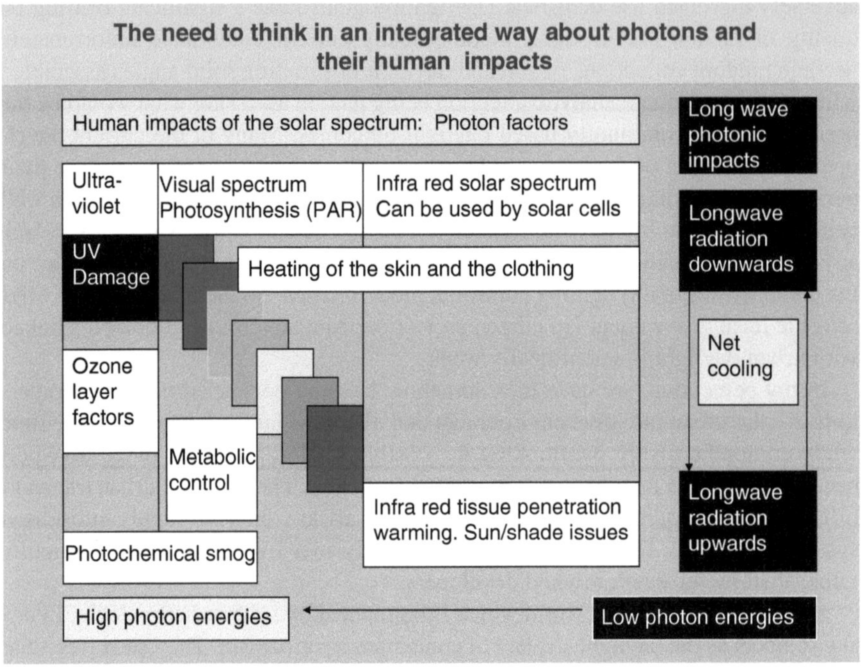

Fig. 1 The radiative climate of the cities we live in. How much do we really know? Why do we study it so little? Source John Page "Bringing together the urban climate research agenda being pursued in different international research organisations to deliver a more effective science based approach to urban and building climatology" Unpublished Paper given to 2nd International Solar Cities Congress Oxford 3–6 April 2006

to develop better technologically based methodologies for daylight design which are more suitable for hot climates. There are many important interactive climatic issues. For example, in desert climates where ground reflectances are typically high, the ground often becomes the main source of illumination for buildings because, at low latitudes, the altitude of the sun is high for most of the day. The roof becomes the dominant beam irradiated area. The walls receive a lot of ground reflected daylight. It follows in such areas the geography of ground reflectance becomes an important daylight design topic. In contrast, in hot humid climates, the luminance of clouds obstructing the direct view of the beam from the high sun become very high and the ground is usually covered with dense dark green vegetation, so sky glare control becomes a dominant factor in window daylighting design in such areas.. While this book points towards achieving better sky glare assessment techniques but does not ponder the precise effective window design solutions suitable for low latitudes.

Readers should be grateful to the authors of this book who have provided us with a new advanced tool box of scientific knowledge providing algorithms that can be applied anywhere in the world. It is now up to the World Daylighting Community to use the well organised tools in this book to create science based local daylight climatologies. They will need to work closely with their National Meteorological Services who hold the local long term climate records. The sustainable climate challenge in daylighting design is to extract data from the local daylight climatology with the aid of National Meteorological Services and then use the compiled information to produce daylighting and artificial light designs that respect sustainable daylight and artificial lighting goals. It is a goal based on the daylight climatology of specific places. In reaching for this goal readers should make themselves aware of how much progress has already been achieved in developing the global climatological data bases needed to support the design of solar energy systems. For example, both the European Union and NASA have made substantial advances in developing mapped climatologies for solar energy design. More and more, these climatologies are drawing on the data streams that can be generated from satellite imagery. Given the existence of methods of estimating luminous efficacies of beam energy and diffuse energy described in this book, it is not a big task to link up with these existing solar radiation data bases to provide a daylighting geography. The book thus lays the foundation for achieving an exciting and relevant future for sustainable daylighting design.

This book provides all the basic methodologies needed to advance international progress on sustainable daylight design on a sound scientific foundation. I strongly recommend it to readers. Remember though tools are only useful if you use them for the purposes for which they are designed.

John Page
Emeritus Professor of Building Science,
University of Sheffield, UK
Email: johnpage@nivshef.freseerve.co.uk

Contents

Chapter 1
Introduction

This is the place. Stand still, my steed,
Let me review the scene,
And summon from the shadowy Past
The forms that once have been.

The Past and Present here unite
Beneath Time's flowing tide,
Like footprints hidden by a brook,
But seen on either side...

And over me unrolls the high
The splendid scenery of the sky,
Where through a sapphire sea the sun
Sails like a golden galleon...

Through the cloud blinds the golden sun
Poured in a dusty beam,
Like the celestial ladder seen
By Jacob in his dream...

This memory brightens o'er the Past
As when the sun, concealed
Behind some cloud that near us hangs,
Shines on the distant field.

Rhymes extracted from poems by H.W. Longfellow:
The Poetical Works of Henry Wadsworth Longfellow
(Edited by W.H. Rossetti), Ward, Lock and Co. Ltd.,
London, New York, and Melbourne (1906)

The ancient humans, perhaps even more than humans today, realized the importance of the sun and sky like mighty nurturers of the environment, enabling preservation of life, health, vision, and the well-being of a habitat which nurtures humankind, living in fruitful harmony with nature and aware of the dangers of its sometimes terrible power. Philosophers and scientists have observed, wondered, and tried to evaluate the manifold effects of changes of sunlight and daylight. Technologists have utilized daily sun paths for the orientation of objects in space, measured time

R. Kittler et al., *Daylight Science and Daylighting Technology*,
DOI 10.1007/978-1-4419-8816-4_1,

with sundials, and built openings in houses to ease daylight and sunshine into interiors.

In physics, especially in optics, radiometry, and photometry the old sunbeam and flame sources evaluated by humans kept researchers and inventors busy for many centuries until, by the end of the second millennium, daylighting technology had developed precise sensors, instrumentation, and methods of rapid data collection and analysis. These advances stimulated the international formation and acceptance of quality standards and measures to regulate the characteristics of such sensors, instrumentation, and computer-aided trackers and scanners. In the recent trend, from about 1970, to utilize solar energy and daylight in the "passive design" of buildings the sun and sky are too often perceived as new "green," free energy resources discovered in building design and operation only recently. However, it should be realized that the first knowledge of solar geometry was applied in sundials, the orientation of buildings, including those in ancient towns, and, indeed, the pyramids. Further, human curiosity in that period of ancient antiquity stimulated many developments in mathematics, trigonometry, astronomy, and physics. The history of architecture also often forgets that human visual tasks moved from outdoor to interior spaces and enclosures that needed illumination; hence, window design became a priority task, stimulated by both necessity and amenity.

In any encyclopedic dictionary under "photometry" there is an explanation that it is a field of physical optics specialized in the measurement of light quantities using human eyes and light-sensitive photocells. However, historically, in parallel with the measurement interests of photometry there has also existed the science of theoretical photometry (photology), containing also calculation methods for the prediction of light propagation in space. In this broad sense is meant the science and photometry of daylight, including, in human-timescale perception, the everlasting influence of sun radiation and sunlight in nature and their extraterrestrial outdoor and indoor parameterization and prediction linked with measured local daylight climate or measured illuminance in building interiors together with their viewed environments. This wide range of science and technology is connected with daylight measurements, solar geometry and mathematics, architecture, building science, illuminating engineering, energy conservation technology, and human visual environments with their health and welfare consequences. Therefore, the interdisciplinary chain of interactions which must be respected embraces the influences of sun, Earth, and global location, sun and sky exterior daylight conditions, building apertures as daylight sources, interior illuminance prediction, and visual environment with its impact on health, activities, and the comfort of users.

Eon-long development of all living creatures with their progress in activities and civilized societies reveals the basic influence of sunlight and skylight on their daily rhythm, health, and visual abilities. Especially in humankind, this prehistoric heritage has to be respected when novel technology enables the production of electric and other sources of light. In the long history of daylight research and photometry, sunbeams and sky luminance were once compared to flame sources to obtain the basic knowledge of the character of light and its propagation.

That history must not be forgotten for these first achievements were necessary for the twentieth century's modern photometry with sophisticated detectors, novel tracking and scanning equipment, and computer-recording facilities. Only a few decades ago all architects and builders still used the ancient "stick and rope" measuring and drawing methods to design and realize buildings, although in a newer form with drawing boards, pencils, rulers, triangles, and a pair of compasses. Even then, advanced urban and architectural designs and creations had to solve already contemporary investment requirements and the latest fashion and status needs of inhabitants. However, they also had to respect the principal or traditional broad knowledge and facts, including environmental conditions and material properties to be applied in optimal and reasonable solutions of projects.

In the current era, ever-advancing computer technology is bringing new possibilities to solve very complex and previously extremely tiresome tasks in a few seconds. Once the "science" is understood, a mathematical model and defensible design or computer simulation system can be developed. Because of the wide discipline base of the audience which could be interested in reading and studying this book, we have tried to explain all problems, interrelations, and available literature in a simple and logical approach without expanding all of the basic assumptions or sometimes complex mathematical expressions and formulae necessary to solve the task. The historical context of the science and technology evolution is important for teachers, inventors, or researchers in order to find new ideas and solve the challenges they create and to formulate trustworthy and precise expressions and laws useful for application in practice. However, for those readers who seek only an "overview" of the route to a solution, are interested especially in practical applications, or want to check a design solution, the simple computer tools are also mentioned where they are available on the Web or via references.

Novel daylight systems such as light shelves and anidolic or tubular hollow light guides require special skills to apply them in computer-aided calculation methods or user-friendly tools to offer an advantage to design offices. The utilization of sunlight and skylight must be considered to be an ever-evolving set of issues in all spheres of science, technology, and practice which address the state of the art.

In such context, this book will try to explain basic historical development trends in daylight science and daylighting technology and convey relevant knowledge of applied photometry and computed prediction methods that can be used. However, such an ambitious aim has undoubtedly been aided by many uncertain and almost forgotten historical achievements by now unknown men and women who contributed in various civilization or cultural centers to the present sophisticated knowledge in various seemingly unrelated special fields. Hence, we wish to acknowledge and record our debt to both the known and forgotten visionaries of the past who laid our present foundations. Our tribute and acknowledgement is extended to every teacher and researcher who has contributed to the science and photometry of daylight even if his or her name is not mentioned in the considerable lists of chapter references.

In spite of our earnest efforts to cover the worldwide problems and information available, we apologize that, owing to authorship copyrights and difficult permission procedures, in many figures only Bratislava International Daylight Measurement Programme station data could be used to document the generally valid interrelations or represent illustrative examples. However, some interesting and instructive information from prominent foreign authors is schematically redrawn with reference to introduce, in a simplified version, their basic ideas or lessons to be learned.

Chapters 1, 2, 3, 5, 7, 8, 9 and 10 were written by Richard Kittler, while Chapter 4 was added by Miroslav Kocifaj and Chapter 6 and partly 3 were written by Stanislav Darula. The author of all eleven appendices is Richard Kittler. In preparing this book, all authors were assisted by David MacGowan, who kindly edited and corrected our English text and as a guest author wrote Chaps. 11 and 12. We owe a debt of gratitude for his patience and friendly cooperation. For Chap. 11, we appreciate also the cooperation of Janos Schanda, who helped to add interesting notes on mesopic vision. Furthermore, with thanks we appreciate the cooperation in drawing some figures of Peter Oberman, who was always so eager to help. With special thanks for the Foreward and valuable comments and checks of our manuscript we would like to express our gratitude to Prof. John Page.

Please enjoy reading this book and if you have any comments, please contact the authors.

Chapter 2
Short Historical Review of Daylight Utilization by Living Creatures

2.1 Solar Radiation and Light Helped To Create and Nurture Life

When the simplest forms of life began evolving in shallow waters of lakes and oceans, they started to grow and survive by utilizing the vital sunlight and solar energy. Survivable living conditions were also created by oxygen produced by blue–green algae roughly 2,000 million years ago (MYA). Owing to available sunlight, in a time span of another 500 million years these algae formed huge reefs of stomata with excess oxygen production. That evolution influenced the composition of the near-to-ground atmospheric layers which changed the ferric oxide component of the ocean floor. The slowly changing original $N_2 + CO_2$ filter of the atmosphere reduced the deadly UV part of the solar radiation spectrum to a $N_2 + O_2$ atmospheric layer around roughly 1,000 MYA. That evolution stimulated the development of the first green algae on ground (so-called terapoda). Probably from that time on, Earth's atmosphere spectrally attenuating solar radiation in the deadly UV region enabled, as today, the development, existence, and survival of various forms of life not only in water but also on land (Fig. 2.1). Hence, the first greenery evolved with a more effective chlorophyll-based photosynthesis process. In photosynthesis, rapidly growing flora utilized especially the blue and red parts of the visible solar spectrum using chlorophyll, water, and carbon dioxide to release oxygen and produce organic matter from 700 MYA. It is interesting that also today very primitive plants and animals live in seas, lakes, and marshes. Protozoa and flagellates in small colonies, the latter having color-assimilation pigment-bearing cells, "chromatophores," are quasi-animals that utilize sunlight for nutrition like green plants. The breathing and feeding needs of primitive fauna could be satisfied either in water or on land. At about the same time, roughly around 650 MYA light-sensitive cells with a photobearing protein pigment, now called opsin, started to operate in all fauna eyes as their biochemical substance and the inheritable control gene coded Pax6 transmitted its capacity to future generations (Gerinec 2007). The vast extent of time which elapsed before the first complex organisms started to

R. Kittler et al., *Daylight Science and Daylighting Technology*,
DOI 10.1007/978-1-4419-8816-4_2,

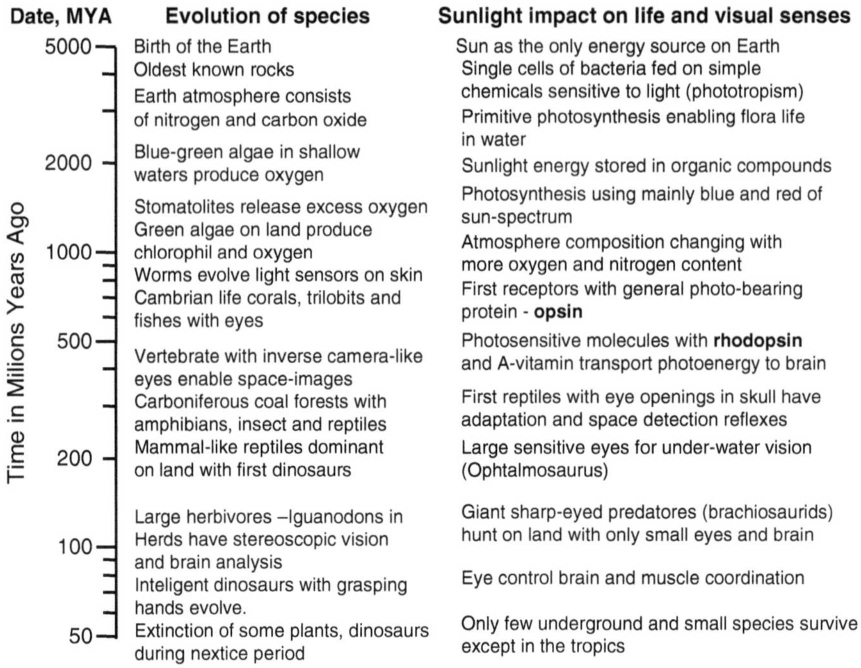

Fig. 2.1 The oldest dependence of life on sunlight

appear, in the Precambrian, ended about 600 MYA. However, during that time some Precambrian worms developed sparsely distributed light-sensitive cells on their skin. A few invertebrate marine animals had epithelium depressions lined with retina-like light sensors and visual nerve contacts to the brain such as the limpet, whereas others could also close their spherical eyeball, having a pinhole pupil as the cuttlefish *Nautilus* (Gregory 1966). Probably that stage of evolution heralded the still prevailing favorable sharing of the sunlight spectral components between most plant types and the visual organs of major groups of animals, including current humankind. Flora utilized mainly blue and red spectral bands for their needs. Fauna were sensitive to the whole luminous spectral range, some fauna, humankind, with maximum daytime visual sensitivity in the yellow–green region of the solar spectrum (Fig. 2.2).

As the main tasks of vision in the image-forming eye were distance, color, and movement discrimination, already the invertebrate animals (i.e., mollusks, especially snails or squids) had developed a form of lens for accommodation and eye micromovements. The first animals with a sophisticated nervous system were chordates, which had visual nerves for transporting and controlling peripheral signals from exterior detectors; these primitive eyes were equipped with lens attachments that formed the environmental image on a mosaic of light-sensitive, retina-type, receptors.

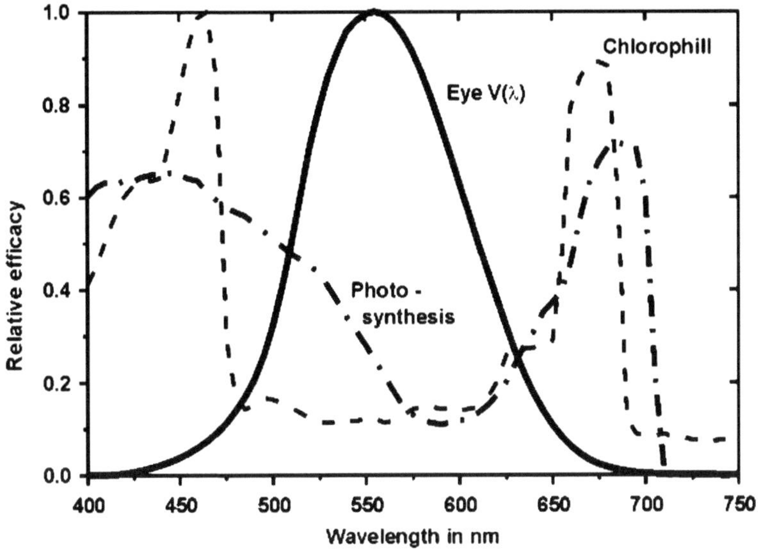

Fig. 2.2 Sharing daylight between greenery and human vision

All mobile animals needed some space orientation to find food, shelter, and favorable breeding and living environmental conditions. Thus, visual attributes also contributed to the Cambrian life explosion during the period 550–530 MYA. Probably then the rhodopsin pigment containing opsin and carotenoid (a vitamin A derivative) created photosensitive molecules that transformed light energy into impulses transportable by specialized visual nerves to brain centers.

The natural environment with fluent changes in brightness and luminance patterns during daytime and nighttime also influenced the further eye developments for necessary adaptation in a very wide range of grey-shade and colorful situations. The original primitive eyes evolved in two different ways:

1. From 500 MYA invertebrates (anthropods, trilobites, and insects) developed compound eyes with many single receptor elements each under corneal and cylindrical lenses placed on a luminance-meter-like microcylinder equipped with sense cells. Such compound eyes had over 1,000 facets that signaled the presence of luminance in tiny solid angles from a direction immediately in front of each facet and, probably, these signals combined in a pattern reproduced as a single environmental image in a small brain.
2. Vertebrates developed the image-forming eyes as camera-like cavities shaped in the form of spherical depressions closed either by a small pupil or by a lens operated by eye muscles. Receptor cells lined the inner retina surface and these were connected by visual nerves to a larger brain.

Epochs of abundant coal forests and the evolution of reptile and dinosaur families as well as later mammals occurred as various diverse life forms evolved,

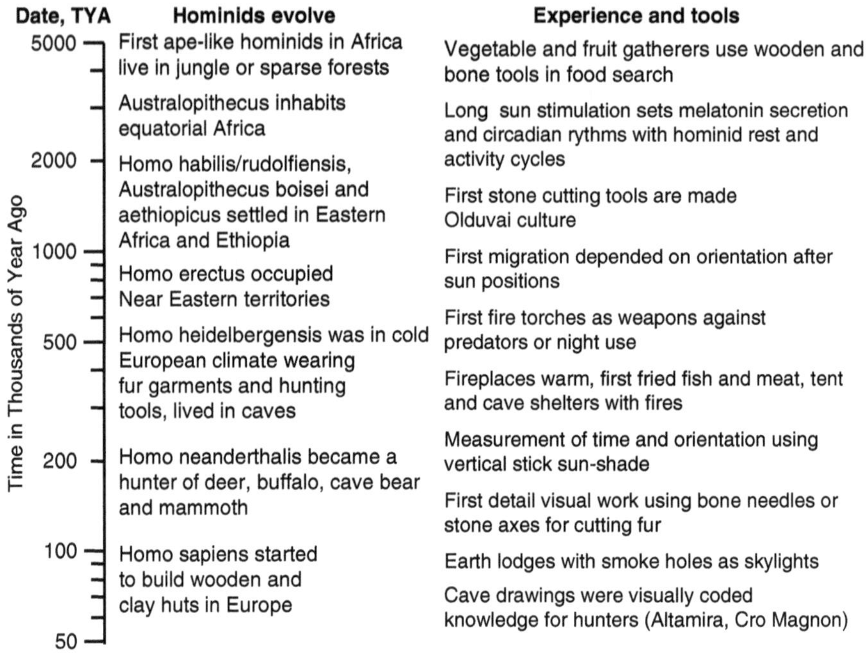

Fig. 2.3 The evolvement period of the human race

developed, and partly disappeared while daylight utilization was gradually more and more effective. Therefore, sunlight and skylight changes stimulated:

- The periodic general and metabolic activity in day and night cycles that influenced the development of all flora and fauna whether living in water or living on land.
- The space movement behavior whether local or migratory with settlement in different climate zones, where adaptation and further evolution predisposed different species to develop with more appropriate body structures, feeding habits, protection qualities, and visual organs.
- The distribution of light-absorbing pigments in respect to their utilization or against their overflow from simple skin light receptors to spherical eye protrusions for camera-like space-forming images.
- The light signals for sexual behavior of different species adjusting seasonal growing, flowering, and seed ripening as well as breeding habits in favor of life reproduction.

Gradual changes and specific evolution alternatives were developed over the eons of time summarized in Fig. 2.3. Small vertebrate animals, e.g., small shellfish and cuttlefish or octopus, had ocelli, image-forming eyes, on either their tentacles or their heads. Some flies, bees, or wasps developed compound eyes with numerous and densely packed sensors. Vertebrate eyes are characterized by eye bulbs with a complex structure behind the pupil aperture with a controllable lens projecting the luminance image onto a sensitive layer of ganglion and receptor cells connected to the central nervous system of the body.

In the Carboniferous, about 300–280 MYA, some lizard-like amphibians already had eye cavities in their skull; these gradually developed into a larger reptile family tree of about 16 orders, of which only crocodiles, lizards, and snakes have survived to the present. Probably, the longest surviving members of the oldest ancestral reptile family of Carboniferous animals survived by evolving skull openings with hidden and shaded eyes to enable distance orientation and searching for food as well as the nerve and vein contact of the brain and muscular body for action. Anapsid skulls of the most primitive reptiles (resembling turtles) already had two smaller eye openings, whereas synapsid skulls of the mammal-like reptiles had eye–brain and ear–brain links. In the late Permian and early Triassic, these mammal-like reptiles were the most abundant terrestrial vertebrates. Their skull openings also provided areas for effective muscles for flexible and manageable eye movements and involuntary reflex actions. Also in the same time period, eyes as sensors to detect luminance patterns and orientation organs to utilize daylight, twilight, or night scenes evolved in different land animal groups to meet their needs.

These different light receptors were adjusted to their life in water or air, their predator or defensive needs, and daytime or nighttime life conditions. Thus, some most efficient eyes of different qualities evolved, e.g.:

- Some arthropodan compound eyes have preferred vision in different spectral ranges, e.g., whereas bee eyes are extremely sensitive to near-UV light owing to their special violet, blue, and yellow receptors, butterfly eyes are more sensitive to blue–green, yellow, and red light (Land and Nilsson 2005).
- Some fishes have big eyes with large spherical lenses and rod-enhanced retina layers which are very sensitive at lower illuminance levels; hence, the life in bigger and deeper seas.
- The most sophisticated vertebrate eyes have excellent image-forming space vision with perfect brain analysis and memory, however, e.g.,
  - In predator birds, daytime hunters such as eagles and vultures have special forward detail daytime (photopic) vision with extreme sharpness at long or short distances, whereas nocturnal hunters, such as owls, have wider iris openings with a spherical visual field and good nighttime (scotopic) vision with many rod receptors.
  - Most victim's eyes are placed on opposite sides of the head symmetrically, and their visual field is thus covered by each eye fully horizontally and vertically to spot possible danger, e.g., the visual field of a rabbit for each eye is 180.5° with the overlaying 10° forward binocular space vision.

In the Mesozoic, as a consequence of the larger-mammal predators, especially, carnosaurs and tyrannosaurids on land and from above *Archaeopteryx*, the first bird-like creatures, many small rodents and vertebrates sought underground and cave shelters and developed arrhythmic behavior, including eye adaption to 24-h use by a special structured rod and cone retina. Hence, their night activities enabled them to gain more feeding time and possibilities under the protection of darkness. The arrhythmic eye soon acquired a contractible iris and a rhodopsin-controlled photochemical process of sophisticated adaption within broad illumination levels. All diurnal animals as well as later hominids had a thicker layer of ganglion cells controlling melatonin secretion and a prevailing number of cone/photopic sensors in the most sensitive part of the retina. For diurnal mammals, besides adaptation improvements, their accommodation ability is also characteristic, i.e., sharp discrimination of objects at different distances from the eyes. At about the same time, better three-dimensional object location used binocular vision owing to the placement of both eyes in the front of the skull, resulting in partly overlapping fields of view. These improvements in stereoscopy meant not only a better space discrimination of objects but also the recognition of special position, size, and form in that stereoscopic image-forming perception.

Whatever caused the extinction of the dinosaurs during a catastrophic era around roughly 60 MYA, it is certain that major volcanic eruptions with devastating floods and protracted reduction of solar transmittance very badly influenced living conditions and killed many plants and dinosaur families. Also, owing to these catastrophic environmental changes followed by ice cover of entire territories, only a few species with the best adaptability and luck survived. In such a forced, but still natural selection some surviving species found food and shelter in tropical and equatorial regions where floral transformation had taken place.

New flowering and seed plants, fruit trees, and palms brought a healthier diet to theropod creatures, including a few smaller dinosaurs such as *Saurornithoides* (small carnivore) and *Stenonychosaurus* (slender clawed reptile). The latter had a relatively large brain and highly developed senses and forward-pointing large eyes enabling reasonable stereoscopic vision. Together, fine control of limbs and hand-like arms with slender flexible fingers indicate a quick-running and manipulative creature with a relatively fine intelligence, a capable predator outwitting its prey and perhaps even using simple weapons such as stones and sticks. Owing to these qualities, some scientists suggested its possible further development to a pre-ape creature called dinosauroid (Norman 1985, p. 55). However, it is still conjecture how the smaller creatures survived that catastrophic era. An almost whole lemur-like skeleton dated to 45 MYA found in a German quarry in Messel near Darmstadt suggested it was an ancestor of both apes and humans, but that too remains conjecture.

Probably somewhere in the different phylogenetic process of mammals their DNA gradually changed, with more and further eye genes predetermining the ape-like visual and brain qualities. These were recently designated ABCA4, Sox1, and Sox3 and added to the key Rx gene of all vertebrate species and the general opsin archetype characteristic for the eyes of all fauna. These inheritable predispositions were already given to the hominid "great grandfather" *Aegyptopithecus* dated to 35 MYA and later to the "grandfather" *Dryopithecus*, sometimes called *Proconsul*, who is thought to have lived around 23 MYA. Both of these followed the great spread of mammals on Earth between 50 and 40 MYA. The oldest hominid fossil skull, from *Sahelanthropus* and dated to 7 MYA, probably documents a milestone at the evolutionary crossroads of hominid's final departure from the ape/chimpanzee branch of evolution.

After many million years of step-by-step evolution and adaptation changes, practically all plants utilize solar radiation for their vital photosynthetic food, whereas animals use their eyes to investigate their environment and living space. In fact, it is clear that flora and fauna survive and evolve, or perish, by their success in voluntary or involuntary adjustment to the forces of environmental imperatives.

2.2 The Hominid Eye Evolved in an Equatorial Environment

Recent studies on the phylogeny of living organisms encompass the evolvement of many qualities that are stored in the DNA sequences, which are transmitted by cells and hemoglobin genes for endless generations throughout the ages of adaptation to their environment. The DNA molecule, although first discovered by Freidrich Miescher in the 1860s (Wolf 2003), was first modeled as a three-dimensional double helix by Watson and Crick (1953). DNA is stored inside the nucleus of a cell and contains the genetic information which has aided flora and fauna to grow, reproduce, evolve, diversify, and function over the four billion year history of life on Earth. In fact, probably the age-long evolution of eyes is repeated in the embryologic prenatal process documented in the six typical stages in Fig. 2.4. It seems incredible, but every newborn has already passed the million year evolution process transforming eyes from ancestral primitive stages to the sophisticated and well-equipped visual sense organs already equipped to learn the myriad of modern life's visual tasks.

One of the specialized organs detecting and utilizing sunlight and skylight is photosensitive cells; those cells enable photosynthesis in plants and vision through the eyes of animals and humans. Retinoic acid, the acid form of vitamin A, influences the early eye and photoreceptor differentiation (Hyatt and Dowling 1997), including the development of original optical cups and lens pits in hominids and present humans as early as 29–35 days after fertilization. In the fetal period, in the gestational 10–12th weeks, the eyelids are more developed and will close the eyeballs until the 28th week to enable all inner-eye components to evolve. Also in that time eyebrows and eyelashes appear and the brain grows rapidly. Then, when the eyelids can open and close, thalamic brain connections can mediate sensory inputs and control eye muscles which operate the lens, pupil, and eyelids. During that embryonic development, retinoic acid levels are highest in the ventral retina, in the photoreceptor cell layer forming cones and rods, and lowest cells in the ganglion cell level of the inner eyeballs.

Only a long-lasting influence could have developed such a complex eye–brain system to fit to the wide ranging stimuli of sunlit to night scenes and to be able to detect spatial objects of various sizes, texture, or color in depth and distance.

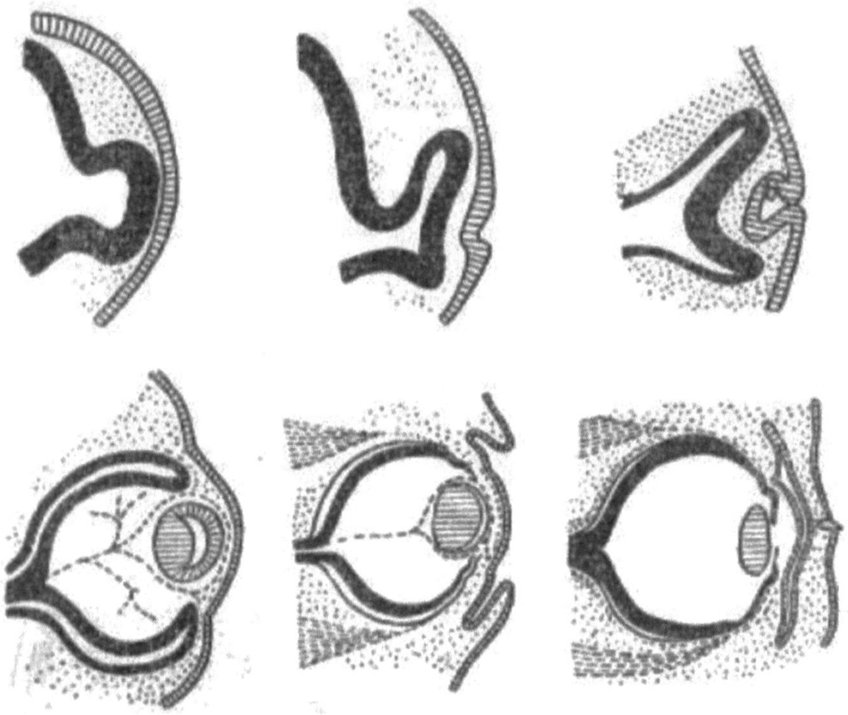

Fig. 2.4 Evolution stages of the prenatal development of the human eye. (After Kravkov 1950)

Hominids as immediate human ancestors or their related extinct forms of two-footed primate mammals inherited eyes that:

- Were placed in the front part of their skull similarly to predator eyes in animals suitable for identifying danger, for overall vision, and for searching for food needed to sustain growth, vital processes, and furnish energy.
- Were equipped with the accommodation mechanism (i.e., the pupil and lens adjustments by ciliary muscles) for detailed near and far vision in the forward binocular visual field (Fig. 2.5) focusing detailed images by the body or eye movements onto the "yellow spot" of the fovea, which has the greatest concentration of photoreceptors.
- Learned to respect the presence or absence of the sun and identify or ignore glaring objects outside the direction of the visual target with accepting the reality in usual environmental situations.
- Owing to movement, color, and luminance changes within the natural scenery in either the central or the peripheral visual field, through the eye search and pulsing reflexes which stimulate the turning attraction of the whole head or body to achieve detail detection in the eye focus and light-implied behavior closely linked to brain and muscle activities.

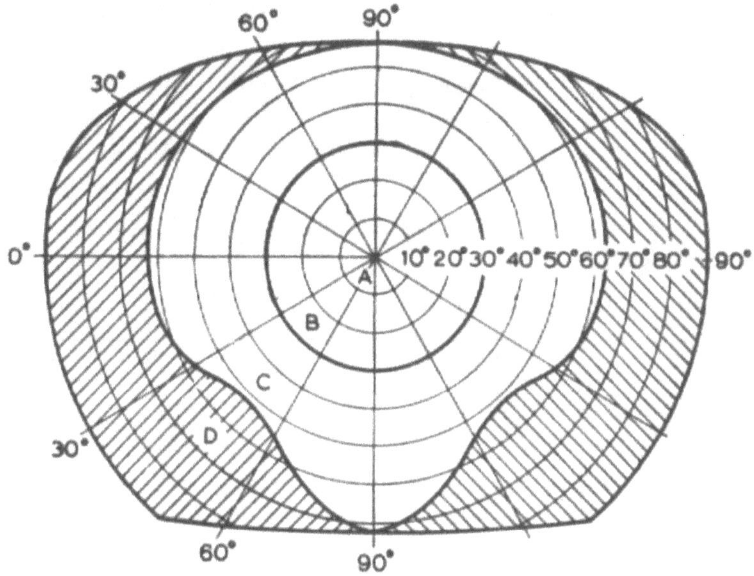

Fig. 2.5 The visual field of a human eye with the foveal zone A, the visual cone B, binocular field C, and the peripheral zone D

- Could adapt their sensitivity from feeble nightlight (i.e., dark colorless scotopic vision mainly by rods) to full daylight vision during daytime (i.e., photopic or color vision mainly by cones based on photochemical doses induced by daylight involving rhodopsin, retinal vitamin A, and protein). Such a sensitivity shift is known as the Purkinje effect.
- Advantageously share the spectral composition of daylight, i.e., whereas plants utilize mainly the bluish and reddish parts of the daylight spectrum, human eyes have maximum sensitivity in the yellow–green part, which is mainly reflected by plants (Fig. 2.2).
- Ganglion cells have also a regulating so-called circadian rhythm in humans, inducing the production and secretion of melatonin, i.e., a natural sleeping dose (Sahelian 1995), starting about 1 h after sunset or with an illuminance level under 250 lx and reaching the peak effect at about 2 a.m. to 3 a.m. during the night, whereas the hormone melatonin is absent or its level is undetectably low in the blood or saliva during daytime (Fig. 2.6).
- Natural year-round cycles in the development, flowering, and ripening of plants, birth and migration trends, and the search for fresh food by herbivores and their predators indicated the seasonal relation to yearly sun-path changes. In equatorial Africa, sunrise permanently happened at what is now known as the cardinal East point, the sun culminated overhead at noon at the equinoxes, whereas the sunset defined the West direction (Fig. 2.7). Hominids were exposed under the equatorial sun to extreme solar radiation and had to cope with its germicidal,

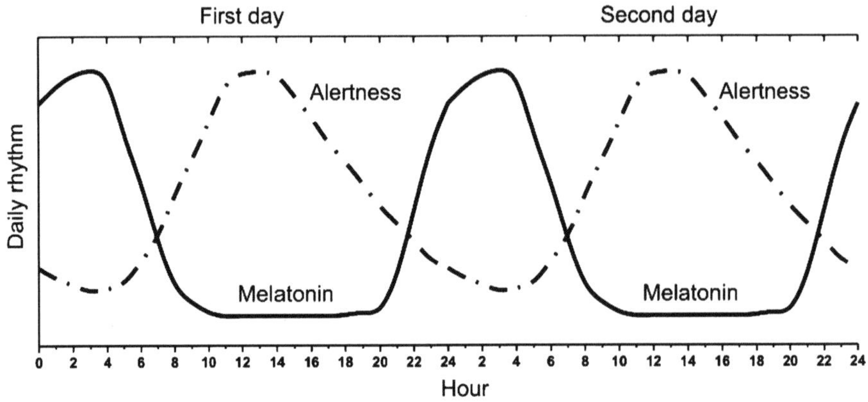

Fig. 2.6 Circadian rhythms influenced by daylight changes

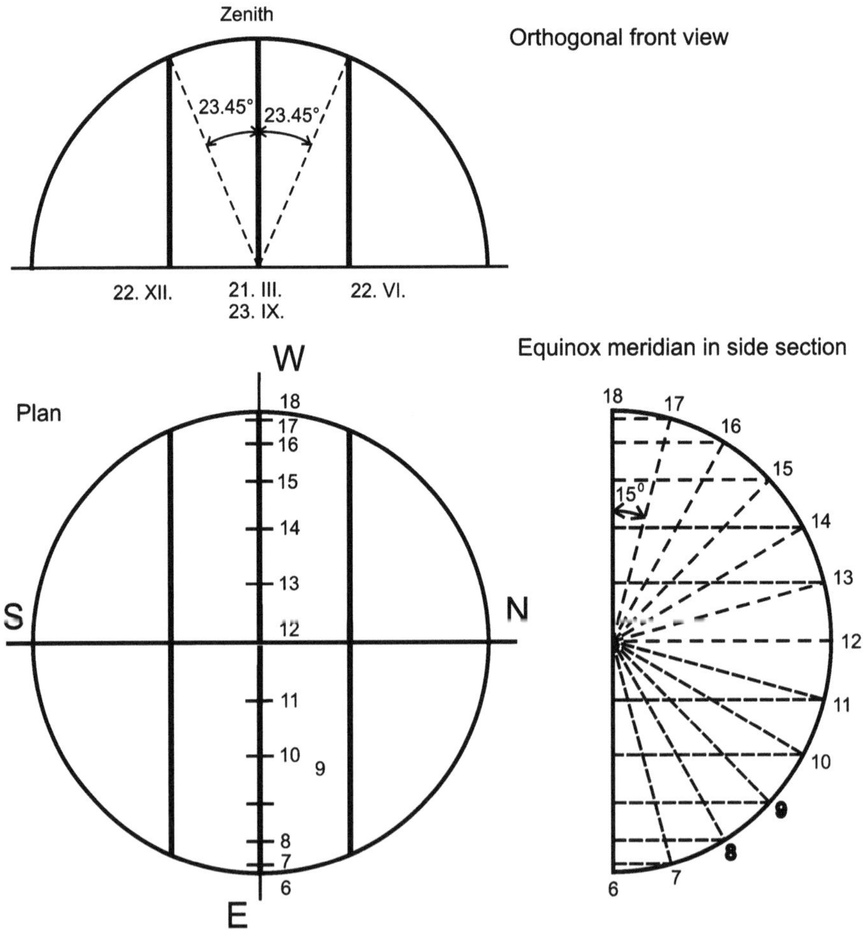

Fig. 2.7 Sun paths for equinox, winter and summer solstice days in equatorial locations

Fig. 2.8 Natural environment of hominids

antirachitic, and erythemal effects. Also, the regular rainy seasons with consecutive floods, e.g., in Central Africa or in Egypt, were sensed as yearly rhythms effecting hominid and human life with the need for a sleeping shelter in a cave, under a cliff overhang, or under a tree (Fig. 2.8). Better enclosing body insulation was required when groups migrated to Europe and northern Asia because of the even more severe influences of fierce winds, frigid air, freezing cold, and snowstorms.

- Are able to search in space either by the longer visual wide-angle images of the environment, including head turning, or by quick saccadic eye movements with concentration on an interesting object and its detail within the most sensitive fovea focus called the yellow spot.
- Mediate reflexes, selective orientation attractions to check and analyze or ignore luminance changes, or movements in the quest for spotting danger, prey, interesting objects, avoiding glare, etc. and thus stimulate further brain processes and muscle actions.
- Besides providing actual visual sensations and reflexes enabled cognitive links and memory relations to be formed, i.e., the ways that individuals characteristically organize their activities and thought processes, which are not direct reflections of the nature of the organism but rather represent preferred memory-biased strategies or experienced opinions the individual has developed over a long period of time.

The everlasting regular daily sun path with 12 h daytime and 12 h nighttime in the equatorial region determined the circadian rhythm, but hominids certainly realized also the sun position as a safe orientation point. Of course, the absolute sun culmination at the zenith is reached only on two "magic" days per year, which are called "equinox days," when the sunshade of a vertical stick shows during the whole day exactly the East–West direction. Owing to very high solar altitudes, the equatorial climate is characterized by stable high average daily temperatures and solar radiation as well as illuminance levels. Therefore, in addition to the beneficial influences of daylight, the sun, especially in direct view, was considered a dangerous environmental element, equated with lightning, thunderstorms, earthquakes, and volcanic disasters.

2.3 Fire as the First Artificial Source of Light and Heat

As early as 1.5 MYA *Homo erectus* (the "upright man") was able to perform handwork at arms reach associated with detail visual control. Primitive bone and stone tools as stone cutters and knives or fine bone needles were the first cultural remnants but other destroyable wooden tools and weapons were probably lost or deteriorated during the ages, e.g., clubs, sticks, branch and leaf beds or shelters, and garments. The great invention was the adoption of fire. Owing to the utilization of fire, *Homo erectus* could start to migrate to colder far-away territories. To keep the fire alight and move the fire, torches were probably used, as recently discovered in the Jordan river valley, where the fire ashes were dated as 79,000 years old.

Owing to cold rainy and windy seasons, these groups had to find shelter inside caves, whereas during fine weather in summer and autumn they enjoyed living at the banks of rivers or streams and probably lived in tents. *Homo sapiens neanderthalis* people are well known as cave dwellers, famous for their remarkable drawings and paintings in French and Spanish caves. These probably originated in winter in deep cave interiors illuminated by torches or the first stone lamps holding oil or animal fat soaking a burning wick. Such lamps were found in the French caves Lascaux and La Mouthe and were dated as 100,000–25,000 years old. Cave wall paintings represent the important know-how of the hunters of buffalos, mammoths, and deer as pictorial guides of their prey with sometimes drawings of their heart or organs as targets for spears or as warnings of their horns. Such "winter schoolrooms" are often interpreted as shows of artistic needs felt by intelligent *Homo sapiens* individuals. However, there was no other way to store and teach future hunters how to face and kill these huge and frightening beasts with bravery and inherited experience in the traditional caveman habitat (Fig. 2.9). The visually depicted dangerous tasks enabled the hunting group to survive, to hunt and store meat for winter, and to provide furs, oil, and bones for tools, all so useful in the cave shelter. Visually coded information, whether mediated by abstract simple drawings and sketches or cut-into-the-rock contours, compensated for the lack of writing and illustrated the important word-delivered knowledge. Of course, during warmer weather, especially during summer, branches, twigs, and skin covers were utilized to build tents and movable provisional shelters (Fig. 2.10).

Fig. 2.9 Cave shelter and fire enabled a better life

Fig. 2.10 Moveable or seasonal shelter in tents

2.4 New Challenges and Progress During the Dawn and Development of Civilization

Whereas *Homo habilis*, a food gatherer, needed mainly visual space contact and orientation, *Homo erectus* used his eyes to control his handcraft in adjusting tools and *Homo sapiens*, the hunter, utilized the visual game image in drawings to store, teach, and apply his knowledge and to express his intentions and the tasks needed for survival in the new vast lands with their strange climate, flora, and fauna (Gore 1997).

Homo sapiens sapiens, the settled and civilized man, felt the influence of the sunshine and seasonal climate changes on his crops and water resources, when in several territories crops were planted and collected as early as 10,000 to 5,000 years ago. Such new food resources stimulated permanent settlement and so the environmental conditions were even more pressing and vital. The new era of taming cattle, raising cultivated cereals, and building shelters meant a revolution in lifestyle and work tasks under a new social organization, work sharing with specified knowledge in larger communities. The old inherited knowledge of space orientation was utilized in longer journeys owing to migration, trade, and war expansions on land and at sea.

The new social organization and civilization trends were expressed also in the architectural concentration in town kingdoms, where the street urban structure meant a rather significant obstruction of the sky. The logical solution in a warm and dry climate was either the shift of manual work onto the top of roofs, which was possible especially in the pueblo building type, or the concentration of life in an interior yard usually with a water pool. The privileged casts or groups, craftsmen, or tradesmen connected to the noble atrium another private court, a garden or a yard with rooms for slaves and stables and sheds for domesticated animals, etc. Standard production of bricks and tiles for builders and clay tablets, papyrus, and parchment utensils for scripts was achieved. The civilization era is summarized with milestones in Fig. 2.11.

Almost all encyclopedias place the first civilizations around 3000 BC (i.e., about 5,000 years ago), were established in Mesopotamia town states in Ur, Uruk, Lagash, and Kish. All were built with clay walls and formed the first urban settlements (Fig. 2.12). Clay cylindrical seals as signatures or ownership stamps and trade pictograms were used, but were soon replaced by the first cuneiform scripts. At the same time, in Egypt the unification of the country by the pharaoh Menes brought the introduction of hieroglyphs. Later, in the town kingdom of Ugarit with a Mediterranean port (now in Syria), this crossroad trade center had to cope with communication in many languages and therefore the first alphabet with 30 sound symbols, i.e., letters, was invented in the seventeenth century BC for their transcription. The famous Phoenician sea traders inspired by the number of basic Egyptian hieroglyphs used 22 letters on papyrus scrolls and spread those to many European ports probably in the twelfth to tenth centuries BC. The diversion from clumsy and heavy clay tablets to papyrus and parchment scrolls enabled better merging of different knowledge from the various civilization centers, town kingdoms, and regional provinces (Anon 2002).

Date, YA	Historical event	New inventions or acts
5000 2990 BC	Imhotep built Dzoser´s step pyramid in Egypt	Sexigesimal time/angle units Sumer houses with unglazed windows and atrium houses
	Ancient Greek and Roman culture	Foundation orientation after equinox sunpath shades
2000 10 AD	Vitruvius writes the first textbook for architects with environmental advice	Sundial and orientation after the analemma rule
		Roman law "Right of Light„
	Roman/Byzantine split	Pantheon in Rome with a 9m unglazed skylight
1000 1010 AD	Romanesque period	Glazed panes in windows of medieval castles and houses
	Gothic culture	Cathedrals with clerestory windows and colour glazing
500 1510 AD	Renaissance culture	
	Industrial revolution	Sawtooth and clerestory day-lighting of industrial buildings,
200 1810 AD	Photometry basis by Bouguer and Lambert	Bouguer defined optical air mass of the atmosphere and clear sky luminance pattern Lambert´s uniform sky
100 1910 AD	Paxton´s Crystal Palace	Glass industry develops, gas lighting of interiors, skylight roof in halls and railway stations Edison´s bulb started artificial lighting of interiors Kähler defined overcast sky luminance gradation
50 1960 AD	Modern/High Tech architecture and building	Artificial sun/sky model studies Glass light guides invented

Fig. 2.11 Developments and civilization achievements of the human race and knowledge

Fig. 2.12 Mesopotamian settlements and urban environment developed civilization trends

The new powerful and religiously supported leaders organized united societies for better labor specialization and cooperation in production and trade links. New agricultural products from irrigated fields, inventions of glass, bronze, and papyrus, excavation of gold and silver on Nubian territory, and town and trade centers with connecting roads for wheel carts enhanced the wealth of society and cultural progress. Soon, planned centers (such as Ur, Nippur, and Mennofer) had better-structured buildings around stone-clad squares and communications, water-supplying and waste channels (the oldest dated about 3000 BC in Nippur), living quarters with community buildings, ziggurats, and atrium houses with workshops, pools, and gardens. In the latter were available not only food products (in shops and bakeries) but also bronze weapons, tools and other utensils, gold and silver jewelry, etc. The handcraft work either in outdoor atria or indoors needed sufficient daylighting, especially for goldsmiths, jewelers, and stone or clay tablet scribes. In religious centers also observations of sun and star paths occurred.

The original vertical stick and rope drawn circles on the ground were still used in several new tasks:

- In addition to rough east–west and south–north directions, a more precise division of the circle to measure actual sunrise and sunset in locales with different latitudes was now needed to determine the sun path year-round.
- To orient the entrance into a tent and a circular clay hut or a rectangular house, mastaba, ziggurat, or pyramid, their design sketches for final building constructions were used.
- To set lots for planting and sowing fields and to orient streets, town walls, and gates, etc., urban planning started.
- To measure distances, right angles, and slopes needed simple geometry rules.
- The visual comparison and counting of things, e.g., grains, tools, animals, or persons, to name and number them as well as to draw their abstract forms in the first scripts were developed.

All these new inventions were needed to pronounce status through the belongings and wealth of individuals in the differentiated society. Many findings of seals and personal signatures on clay tablets or in stone inscriptions testify this trend in Egypt, Mesopotamia, and other early Asian civilization centers whether in India or in China.

The straight line, i.e., the shortest distance, between two points as drawn by the stretched rope and the average forearm was taken as the oldest unit for length, the so-called cubit, inherited from the Sumerian system as kush was equal to 0.5 m. Its two alternatives in Egypt were the small cubit, equivalent to 6 palms (0.45 m), and the royal cubit, which was applied especially in architecture and which was equivalent to 7 palms (0.525 m). For longer distances 100 cubits (52.5 m) or the river measure 20,000 cubits (10.5 km) was used. As indicated, suitable mathematical operations such as adding, multiplying, and dividing were applied respectively.

The solar year respecting moon image changes was used for measuring time periods. The old Sumerian time units, standardized as long ago as 2650 BC using the sexagesimal system, is still valid with minor corrections, i.e., 1 year is 12 months, which is 12 times 30 days, and had 360 days with 2 times 12 h a day and 60 min per hour. The same sexagesimal rule was used to measure the circle, divided into 360° with 90° among cardinal points with a circular turn of 360 days within a year and a 30° turn per double-hour within a day. The units of some distance and area measures on Babylonian tablets with cuneiform texts have been analyzed (e.g., by Neugebauer 1935).

Owing to the old tradition, every year started on the day of the spring equinox (i.e., 21 March) and ended with an additional five or six holiday days to compensate for the "irregular solar year of 365.25 days." The prehistoric tradition of worshipping motherhood (documented by many excavated statuettes of Venus) was repeated by ceremonial feasts to the goddess Inanna in Sumer and Maat festivities in Egypt. The perfect east–west direction indicated by the vertical stick shadow during the equinox day served as a rule to orient buildings, e.g., Egyptian pyramids and graves (Kittler and Darula 2008a). The Egyptian title "Keeper of the Secrets" indicates that only the few initiated knew about the restricted knowledge behind the pyramid foundation process, including the procedure of "stretching the cords." From the solar geometry based on the vertical stick (gnomon) shadows, the right-angle triangles (triplets) were derived and probably besides the triangle with sides in the ratio 3:4:5 units and a slope of 53° (later well-known as Pythagorean) were found several more such as those with ratios of 20:21:29 (slope 43°36′) and 8:15:17 (slope 62°), all of these being used in designing and building pyramids with different slopes (Rossi 2007).

Many temples of the Middle Eastern tradition were oriented along the north–south axis, and Solomon's palace in Jerusalem had the same orientation. Popular religious ceremonies were performed in the open courtyard in front of the southern temple entrance. Also, in Chinese urban planning and architecture, the main streets, town gates, and palaces in the rectangular oblong or square plan of the capital Ch'ang-an, now Xi'an, were designed and built with the traditional sun-based southern orientation in the seventh century (Anon 2002). It is interesting that the tradition to orient building fronts to cardinal points lasted also into the historical periods, e.g., in the White Tower of the Tower of London begun about 1078 by Bishop Gundulf, the architect and also builder of Rochester Cathedral, oriented west–east, and in many palaces and temples.

Ceramic containers and pots used for storage or cooking were handmade in prehistoric cultures on horizontal boards and dried by sun probably as long as 5,000 years ago. However, the board rotation enabled good round pottery to be produced probably after the invention of wheels. In Mesopotamia, where stone and wood resources were scarce, either reed or the first primitive clay bricks were produced to build dwellings, town fortification walls, and ziggurats.

The oldest Sumer clay and Egyptian stone buildings document the trials to realize safe and convenient interiors for living and work. With regard to utilizing daylight, two primary house types are specific:

1. Originally, the winter partly underground hut or early earth lodge developed in a semiground plateau house with a larger smoke hole which served for zenith

lighting and as the entrance down the log notched ladder and later also in pueblo-like structures in which roof apertures were used as doors, chimneys, and roof-light openings.

2. In arid lands very condensed living quarters without a road were built with flat roofs in different heights to allow small windows to be used also as entrances at the roof level as found in Çatal Hüyük (approximately from 6000 BC, after Anon 2002) or in the Indus valley (approximately from 3000 to 2500 BC).
3. The trend to fortify clay houses introduced the rectangular atrium structure, where the inner courtyard was used as a private safe space for children and for work under good daylight, whereas street walls were often without windows and all interiors were usually illuminated from atrium doors. To hinder the entrance of animals, a sill was built into these doors and they were blocked by wooden or fabric shutters at night. Such first windows were found in Sumer houses around 2350 BC (Paturi 1988). However, the first Sumer towns such as Uruk built around 3000 BC and Ur built as a planned capital a few years later probably had buildings with only unglazed windows.

After that time, vertical rectangular window apertures became the most usual planar daylight sources in interiors, with opening covered by occasional shutters. In Egyptian and Crete noble houses a few windows were permanently covered by translucent animal membranes or parchments. From the first century BC in Rome also glazed windowpanes were discovered with bronze frames or small glass pieces in lead frames. In the Roman law system (Hunter 1903) the right to exploit skylight was defined by the rule to consider the height of adjacent buildings when new construction was designed.

Vitruvius (13 BC, 1487) in chapter 6 of book VI of his *De Architectura*, written around 20 BC, stressed the need for interiors to be day-lit properly because obstructions of windows had occurred in towns. In that case his advice was to imagine a straight line from the upper side of the obstructing wall to the illuminated place and then "when viewing along this line a considerable part of the sky is seen, the illumination of that place is not disturbed." This advice was probably influenced by the Roman law justifying the right to obtain unobstructed daylight (Kittler and Darula 2008b). Furthermore, his analemma rule (Kittler and Darula 2004) defining the geometry of the sun path in the main locations enabled sundials to be designed and the insolation of towns, buildings, or interiors to be studied. So, applying insolation conditions, Vitruvius also recommended optimal orientation of windows in particular rooms in chapter 4. However, in book IX of his *De Architectura*, he mentioned that Berossos, a Chaldean priest, had studied Mesopotamian sun-path geometry in 253 BC and therefore this knowledge was quite old.

During the Roman Empire, thinner window glass was made in rectangular forms into which blown glass bubbles were pressed. In Pompey and Herculaneum, which were destroyed in AD 79 were found large bronze window frames measuring 54 cm $\times$72 cm or 100 cm $\times$70 cm into which smaller glass bits or panes with lead fixing bars were inset. The atrium building form in traditional Greek and Roman buildings (Fig. 2.13) was unsuitable in the colder climate of western or central Europe, where

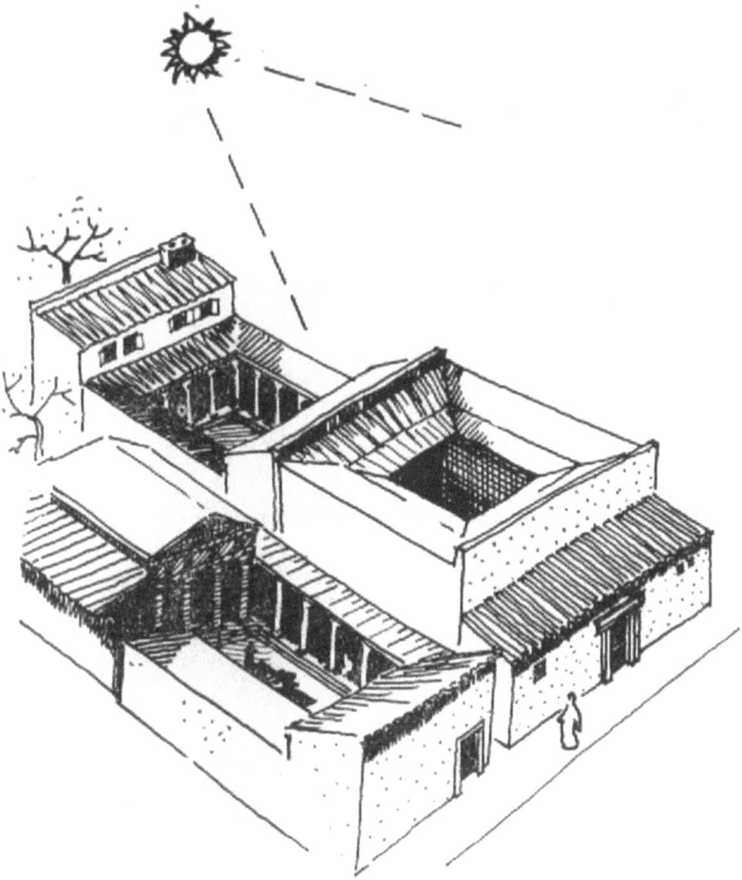

Fig. 2.13 Greek and Roman atrium yards served as safe spaces and work spaces

individual Romanesque houses or churches had smaller windows in exterior walls because of the weather and also for safety reasons. Fortifications embracing larger towns concentrated trade and manufacturing workers into interiors needing more daylight (Fig. 2.14). Window shutters of different materials or glass mosaics in stone frames were developed. Gothic cathedrals had rose-color windows and three-storey, i.e., aisle, triforium, and clerestory, windows or glazed rich broken wall arcading behind altars, all applied the typical Gothic groined vault structure. The medieval house usually had smaller windows on the first floor, where work and living were concentrated, whereas on the ground floor there were horse stables or shops with walking arcades (Fig. 2.15).

The influence of Roman culture and tradition was quite strong as almost all scripts were written in Latin until the eighteenth century, although many old manually written texts were destroyed or were kept forgotten in monastery libraries. The first architectural text-book by Vitruvius was found in 1415 in the Swiss

Fig. 2.14 Compact towns protected the good-producing society

Fig. 2.15 Medieval houses with smaller windows

monastery at St. Gallen and was published in 1487 as one of the first prints. It became the basis of the new Renaissance architecture as the revival of ancient Greco-Roman building style and knowledge. However, the windows had a half-circular upper rim or new rectangular windows as designed by Andrea Palladio in all rooms of the Valmarana Palace in 1566 and the Chiericati Palace in 1570, both in Vicenza, were fashionable. These were soon very popular in European architecture, ever since respecting the progress of technology. Although at first for safety the ground-floor windows were kept smaller and square, in higher storeys the oblong and higher windows were placed in noble rooms on high sills (e.g., in the Medici-Riccardi Palace, Florence, the sill was 165 cm high).

The great Italian architect Filippo Brunelleschi designed and almost finished before his death in 1446 an excellent 39.5-m cupola on the Santa Maria del Fiore Cathedral in Florence. This double-skin cupola was inspired on the exterior by the pointed Gothic vaulting, whereas a hemispherical interior shape resembled the Roman Pantheon inside illuminated by a shaft of light guided from the top lantern. The novel cupola geometry with a double-skin-like encasement, the use of the so-called Roman plaster mix, and armoring inset iron chains enabled to realize a unique three-dimensional construction. This design was the result of his studies of vault and cupola buildings applying current building statics and stress analysis before the design and start of building in 1420. His spatial imagination and designer needs to interpret it in a planar picture made him invent basic principles of perspective drawings together with Lorenzo Ghiberti, his close coworker. Both used the perspective technique in their artistic activities, e.g., in their pictures and bronze reliefs. Brunelleschi's experience and building techniques were later used by Michelangelo Buonarroti in his cupola concept and design for Saint Peter's Cathedral in Rome when he was asked to take over Bramante's work after his death in 1514.

At about the same time, Leonardo da Vinci studied the light spectrum and wondered how it is conceived within the eye. He operated secretly on eyeballs in morgues and discovered nerve connections to the brain (Nardini 1974). Francesco Maurolyco (1575) explained the illuminance from a candle source imagining the light flow within a pyramid to the illuminated plane which presented its solid angle. Candles made of resin, wax, or suet were used from 750 BC during Greek religious ceremonies or in Roman churches around AD 100 and were used as a sufficiently stable and long-burning artificial light source. Later wax candles were cast in France from the eleventh century onwards massively in almost standard forms.

Renaissance research progress in geometry was followed also by progress in mathematics after studies of right-angle triangles and the ratio relations of their sides. Johannes Müller Regiomontanus (1461) introduced trigonometric functions. In 1551, Georg J. von Lauchen (Rheticus) worked out precise trigonometry (sine, cosine, tangent, and cotangent as well as secant and cosecant) tables, which were published by his pupil Valentin Oth in 1596 under the title *Opus Palatinum de Triangulis*. Soon trigonometric functions were applied in astronomy and also the original geometry of the sun path (analemma rule) was expressed by trigonometric formulae for solar altitude and azimuth angles.

Further considerable advance in the eighteenth century resulted in the invention of the subjective luminance meter by Marie (1700) and its use by Bouguer (1729, 1760). The latter measured the optical air mass of the atmosphere in November 1726 using the low luminance of the moon comparable to his candle source under the assumption of luminance contrast thresholds. This was a considerable step to calculate sunlight illuminance at ground level by applying Bouguer's exponential law. Later he measured the luminance pattern of a clear sky, but unfortunately his notes with luminance readings were lost when his posthumous work was submitted for print in 1760. At about the same time, Lambert (1760) based his photometry calculations on his cosine law. He also defined formulae for the illuminance distribution from window apertures and for fluxes from plane to plane on the assumption of a unity uniform luminance of the rectangular light source.

Marie's and Bouguer's subjective measurements were based on the assumption that the observed brightness and physical luminance of two close-by diffuse paper or glass patches can be identified by a human observer as equal; thus, the adjusted luminance from the replaceable candle can be calculated. On the basis of this idea many subjective photometers were made and the optics were improved, and inset incandescent lamps were used even 200 years later (Uppenborn and Monash 1912). However, the sources of the radiation and light output were soon measured in physical terms and units and the human response was questionable.

At the beginning of the nineteenth century, Young postulated the existence of three photoreceptors in the eye and proposed a three-receptor theory of color vision, later Purkinje (Purkyně 1825) introduced the spectral luminosity shift different for daytime (photopic) and dark (scotopic) adaptation of human eyes after Fraunhofer (1821) had predicted the monochromatic colors determined by wavelength after light diffraction studies. König and Dieterici (1886) and König (1891) studied and defined the spectral eye sensitivity under different wavelengths, now known as a $V(\lambda)$ curve (CIE 1926), which was foreseen in König's book *Gesammelte Abhandlungen zur Physiologischen Optik* (König 1903).

Although subjective photometry was based on the equalization of two adjacent brightnesses, E.H. Weber (1795–1878) started to study the human response (e.g., brightness) to a physical stimulus (luminance) in a quantitative fashion, i.e., what is the smallest noticeable difference in luminance L_d that can be identified in the visual field having the threshold or background/adaptation level L_o. His colleague Fechner (1860) offered an elaborate theoretical interpretation, the logarithmic Weber–Fechner law defining the perceived intensity of the human sensation as proportional to the logarithm of the ratio L_d/L_o. Almost 100 years later Stevens (1957, 1961, 1963) changed it to a power law applicable in general psychophysics. Applying his psychophysical knowledge, Fechner (1876) also tried to formulate the basis of esthetic feelings expressing the thirst for enjoyment and happiness via sensual impressions.

Studies of the visual field had been known for a long time, but the tiny eye movements were studied first in the nineteenth century (Müller 1826), and later with more sophisticated instruments, less than 0.001-s rapid photography or photoelectric registration with eye-tracking hardware (Yarbus 1965, 1967). Visual

Fig. 2.16 Typical light sensitivity of human eyes depending on age under photopic and scotopic vision

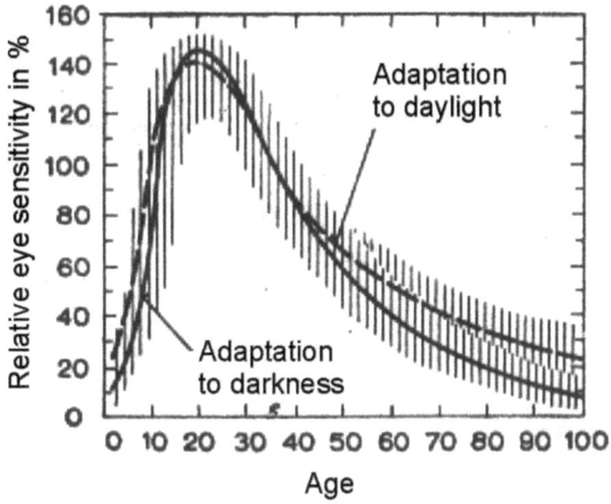

attraction to, search for, or orientation to stationary or moving objects in the visual environment depends on many voluntary or involuntary visual reflexes stimulating search following the brightest targets or scenes, attracted by moving or noisy objects, by familiar color or shape patterns, etc. Very quick unconscious microsaccadic reflexes "touch" and cover any object seen and its details in a sequence of jerk-like involuntary glances to identify and comprehend a specific scene or event. Binocular reflexes help also to determine the depth, height, or distance of any objects seen.

More research results concerning the visual qualities of human eyes were achieved in the late nineteenth century and during the twentieth century. The 24-h living rhythm as a natural clock certainly influences humans as an inherited biochemical consequence of the million-year life under equatorial sun-path conditions. The eye gland secretion of melatonin and daily cycles of daylight stimuli to activity and rest are now certain and are presented as circadian rhythm (Fig. 2.6).

Further investigations were conducted on eye sensitivity (Fig. 2.16), accommodation, and adaption abilities in accordance with the age of the individual. A child can discriminate details at a closer distance between the object and the eyes than an adult (Fig. 2.16). The focal accommodation range (Fig. 2.17) decreases very considerably with age. At the same time, the adaptation sensitivity to daylight and twilight increases with age until young adulthood and then slowly drops (Kittler and Kittlerová 1968, 1975). Thus, in most visual tasks it is important to determine task and work area luminances and luminance contrasts (Fig. 2.18).

Several innovations and new inventions in the production of window glass were accomplished by the continual drawing and rolling technique invented in 1844 and realized by Robert L. Chance in England. This made larger and cheaper window sizes available in mass quantities as documented in the Crystal Palace built for the Great Exhibition of 1851. Such industrial glass production in the twentieth century facilitated the development of modern architecture in multistorey buildings with larger glazed windows and "glass curtain facades" (Fig. 2.19).

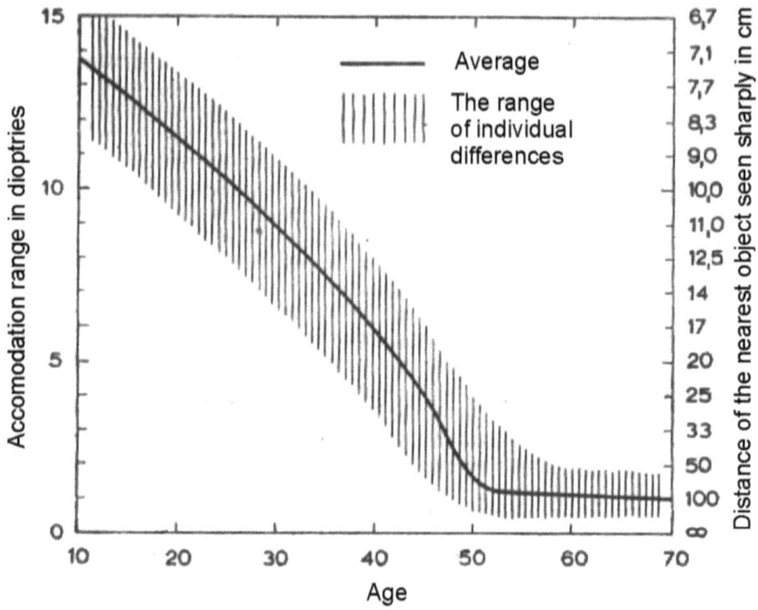

Fig. 2.17 Accommodation ability of human eyes changes with age

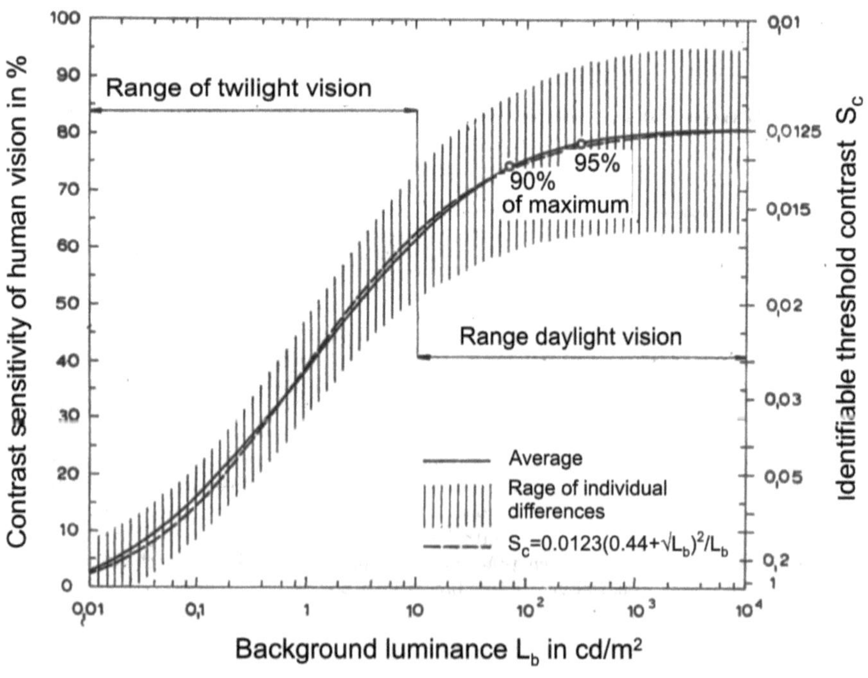

Fig. 2.18 Contrast sensitivity of the human eye during twilight and daylight

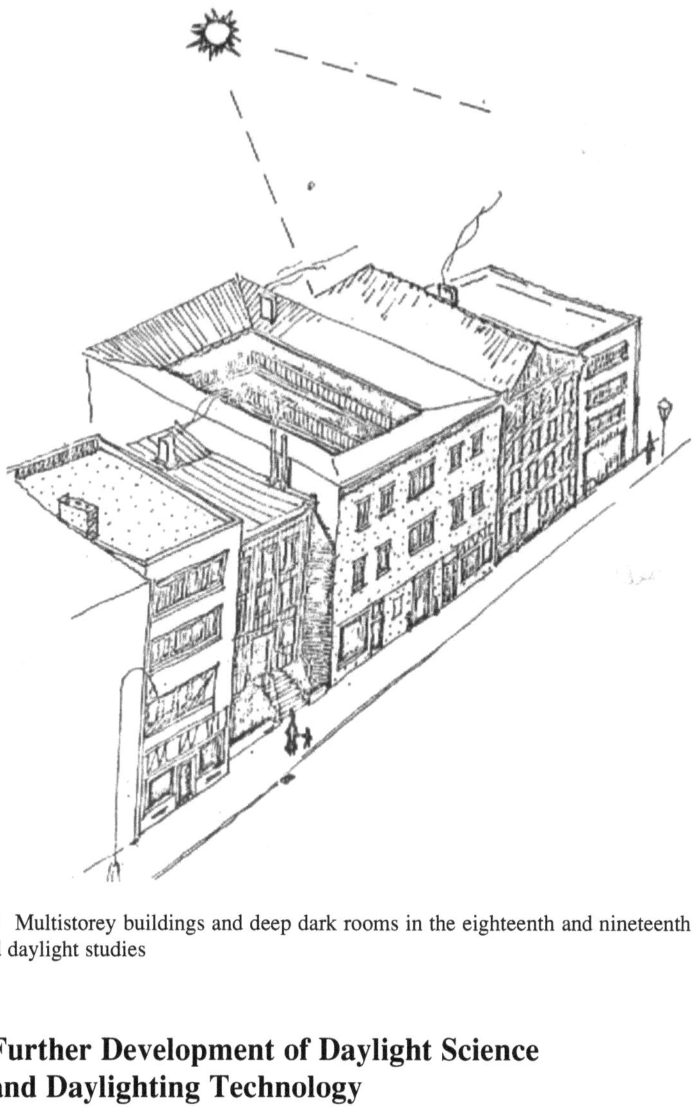

Fig. 2.19 Multistorey buildings and deep dark rooms in the eighteenth and nineteenth centuries provoked daylight studies

2.5 Further Development of Daylight Science and Daylighting Technology

The quantitative aspects of atmospheric extinction or attenuation of sunlight were expressed by Bouguer, but the atmospheric extinction coefficient and the extraterrestrial solar flux were uncertain.

Rayleigh (1899) studied the extinction and scattering within the clean clear sky and found that the extinction varies with wavelength λ and for a blue sky is proportional to $1/\lambda^4$; however, later its dependence on the atmospheric air mass was determined too. The sun radiation at ground level prevented estimation of the extraterrestrial solar flux owing to difficulties with insensitive instruments and changing atmospheric conditions. The idea of a solar constant was introduced in

France by Pouillet (1838), who developed a water calorimeter and obtained the first approximate value of 1.76 cal/cm^2 min (Henderson 1970), equivalent to 1,228 W/m^2. After recent satellite measurements at the border of the atmosphere, the exact extraterrestrial solar spectrum was determined and integrated to define the solar constant $1,366.1$ W/m^2 (Gueymard 2004) and the luminous solar constant is approximately 133.8 klx (Darula et al. 2005).

Weber (1885) after producing his milk-glass photometer and the solid angle meter started to measure daylight illuminance, and his colleagues Schramm (1901) and Kähler (1908) characterized the luminance pattern of overcast skies with the sine gradation function dependent on the elevation angle above the horizon normalized by the zenith luminance.

The revolutionary change from subjective photometry, i.e., visually operated equipment, to objective photometry started with the investigation of selenium layers producing small electric currents after their exposure to light (Smith 1873, Adams and Day 1876), with improvements by Fitts (1883) in his cells. These were corrected to the spectral sensitivity of human eyes by yellow–green filters later (Dresler 1933) as well as for the cosine error (Barnard 1935). The measurement instruments were used either in particular buildings or in models of buildings or interiors under artificial skies (Kittler 1960).

After a 40-mm-high flame had been produced in 1883 by Hefner-Alteneck (1896), German experts started to question the candle as a unit of luminous intensity. As a consequence, international cooperation was successful in standardizing photometric units when Blondel (1896) proposed the adoption of the international standardization of a photometric system of units for the light intensity (cd), luminance (cd/m^2), light flux (lm), and illuminance (lux). After that, agreement was reached by the American Bureau of Standards, the National Physical Laboratory in England, and the Laboratoire Central d'Electricité in Paris that these units, based on the so-called international candle (cd), be accepted worldwide. The International Commission on Illumination (CIE) was founded in 1913. In 1942, the new candle was defined as the light intensity of a blackbody radiating at the temperature of melting platinum, which is 2,046 K. Thereafter, international agreements standardized lighting and photometry issues, such as the CIE (1924, 1983) photopic luminosity function and the $V(\lambda)$ function, which represents human sensitivity to radiation in the visible spectrum.

Although the old Lambertian assumption of uniform and unity sky luminance was used as the design standard for a long time, during which time several graphical tools were developed and used in practice to determine the daylighting of buildings, a uniformly luminous sky really did not reflect reality; the science and daylighting technology developed from old Lambertian assumptions needed to be revisited. A new daylight factor (DF) was accepted internationally as a basic criterion, with a later innovation adopted as the CIE (1955) overcast sky with luminance gradation. The early daylighting design tools were called sky factor assessment aids. Their development took into account only the projected solid angle, as described by Wiener (1884) and later used and developed by Waldram and Waldram (1923) in their "rectangular diagram" with refinements by contour droop lines (1950). In the

USSR and in central Europe "angular charts" by Danilyuk (1931, 1935) were used to predict the sky component of the DF, whereas in western Europe similar Building Research Station protractors were published by Dufton (1946), which were quite popular in Great Britain for a long time (Walsh 1961; Hopkinson et al. 1966).

However, realizing that daylight year-round variability and availability change considerably, in 1983 a CIE Daylight Technical Committee, TC 3-08, initiated the so-called International Daylight Measurement Programme (IDMP). That program was officially launched by the CIE president Bodmann (1991). Thereafter, IDMP stations were established in several global geographical locations, where regular sunlight and skylight data were recorded in 1-min time steps containing:

- Global (sun plus sky) illuminance and irradiance values on a horizontal surface
- Diffuse (sky) illuminance and irradiance values on a horizontal surface
- Zenith luminance data
- Illuminance on vertical planes with east, south, west, and north orientations

A few research stations are also scanning the sky measuring usually in 15-min steps luminance distribution in 145 directions covering the whole sky vault.

These data collected over several years enable one to determine the long-term characteristics of the local daylight conditions in detail, i.e., the probability of occurrence of different typical sky patterns, sunshine duration, solar and sky luminous efficacy, etc., in daily and seasonal changes. Such sunlight and skylight availability information is needed to evaluate, control, and monitor different energy-saving measures or means to utilize better solar energy facilities or novel solutions, e.g., photovoltaic panels, hollow light guides, window shelves, or other innovative daylight designs.

Currently, it is felt that the past rigid DF system is obsolete and sunlight and skylight availability in different regions has to be represented worldwide in a new evaluation system expressed in SI units to be comparable when combined with or supplemented by artificial lighting systems (Kittler 2007).

Historical advances in applying various laws, rules, and criteria for daylight evaluation and design are briefly summarized in the appendix.

2.6 Partial Conclusions

Only recently have all humans realized that the sun is a limitless, ubiquitous, and clean energy and light source which created and formed all creatures on Earth. Basic solar influences on adaptation and utilization processes that formed differently various species were inherited and stored in genetic markers and constitute their history book with many consequences that affect their life and health, activities and creativity, comfort and well-being. The reaction to sunlight started 3,500 MYA in single cells of water bacteria, followed by photosynthesis in algae 1,000 MYA to the evolvement of light-sensitive cells in the first Precambrian worms, corals, and fish around 500 MYA. Reptiles, amphibians, birds, and mammals have had skull-embedded eyes as the most sophisticated sense in phylogeny heritage since 300 MYA. The multiplication of photoreceptors in the retina of the eye as the neuro-ectodermal extension cord of the frontal part of the brain and the pupil with a crystalline lens operated by eye muscles in each eyeball form the optical instrument giving a minute inverted image of the outer space. The dense mosaic of rods and cones convert this pattern into fine electrical impulses transported by opsin or rhodopsin into the brain, where they are analyzed, read, and stored. Such a visual system is characteristic of many vertebrate animals having a body with a head and brain connected to a spine, including humans. Human eyes have experienced slow painful phylogeny trials and errors that shaped them, and human tears still document a recreation of the primordial ocean water which bathed and washed the first eyes.

In this short historical review, the focus was on the complex chain system connected with:

- Sunlight and skylight as basic light sources in exterior and interior space influencing the environment quality
- Especially the human environment with outdoor and indoor daylight for any activity
- Human visual requirements, performance, and health
- Civilization and cultural heritage, including development of daylight research and building and technology progress
- Hygienic, energy-saving and economy requirements

This complex system can be historically summarized in three capital eras:

1. The evolvement of life on Earth in different flora and fauna species in a very slow adaptation process comprising conditions of water and air environments, solar radiation, and sunlight respecting the need of behavioral requirements to receive visual information from the living space. The successful trials and tragic errors often resulted in the survival or extinction of individuals or whole species under favorable or dramatically changing environmental situations. This era lasted roughly from 5,000 to 5 MYA (schematically shown in Fig. 2.1).
2. The origin and development of hominids and humankind encompasses different stages of the development of the abilities of hominids to improve quick movements (from quadrupeds in trees to bipeds on the ground from 4.4 TYA), the eye–brain–hand coordination in the favorable equatorial region, mastering hand tools (from 2.6 MTYA), food gathering, and hunting skills, their increasing reproduction, and migration to vast worldwide territories. This era lasted roughly from 5 MTYA to 5,000 years ago (schematically shown in Fig. 2.3).
3. The human civilization era contains the development of observations, research, and knowledge, i.e., learning and communication means for the utilization of natural resources and cultural progress. This era lasted roughly from 5,000 years ago to the present (schematically shown in Fig. 2.11).

In all these eras several aspects and trends can be shown, e.g.:

1. The basic dependence of life on primary solar energy and daylight influences causing the evolution of all beings, with adaptation to spatial and environmental conditions in different worldwide regions.
2. The million-year influence is deeply impressed in the DNA-coded body structure and sense genotypes, and in health, well-being, and activity requirements of all species, including humans.
3. The sun-path-defined daytime and seasonal changes are impressed in human circadian rhythms of everyday activity and relaxation, preparedness, creativity and performance efficiency, the measurement of time, orientation, and angular relations leading to descriptive geometry or trigonometry.
4. The respect for natural environmental resources have to be respected to enable all human activities and new inventions to achieve healthier and more comfortable improvements in environmental conditions during present and future life. So, for instance, the optimal utilization of sunlight and skylight in all building interiors has to be preferred because of their time, directional, and spectral variability stimulating and training the accommodation and adaptation qualities of human eyes and bodies. Nowadays, human eyes have to cope with informations in novel forms and on screens with projected or reproduced pictures or texts and many detailed visual tasks to identify shapes, color, or contrast differences and recognize familiar or virtual objects and scenes.
5. The era-long exposure to daylight influenced different animal species to acquire alternative actinic, adaptation, and accommodation abilities inherited in complex body and visual qualities and mechanisms especially in their visual systems. These human systems can be basically characterized by:

   - Stimulated actinism influencing photochemical changes (e.g., bactericidal effects purifying the human environment and stimulating vitamin D creation and circadian rhythms by photostimulation of pineal glands producing melatonin).
   - Visual adaptation, which encompasses the possible modification of the visual system to exposed light with overall different levels, contrast and spectral distribution, or angular relations.
   - Visual accommodation, which is an unconditional reflex of the sharp vision of the eye achieved by projecting the image of the object at any distance onto the fovea of the eye by adjusting the binocular dioptric range of both crystalline lenses to focus it on the retina or the yellow spot of the fovea.

Daylight complexity and everlasting changes in daylight climate regions are challenges for humans that also present opportunities for the science and photometry of daylight as humans face an exacting and ever-confounding task to uncover elegant solutions for sophisticated design aspirations to achieve energy savings and to nurture the health and well-being of humankind; in fact, to live in harmony with nature.

Appendix 2

Comparison of Historical Daylight Rules and Standards

Owing to old religious beliefs in ancient Egypt, where the sun was worshiped like the mightiest god of life and death, many burial places and pyramids were oriented to cardinal points with huge-gathering courtyards facing their southern side. Probably, long experiences with the conservation of buried bodies and mummies exposed to solar drying and favorable space magnetism supported the orientation of pyramids, applied for the first time in Djoser's step pyramid designed by its main architect Imhotep. The foundation of pyramids by pharaohs used already known simple methods of grave orientation, but the festive procession on chosen days indicates that equinox days were respected because of the best conditions (Kittler and Darula 2008a, b). Noble houses were also often oriented toward the south with front gates in a high wall with a palm-shaded front garden, whereas the water pool, stables, and vegetable garden were behind the residence on the north side. Time orientation after sundials, benefits of insolation, and sun shading were the original reasons to study the sun path with its daily and yearly changes. Sun-path and sunshade studies inevitably used a vertical stick, in ancient Greece called a "gnomon," and angular relations were expressed in fractions of the sides of a right-angle triangle. The Chaldean priest Bel-re'ushunu (mentioned by Vitruvius in 13 BC as Berossos) compiled the old knowledge of the sun-path diagram from Sumerian and Babylonian tablets, having at his disposal the archives of the church of Esagila in Babylon probably during 258–253 BC. The geometrical principle for drawing the sun-path diagram was called the analemma rule by Vitruvius and was proposed not only for sundials but also for designing the orientation of buildings and towns (Kittler and Darula 2004a, 2004b).

Atria used as workplaces in private houses in Mesopotamia, Greece, and Rome soon indicated some problems with the small patches of the sky owing to their design and obstructions by the court roof cover usually directing rainwater to the inner pool. With regard to safety reasons, the atria houses did not have larger apertures facing the open space or street outside. Some shops and provisional selling places in front of the houses as well as small windows on the first floor needed no daylighting rules or standards. However, in book VI of his *De Architectura*, Vitruvius (13 BC) mentioned the importance of daylight in interiors and considered the so-called no-sky line, i.e., the place from which not even a small patch or strip of the sky can be seen from the working place.

The first "rule of no-sky line" documenting the right to use skylight in any private adjacent houses was proclaimed in ancient Rome by Augustus (27 BC to AD 14) in a set of six easements called originally servitudes concerning the "easement of light" (Swarbrick 1933). The term "easement of light" meant the right of an owner or occupier of a building to prevent diminution of daylight within the interior by a new building on the neighboring site considering the situation acquired by long enjoyment. Later, these rules became Roman laws in the section *jus urbanorum praediorum* (i.e., applied to land as only limited rights), which also contained:

- *Jus luminus immitendi*, i.e., law to obtain skylight from common or neighboring land through a window in its direction.
- *Jus officiendi luminibus vicini*, i.e., law of reobtaining daylight by the neighbor to share the light with neighbors.
- *Jus altius tollendi*, i.e., right to design a tolerably higher building.
- *Jus altius non tollendi*, i.e., right to protest against and ban an intolerable height of the designed neighboring building.
- *Jus ne prospectui officiator*, i.e., right to prohibit the reduction of the view outside.
- *Jus ne luminibus officiator*, i.e., right to prohibit the reduction or shading of skylight falling on a window already used.

The relatively fair and impartial although disputable Roman right of daylight was valid probably to the fall of the Roman Empire and Justinian I during his reign in Byzantium (the Eastern Roman Empire) is famous for his codification of the Roman law known as *Corpus Iuris Civilis* in AD 533–534. These encyclopedic 62 books plus *Novellae* were studied in the Italian University of Bologna from the eleventh century and soon were considered as a universal basis especially for town laws in the Roman-German Imperial Law System for so called "Stadtrechte und Landrechte." In the medieval period, these laws were probably taken as "the time immemorial easements" and were respected also in England from the reign of Richard I (1189) as mentioned in the introduction to the Prescription Act 1832 with a regret that "the title to matters that have been long enjoyed is sometimes defeated by showing the commencement of such enjoyment, which is in many cases productive of inconvenience and injustice." The right to light as a form of easement in English law was most usually acquired officially under the Prescription Act of UK Statute Law (1832), where the claim to use skylight enjoyed for 20 years without interruption was declared to be deemed absolute and indefeasible. It is interesting to note the rise of law-court cases regarding lawsuits of easements of daylight summarized by Swarbrick (1930), where for the 1562–1928 period 572 cases that took place in English courts are mentioned. Almost 70% of those contradictory cases were processed during the nineteenth century when the building boom of high-rise construction was stimulated by investments in town centers.

Unfortunately no such evidence is available from sources in America, where the first skyscrapers were built at the end of the nineteenth century and caused a considerable lack of skylight in the lowest storeys of adjacent houses. In contrast, the Florida Appellate Court in 1959 stated that the "ancient lights" doctrine had been unanimously refused to be accepted in the USA, possibly to avoid financial compensation in the case of serious obstructions.

The main problem in complex urban space with irregular obstructions was the shape of the visible sky, which was quite irregular and therefore either approximate formulae or some graphical methods could not be used in the design process or tested by measuring methods in real situations. The two volumes of Swarbrick (1930, 1933) documented and summarized many methods that could be used, but exact criteria later were questionable or unsettled and still lacking precision (Swarbrick 1953). Kerr (1866) tried to apply imaginative diagrams in circular and square/cylindrical projection to prove the obstruction influences for the jury, in fact to substitute the previous tests using the simple method of critical obstruction angles documented only in sectional drawings. The latter method was used by the French Education Department from 1882, requiring in schools for each student position a sky patch at least 30 cm from the top of the window to be visible (Kerr 1914).

In continental Europe the rough eighteenth century ratio of 1:10 of the window to floor area as well as the oldest "no-sky line" rule was no longer acceptable in extremely obstructed quarters.

In Germany, Cohn (1884) suggested the design of schoolrooms with a window area at least one fifth of the floor area after visual task examinations and daylight measurements in schools. He also recommended 50 lx artificial illuminance, which "can give the same ease and speed of reading as 'good daylight,' but there can never be too much light in a school." Further discussions on daylight illumination (IES 1914) were also oriented to school design codes that gave a good summary of expert opinions from different European countries. It was interesting that the recommendations of the best criteria were those which specified the solid angle and its minimum at each desk defined as 50 reduced square degrees as already proposed in Germany in 1904. Previously, König (1897) in her classic work on the relation between illumination and visual acuity initiated experimental studies related to the demands of visual tasks on interior illuminance. A more sophisticated standardization of daylighting respecting actual levels was introduced after the First World War when critical illuminance levels and sky factors or DFs as code criteria were studied. In England, Taylor and Weston (1926), Taylor (1931), and Lythgoe (1932) followed with more detailed experiments. However, these studies were mainly oriented to artificial illumination in schools, offices, and factories and the findings were additionally applied in daylighting standards in relation to the new interests in applying the rights of skylight. It is interesting that the specialized barrister and expert in the right of light and restrictive covenants affecting freehold land Francis (2000) and in his recent article, Francis (2008) cited the English standard in England and Wales, which requires only the lowest sky factor of 0.2% on at least 50% of the work plane area at 0.85 m above floor level (the 50:50 rule). Furthermore, if the room was already badly lit, i.e., to less than 50%, then a reduction of 10% "will be regarded as actionable." The first article by Francis immediately provoked a discussion started by Pitts (2000), later followed by Chynoweth (2004, 2005, 2009). In fact Waldram (1909a, b) as a chartered surveyor and illuminating engineer, was the first to propose the sky factor as a criterion in ancient lights claims, and later Waldram and Waldram (1923) suggested a sky factor of 0.2% as a "grumble point," i.e., a legal threshold to determine whether a right to skylight had been infringed in a particular obstruction situation. Furthermore, they introduced their diagram as a graphical tool for the identification of the grumble point and line in a side-lit room with obstructed windows. Unity uniform sky was assumed to produce interior illuminance on horizontal surfaces under outdoor 500 foot-candles (5,380 lx). Graphically illustrative cylindrical diagrams were often used with droop-line overlays to depict the rectangular system of obstructions.

A different and more general approach to the daylight standardization was introduced in a German standard (DIN 1935), which assumed the critical exterior horizontal illuminance of 3,000 lx and the DF as the ratio of interior to exterior illuminance followed the prescribed levels in four categories of visual work:

1. Common rough tasks with minimum artificial illuminance $E_{min} = 40$ lx, i.e., corresponding minimum DF = 1.33%.
2. Medium to fine visual tasks with $E_{min} = 80$ lx, i.e., minimum DF = 2.66%.
3. Fine work with $E_{min} = 150$ lx, i.e., minimum DF = 5%.
4. Very fine work with small details with $E_{min} = 300$ lx, i.e., minimum DF = 10%.

Today, all these requirements would be considered by daylight experts as extremely high, but it is important to realize the historical context and assumptions. The outdoor 3,000-lx level was roughly proportional to the outdoor winter level under the uniform overcast sky at around 9 a.m. when the artificial lighting system should be switched off to save energy. After long-term measurements (Darula and Kittler 2004) under the uniform overcast sky with the prevailing $D_v/E_v = 0.22$ and the worst winter conditions have to be expected with the lowest declination $\delta = -23°$; thus:

- For the locations around the geographical latitude 48°N, e.g., on the line Freiburg–Munich–Vienna–Bratislava, at 9 a.m. an outdoor illuminance of 4,260 lx can be expected.
- At locations around 52°N, e.g., London–Hanover–Magdeburg–Berlin–Warsaw, only 2,723 lx could happen.
- Around 56°N, e.g., Glasgow–Edinburgh, the illuminance falls to only 1,176 lx.
- In Scandinavia and northern Russia (Oslo–Stockholm–Helsinki–St. Petersburg) at 9 a.m. in wintertime the sun is under the horizon and therefore the outdoor illuminance is close to zero.

So it seems that the 3,000-lx reference level was an approximate mean value for the German territory. However, after the adoption of the CIE overcast sky standard with the gradation 1:3 (CIE 1955) the exterior horizontal illuminance levels were reduced owing to typically lower D_v/E_v=0.1 under these skies, so at 9 a.m. under such densely overcast sky one could be expected:

- At around 48°N latitude, only 1,936 lx
- At around 52°N latitude, only 1,237 lx
- At around 56°N latitude, only 535 lx

That is, roughly less than half of the originally expected levels. Thus, either a longer morning and afternoon use of artificial lighting had to be supported or lower DF requirements had to be allowed. The latter was adopted in many standard requirements. For example, even earlier than the CIE overcast sky standard there was approved a preliminary proposal for the daylight standard for industrial buildings in Czechoslovakia by Hannauer (1947), who proposed an exterior illuminance of 4,000 lx as well as a standard average DF in four categories of visual work in the ranges over 0.5, 1.25–2.5, 2.5–3.75, and 3.75–7.5%. The actual standard approved in 1955 as ČSN (1956) required the absolute minimum of the DF (1:1) = 1.5, 2.5, 3.8, and 5%, with higher recommended levels 2.5, 3.8, 5, and 7% in four respective categories under the exterior 4,000 lx. The new revised German standard DIN (1957) applied the CIE overcast sky immediately with the strict general requirement of the minimum DF(1:3) value of 1%, with the only exception being for school classes, where at least a DF of 2% was required. All eastern European countries cooperating within the RVHP (Council for Mutual Economic Help) agreed on the exterior level of 5,000 lx under the CIE overcast sky in 1962–1963 (Matoušek 1964, 1965), applying six classes of visual activities with minimum DF(1:3) values in side-lit rooms of 0.25, 0.5, 1, 1.5, 2, and 3.5%, respectively. The conceptual idea was that whenever or wherever the outdoor illuminance dropped under 5,000 lx, the interior artificial lighting system should be turned on. It is evident that in any specific location under the densely overcast sky with 1:3 gradation and $D_v/E_v = 0.1$ the valid momentary solar altitude has to be

$$\gamma_s = \arcsin\left(\frac{5,000}{1,333.4}\right) = \arcsin(0.375) = 22^\circ,$$

which means that such worst outdoor illuminance could be expected during wintertime ($\delta = -23^\circ$) at least during noon in places with geographical latitude $\phi = 180^\circ - 90^\circ - \delta - 22^\circ = 45^\circ$, i.e., south of the line Bordeaux–Turin–Belgrade–Bucharest in Europe, whereas in central and northern Europe on such overcast whole days one has to depend on artificial lighting only.

Today, economic and profit tendencies of investors influence governmental standards to reduce the legal requirements for the inhabitants especially in built-up areas of cites regarding their right for sufficient insolation and daylight, which is essential for their well-being in a healthy and work-efficient human environment. In the European Union, after consultations with member states, mutual essential requirements and harmonized standards were sought (EN 2002). However, these standards should remain voluntary, and for daylighting no exact criteria are given except in the ISO (2002) standard: minimum $DF(1:3) = 1\%$ at 3-m distance from the window in side-lit rooms is required. Therefore, daylight criteria and standards have to be considered with regard to regional daylight climate and agreed with respect to the right for daylight and visual work requirements as well as optimal annual energy-saving policies in particular countries.

References

Adams, W.G. and Day R.E.: The action of light on selenium. Proc. Royal Society **25**, 113–117 (1876)

Anon: Vanished Civilisations. Reader's Digest Corp. Ltd., London (2002)

Barnard, G.P.: Portable photoelectric Daylight Factor meter. J. Sci. Instruments, **13**, 392–403 (1935)

Blondel, A.: Rapport sur les unités photométriques. Congrés International des Electriciens. Geneva (1896)

Bodmann, H.W.: Opening of the CIE Conference by the President. Proc. 22nd Session CIE, Vol.2, 3 (1991)

Bouguer, P.: Essai d'Optique sur la gradation de la lumiére. C. Jobert, Paris (1729, 1921)

Bouguer, P.: Traité d'Optique sur la gradation de la lumiére. Paris (1760), Latin translation by Richtenburg, J.: Bougeri Optice de Diversis Luminis Gradibus Dimetiendis. Vienna-Prague (1762), Russian translation by Tolstoy, N.A. and Feofilova, P.P.: Opticheskiy traktat o gradaciyi sveta, edited and commented by A.A.Gershun, Izd. Acad. Sci. USSR, Moscow (1950), English translation by Knowles Middleton, W. E.: Pierre Bouguer's optical treatise on the gradation of light. University of Toronto Press, Toronto (1961)

CIE – Commission Internationale de l'Éclairage: Principales décisions λ function, CIE Proc. 6th Session 1924 p.67–70 Cambridge University Press (1926)

CIE – Commission Internationale de l'Éclairage: Compte Rendu, p.67 or The basis of physical photometry. Publ. CIE 18.2 (1924, 1983)

CIE – Commission Internationale de l'Éclairage: Natural Daylight, Official recommendation. Compte Rendu CIE 13th Session, 2, part 3.2, 2–4 (1955)

Danilyuk, A.M.: Raschot yestestvennogo osveshcheniya ot pryamougolnych svetoproyomov. (Calculation of daylight from rectangular windows). Trudy II. Vsesoyuznoj svetotekhnicheskoy konf. 4, 168 (1931)

Danilyuk, A.M.: Ein neues graphisches Verfahren zur Berechnung der natürlichen Beleuchtung von Räumen mit recheckigen Lichtöffnungen. Das Licht, **4**, 241–243, **5**, 16–20 (1935)

Darula, S., Kittler, R., Gueymard C.: Reference luminous solar constant and solar luminance for illuminance calculations. Solar Energy, **79**, 559–565 (2005)

Dresler, A.: Über eine neuartige Filterkombination zur benauen Angleichung der spektralen Empfindlichkeit von Photozellen an die Augenempfidlichkeitskurve. Licht, **3**, 41–43 (1933)

Dufton, A.F.: Protractors for the computation of Daylight Factors. D.S.I.R. Build. Res. Techn. Paper 28. H.M.S.O., London (1946)

Fechner, G.T.: Elemente der Psychophysik II. Breitkopf-Hartel Verlag, Leipzig (1860)

Fechner, G.T.: Vorschule der Aesthetik.I.II. Breitkopf-Hartel Verlag, Leipzig (1876)

Fitts, C.E.: A new form of selenium cell. Amer.J. Science, **26**, 465–472 (1883)

Fraunhofer, J.: Neue Modifikation des Lichts durch gegenseitige Einwirkungen und Beugung der Strahlen, und Gezetze derselben. Paper to the Royal Bavarian Academy of Sciences, Munich (1821)
Gerinec, A.: Fylogenéza oka a zraku. (Phylogeny of the eye and sight. In Slovak), Kapos publ., Bratislava (2007)
Gueymard, C.: The sun´s total and spectral irradiance for solar energy applications and solar radiation models. Solar Energy, **76**, 4, 423–453 (2004)
Gore, R.: The dawn of humans: The first steps. National Geographic, **191**, 2, 72–99 (1997)
Gregory, R.L.: Eye and brain. World University Library, London (1966)
Hefner-Alteneck, F.: Die Hefnerkerze. Elektro-Tchnishe Zeitschr., **17**, 754–756 (1896)
Henderson, S.T.: Daylight and its spectrum. A. Hilger Ltd., London (1970)
Hopkinson R.G., Petherbridge P., Longmore J.: Daylighting. Heinemann Ltd, London (1966)
Hunter, W.A.: Roman law. London (1903)
Hyatt G.A. and Dowling J.E.: Retinoic acid: A key molecule for eye and photoreceptor development. Investigative Ophtalm., and Visual Science, **38**, 8, 1471–1475 (1997)
Kähler, K.: Flächenhelligkeit des Himmels und Beleuchtungsstärke in Räumen. Meteorol. Zeitschrift, **25**, 2, 52–57 (1908)
Kittler, R.: An historical review of methods and instrumentation for experimental daylight research by means of models and artificial skies. Compte Rendu CIE Session, B, 319–334. Bureau Central, Paris (1960)
Kittler, R. and Kittlerová, L.: Návrh a hodnotenie denného osvetlenia. (In Slovak. Design and assessment of daylight illumination). Slovak Publ. Techn. Lit., Alfa, Bratislava (1968, 1975)
Kittler, R. and Darula, S.: Analemma, the ancient sketch of fictitious sunpath geometry – Sun, time and history of mathematics. Architectural Science Review, **47**, 2, 141–144 (2004)
Kittler, R.: Daylight prediction and assessment: Theory and practice. Architectural Science Review, **50**, 2, 94–99 (2007)
Kittler, R. and Darula, S.: Applying solar geometry to understand the foundation rituals of 'Old Kingdom' Egyptian pyramids. Architectural Science Review, **51**, 4, 407–412 (2008a)
Kittler, R. and Darula S.: Historical developments in practical predicting of sunlight and skylight availability. Przeglad elektrotechniczny, **84**, 8, 15–18 (2008b)
König, A. and Dieterici C.: Die Grundempfindungen und ihre Intensitäts-Vertheilung im Spectrum. Sitz. Akad.Wissensch. Berlin, 805–829 (1886)
König, A.: Über den Hellikeitswert der Spektralfarben bei verschiedener absoluter Intensität. Beiträge zur Psychologie und Physiologie der Sinnesorgane, (1891)
König, A.: Gesammelte Abhandlungen zur Physiologischen Optik, J.A.Barth, Leipzig (1903)
Kravkov, S.V.: Glaz i yego rabota: Psichofiziologiya zreniya, gigiena osveshcheniya.. (In Rusian, The eye and its work: Psychophysiology of vision, hygiene of illumination). Publ. Acad. Sci. U.S.S.R,, Moscow (1950)
Lambert, J.H.: Photometria sive de mensura et gradibus luminis, colorum et umbrae. Augsburg (1760), German translation by Anding, E.: Lamberts Fotometrie. Klett Publ., Leipzig (1892), English translation and notes by DiLaura, D.L.: Photometry, or on the measure and gradation of light, colors and shade. Publ. IESNA, New York (2001)
Land, M.F. and Nilsson D.E.: Animal eyes. Oxford University Press, Oxford (2005)
Marie, F.: Nouvelle découverte sur la lumiére. Pour la mesurer et en compter les degrés. L. Sevestre, Paris (1700)
Maurolyco, F.: Photismi de lumine. Venezia (1575), English translation by Crew, H.: The Photismi de Lumine of Maurolycus. A Chapter in late medieval optics. Macmillan, New York (1940)
Müller Regiomontanus J.: De triangularis omnimodes libri quinque. Rome (1461)
Müller, J.: Zur vergleichenden Physiologie des Gesichtssinnes des Menschen und der Tiere. Nebst einem Versuch über die Bewegungen der Augen und über den menschlichen Blick. Cnobloch Verlag, Leipzig (1826)

Nardini, B.: Leonardo da Vinci. Centro Intern. del Libro, Florence (1974)
Neugebauer, O.: Mathematische Keilschrift-texte. I.-III. Springer, Berlin (1935)
Norman, D.: The illustrated encyclopedia of dinosaurs. Salamander Book Ltd., London (1985)
Paturi, F.R.: Chronik der Technik. Harenberg Verlag, Dortmund (1988)
Pouillet, C.S.: Compt. Rendu Acad. Sci., Paris **7**, 24–65 (1838)
Purkyně, J.E.: Zur Physiologie der Sinne. Prague (1825)
Rayleigh, J.W.: On the transmission of light through an atmosphere containing small particles in suspension and on the origin of the blue of the sky. Philos. Mag., **47**,5, 375–384 (1899)
Rossi, C.: Architecture and mathematics in ancient Egypt. Cambridge University Press, Cambridge (2007)
Sahelian, R.: Melatonin: Nature's sleeping pill. Be Happier Press (1995)
Schramm, W.: Über die Verteilung des Lichtes in der Atmosphäre. Schriften d. Natur Vereins f. Schl.-Holst., **12**, 1, 81–129 (1901)
Smith, W.: The action of light on selenium. J. Soc. Telegraphic Engineers, **2**, 31–33 (1873)
Stevens, S.S.: On the psychophysical law. Psychological Review, **64**, 3, 153–181 (1957)
Stevens, S.S.: To honor Fechner and repeal his law. Science, **133**, 3446, 80–83 (1961)
Stevens, J.C.and Stevens, S.S.: Brightness function: Effects of adaptation. Journ. Opt. Soc. Amer., **53**, 3, 374–380 (1963)
Uppenborn, F. and Monash, B.: Lehrbuch der Photometrie. Oldenburg Verlag, Munich (1912)
Vitruvius, M.P.: De architectura libri X. Rome (13 BC, 1487), English translation by Granger, F.: Of Architecture, The ten books. Heinemann, London (1970)
Waldram, P.J. and Waldram, J.M.: Window design and the measurement and predetermination of daylight illumination. Illum. Engineer, **16**, 4–5, 90–122 (1923)
Waldram, P.J.: A measuring diagram for daylight illumination. Batsford Ltd. London (1950)
Walsh, J.W.T.: The science of daylight. MacDonald, London (1961)
Watson J. and Crick F.: Molecular structure of nucleic acids, a structure for deoxyribose nucleic acid. Nature, **171,** 4354, 737–8 (1953)
Weber, L.: Intensitätsmessungen des diffusen Tageslicht. Annalen der Physik und Chemie, **26**, 11, 374 (1885)
Wiener, C.: Lehrbuch der darstellenden Geometrie. Taubner Verlag, Leipzig (1884)
Wolf, G. Friedrich Miescher: The man who discovered DNA. Chemical Heritage**, 21**, 10–11, 37–41 (2003)
Yarbus, A.L.: Rol dvizheniy glaz v procese zreniya. (In Rusian: The task of eye movements in the process of vision.) Publ. Nauka, Moscow (1965)
Yarbus, A. L.: Eye Movements and Vision. Plenum Press, N. Y. (1967)

References to Appendix

Cohn, H.: Tageslicht-Messungen in Schulen. Deutsche Medicin. Wochenschrift, **10**, 38, 609–611 (1884)
CIE – Commission Internationale de l'Éclairage: Natural Daylight, Official recommendation. Compte Rendu CIE 13th Session, 2, part 3.2, 2–4 (1955)
ČSN – Československá státní norma (Standard): Denní osvětlení prùmyslových budov. (In Czech. Daylighting in industrial buildings), ČSN 73 0511. Vyd. Úřadu pro normalisaci, Prague (1956)
Chynoweth, P.: Progressing the right to light debate – Part 1:.A review of current practice. Structural Survey, **22**, 3, 131–137 (2004)
Chynoweth, P.: Progressing the right to light debate – Part 2: the grumble point revisited. Structural Survey, **23**, 4, 251–264 (2005)

Chynoweth, P.: Progressing the right to light debate – Part 3: judicial attitudes to current practice. Structural Survey, **27**, 1, 7–19 (2009)
Darula S. and Kittler R.: New trends in daylight theory based on the new ISO/CIE Sky Standard 1. Zenith luminance on overcast skies. Build. Res. Journ., **52**, 3, 181–197 (2004)
Deutsche Beleuchtung Gesellschaft: Die Tagesbeleuchtung von Innenräumen. Berlin (1927)
Deutsche Industrie Normen (DIN): Leitsätze für Tagesbeleutung. DIN 5034, Dinorm, Berlin (1935, 1957)
EN – European standard: Light and lighting – Lighting of work places- Part1: Indoor work places. EN 12464–1:2002, (2002)
Francis, A.: Dealing with rights of light. Structural Survey, **18**, 5, 239–243 (2000)
Francis, A.: Right of lights ahead ! Journ. of Building Appraisals, **4**, 5–13 (2008)
Hannauer, K.: Normy přitozeného osvetlení prùmyslových budov. Předběžný návrh. (In Czech. Standards of natural lighting of industrial buildings. Preliminary draft.), Orbis, Prague (1947)
IES - Illuminating Engineering Society: Some queries on daylight illumination , The Illuminating Engineer, **1**, 27–43, 2, 96–97 (1914)
ISO – International Standardisation Organisation: Lighting of indoor work places. ISO 8995:2002 (2002)
Kerr, J.: Short history of investigations on the natural lighting of schools. The Illuminating. Engineer, **1**, 27–30 (1914)
Kerr, R.: Remarks on the evidence of architects concerning the obstruction of ancient lights, and on the practice of proof by measurement: with reference to recent cases in the Court of Equity. Transac. Institute of Roy. Inst. Brit. Architects, London (1866)
Kittler, R. and Darula, S.: Analemma, the ancient sketch of fictitious sunpath geometry – Sun, time and history of mathematics. Architectural Science Review, **47**, 2, 141–144 (2004a)
Kittler, R. and Darula, S.: Záhada analemy – historický základ využívania slnečného svetla. (In Slovak. The puzzle of Analemma – the historical basis of sunlight utilisation). Světlo, **7**, 2, 62–64 (2004b)
Kittler, R. and Darula, S.: Historická dôležitosť solárnej geometrie pre orientáciu v čase, priestore a v architecture. (In Slovak. Historical importance of the solar geometry for the orientation in time, space and in architecture). Architektúra a urbanizmus, **42**, 3–4, 159–165 (2008a)
Kittler, R., Darula, S.: Applying solar geometry to understand the foundation rituals of 'Old Kingdom' Egyptian pyramids. Architectural Science Review, **51**, 4, 407–412 (2008b)
König, A.: Die Abhängigkeit der Seehschärfe von der Beleuchtungsintensität. Sitzbericht d.k.p. Akad. Wissensch. Berlin (1897)
Lythgoe, R.J.: The measurement of visual acuity. Med. Res. Council Report 173, London (1932)
Matoušek, J.: Denní osvětlení – projektová norma RVHP. (In Czech. Daylighting – the design standard RVHP), Investiční výstavba, **2**, 7, 215–220 (1964)
Matoušek, J.: Projektová norma RVHP o denním osvětlení budov. (In Czech. The design standard of the Council of mutual economic help (RVHP) for daylighting of dwellings), Světelná technika, **5**, 1, 13–18 (1965)
Pitts, M.: The grumble point: is it still worth the candle ? Structural Survey, **18**, 5, 255–258 (2000)
Swarbrick, J.: Easements of light. Vol. I. Modern methods of computing compensation. Batsford Ltd., London, and Wykeham Press, Manchester (1930)
Swarbrick, J.: Easements of light. Vol.II. Optical and photographic methods of assessing depreciation. Batsford Ltd., London, and Wykeham Press, Manchester (1933)
Swarbrick, J.: Daylight, its nature, therapeutic properties, measurement and legal protection. Batsford Ltd. and Wykeham Press, London (1953)
Taylor, A.K.: The daylight illumination required in offices. Deptm. Sci. Industr. Research, Techn. Paper 12, HMSO, London (1931)
Taylor, A.K. and Weston H.C.: The relation between illumination and industrial efficiency. Deptm. Sci. Industr. Research and Med. Res. Council Report, London (1935)
UK Statute Law: Prescription Act 1832, Chapter 71, 3, Revised Statutes Law Database Prescription Act, c.71, http//:www.opsi.gov.uk/RevisedStatutes/Acts/ukpga/1832

Vitruvius, M.P.: De architectura libri X. Rome (13BC, 1487), English translation by Morgan, M. H.: The ten books on architecture. Harvard Univ. Press (1914) and by Granger, F.: Of Architecture, The ten books. Heinemann, London (1970)
Waldram, P.J.: The measurement of illumination, daylight and artificial: with special reference to ancient light disputes. Journ. Soc. of Architects, **3**, 131–140 (1909a)
Waldram, P.J.: A standard of daylight illumination of interiors. The Illuminating Engineer, **3**, 469–472 (1909b)
Waldram, P.J. and Waldram, J.M.: Window design and the measurement and predetermination of daylight illumination. The Illuminating Engineer, **16**, 90–122 (1923)
Weston, H.C.: Illumination levels for building interiors. Proc. Building Research Congress Div.3, Part III, 139–141 (1951)

Chapter 3
Daylight Photometry: History, Principles, and Empirical Development

3.1 The Interrelation of Radiant and Luminous Quantities, Terms, and Units Under Simple Assumptions

The earliest more systematic measuring processes in the first Sumerian and Egyptian civilization centers respected the prehistoric experiences with the length and orientation evaluation tasks. However, the primitive units of length, based on the foot, the number of footsteps, yards, etc., were supplemented by orientation to cardinal points. Such orientation, based on the study of the sun-shading originally conducted in equatorial Africa, was also interlinked with time measurements. In the case of solar light flux, the parallel beams indicated clear differences between insolated and shaded places. Therefore, the radiation and light flux in certain directions could also be clearly identified. And vertical stick's (gnomon) shadows enabled prediction of eastern and western directions or all four cardinal point orientations. The direction of the solar beam was also studied using gnomons to design sundials. This knowledge was summarized in the sexagesimal system established in Sumerian town kingdoms around 2800 BC and included 12 months, 2 times 12 h a day, 60 min, and 60 s as well as 90° and 360° in a circle.

Ever since the days when humans first realized the everlasting influence of the sun's radiation and light on their lives, the sun was worshiped as the mightiest god, with its energy and light benefits and sometimes also destructive powers that could be compared only with fire. All later artificial light sources, such as fire flame, torch, candle, and wick lamps, used until the twentieth century, were temperature sources with spectra radiating especially in the infrared and visible range, i.e., as later named Planckian radiators. However, these were considered as small local point sources with considerably lower temperature in comparison with that of the sun. Although the solar disk seen from ground level also seemed quite small, the parallel sunbeams indicated an extremely large distant radiating source with the utmost brightness. And it was obvious that the farther away the fire or torch, the more diminished was its illumination. Probably Maurolyco (1575) defined this effect by imagining the illumination spread of light from a point source in the form of a

R. Kittler et al., *Daylight Science and Daylighting Technology*,
DOI 10.1007/978-1-4419-8816-4_3, © Springer Science+Business Media, LLC 2012

pyramid with its base on the illuminated plane. This indicated the influence of the solid angle ω, i.e., the ratio of the illuminated normal area A_n to the square distance l^2 from the source expressed in steradians (sr):

$$\omega = \frac{A_n}{l^2} \text{ (sr)}. \quad (3.1)$$

It seemed that to measure illuminance from planar sources it was important to standardize the units for length, distance, and area. Since the Sumerian and Egyptian civilization very different measures had been used, based on local agreements on foot or footsteps, thumb or inch, fathom or yard, etc. So, the history of the international standardization of length units is rather long. During Bouguer's lifetime (1698–1758) in France the royal fathom called toise, i.e., 1.949 m, was used as the length unit established by the French kings Louis XIV and Louis XV and was defined by the half width of the gate of their residence the Louvre in Paris. This "disgraced" royal measure was unacceptable after the French Revolution. Therefore, in 1791 the French Academy of Science (established in 1666) started discussions about a new standard of length and appointed a committee under the chairmanship of P.S. Laplace to deal with the problem. Of course, the English fathom (equal to 6 ft, equivalent to 1.85 m) was also unacceptable. Therefore, in 1792 experimental measurements of the Paris meridian took place between Dunkirk and Barcelona (roughly around the 2°E meridian) and a subsequent calculation of its length from the globe pole to the equator gave ten million units. In 1799, a platinum bar of a standard meter was deposited in the Archives de la République in Paris, which was the first step in the practical definition of the basic meter standard of the present International System of Units (SI). In November 1800, Laplace's committee finally officially announced the definition of the meter as equal to the 10^{-7} part of the Paris meridian mentioned. The last Conférence Générale des Poids et Mesures (CGPM; i.e., General Conference on Weights and Measures) definition adopted in 1983 is based on the speed of light (exactly 299,792,458 m/s) and states: "The meter is the length of the path traveled by light in a vacuum during a time interval of 1/299,792,4588 of a second." Thus, a meter is related also to the astronomical unit of a light year, which is defined as the distance that light in a vacuum will travel in 1 year, i.e., equal to 9.4605×10^{15} m/year.

However, in the case of sun radiation, the density of the radiant and light flux penetrating a fictitious plane perpendicular to sunbeams was suggested as typical for rectilinear propagation of light, and so Kepler (1604) defined the normal illuminance E_{vn} as proportional to the point source intensity I_{vn} and diminishing with the square distance from the light source l,

$$E_{vn} = \frac{I_{vn}}{l^2} \text{ (lx)}, \quad (3.2)$$

now also called the photometric distance law, which is valid for point sources.

In the SI system of units standardized in 1979, the candela was chosen as the basic unit of luminous intensity I_{vn} and is equal to the radiant intensity of a monochromatic source of 1/683 W/sr for the wavelength of 555 nm.

Although Kepler's relation was meant to express the luminous intensity I_{vn} of stars as point sources depending on their distance from Earth or from each other l, it was derived from optical experiments with candles. In the still medieval atmosphere of the Prague castle, full of astrological prophets and peculiar alchemists, nobody could detect the more substantial photometric interrelations as

$$E_{vn} = \frac{\Phi_{vn}}{A_n} = L_{vn}\omega = \frac{I_{vn}}{l^2} \text{ (lx)}, \tag{3.3}$$

where Φ_{vn} is the normal luminous flux falling on an area A_n, in lumen (lm), L_{vn} is the normal luminance of the light source, and illuminance E_{vn} in SI units is in lux.

These interdependences demonstrate that the normal incident illuminance of any planar element is related to the normal luminance of the light source and its solid angle ω.

Since medieval times the candle had been accepted as a standard luminous intensity unit. In the eighteenth century Bouguer (1726) and Euler (1750) studied and tried to define the solar luminance as proportional to a candle flame. However, owing to the inexact flame luminance and incorrect atmospheric transmittance and the comparison made with solar luminance under the 31° solar altitude, their result was rather problematic, i.e., the sun was assumed to be roughly 175,000 times brighter than the candle (Gershun 1958). Furthermore, the extremely high temperature of the sun as well as the subjective sensation of brightness in relation to objective luminance was unknown and not measurable, except in their comparison of equal brightness utilized in subjective photometers. And luminance with its contrasts and distributions in space was considered the cause of the immediate visual stimulus that eyes analyze, seen in shapes, differences, and colors within a scene in the visual field. Luminance distribution or patterns can characterize either a luminous source or any interface, i.e., reflecting surfaces or transmitting materials.

Euler had already noted that the sun luminance cannot be simulated by any ground-based material but as the closest in radiance he suggested utilizing the radiation of melted metals. So in the last century, the temperature of solidifying elemental platinum at its freezing temperature of 2,045 K was chosen as a blackbody emitting a comparable light flux or luminous intensity to be taken as a standard. The molten platinum surface kept in a closed pot radiates vertically with the luminous intensity of 1 cd simultaneously emitting the luminous flux of 1 lm seen from a small circular opening which should have a 1-sr solid angle in the normal direction. Under such circumstances, the normal luminous flux Φ_{vn} in the unit solid angle is 1 lm and the luminous intensity I_{vn} is 1 cd; therefore,

$$I_{vn} = \frac{d\Phi_{vn}}{d\omega} \text{ (cd)}, \tag{3.4}$$

i.e., in SI units 1 cd =1 lm/sr.

Of course, Bouguer's candle was poorly reproducible: therefore, after the invention of carbon filament lamps and later incandescent bulbs and their use on photometric benches better luminous-intensity sources became available. Although the high platinum temperature of 2,045 K was difficult to maintain even in basic metrological laboratories, it could replace the uncertain flame and electric fiber sources. After Blondel's (1896) proposal and several international corrections, the so-called new candela (cd) unit was adopted.

The strict standardizing conditions today assume that the platinum emanating frequency of 540×10^{12} Hz corresponds to a wavelength of about 555 nm with a spectral radiant flux of 1/683 W/sr, or 1.46×10^{-3} W/sr, corresponding to a maximum luminous efficacy of 683 lm/W for 555 nm. Since 1979 the candela has been used in the International System of Units (SI) as the photometric base unit linking together photometric and radiometric quantities (Blewin and Steiner 1975; BIPM 1979). Thus, the candela 191 previously defined for natural white light is now standardized in the monochromatic 192 yellow–green part of the spectrum. However, many laboratory reference instruments and meters for the international comparison of photometric units and methods now use cryogenic absolute radiometers with reference radiometers at a few discrete wavelengths. Today reference photometers with high reflection precision trap detectors, or total luminous flux meters with 2.5-m integrating spheres and test reference lamps and reference lux meters are used on the photometric benches, etc. (Ohno 1996; Toivanen et al. 2000; Hovila et al. 2002).

Although the candela as a primary luminous unit had a long tradition, for several years the luminous flux with its lumen unit or its density ($\mathrm{lm/m^2}$) was also considered basic and more applicable especially in the case of sunlight. It is interesting that in Russia from 1948 a platinum source with a "pinhole" aperture of 0.5305 $\mathrm{mm^2}$ was standardized to measure 1-lm flux (Meshkov 1957) and assuming the acceptance angle of 1 sr, the luminous intensity was equal to 1 cd. The concept shown in Fig. 3.1 is instructive to show the interdependence of all luminous terms and units. The first assumption is that the horizontal platinum surface is parallel to the pot aperture plane as well as the above fictitious plane having just 1 $\mathrm{m^2}$, where the illuminance is 1-lux. Then the vertically directed flux as well as the luminous intensity in a unit solid angle is realized. Thus, a second assumption of a 1-sr solid angle ($\omega = 1$) has to be applied and this solid angle has to maintained by the aperture as well as the upper fictitious plane to link the scalar or spatial (i.e., flux and illuminance) units with the vector or direction (i.e., intensity and luminance). Thus all photometric unit are related and comparable to each other.

The solid angle ω is formed by a fictitious cone of the circular aperture with the space expanse angle projected on a sphere with a cone peak point in its center and diminishing with the square distance l from the light source. If the pinhole aperture area is $A_{\mathrm{np}} = 0.5305\ \mathrm{mm}^2$ and the circular area of the cone base is $A_{\mathrm{np}} = \pi r^2$, a unity solid angle defines the distance of the platinum surface from the aperture.

$$\text{If } \omega = 1 = A_{\mathrm{np}}/l^2, \text{ then } l = \sqrt{0.5305} = 0.72835\ \mathrm{mm},$$

$$r = \sqrt{0.5305/\pi} = 0.41093\ \mathrm{mm} \text{ and } \omega = \pi r^2/l^2 = A_{\mathrm{np}}/l^2 = 1\ \mathrm{sr}.$$

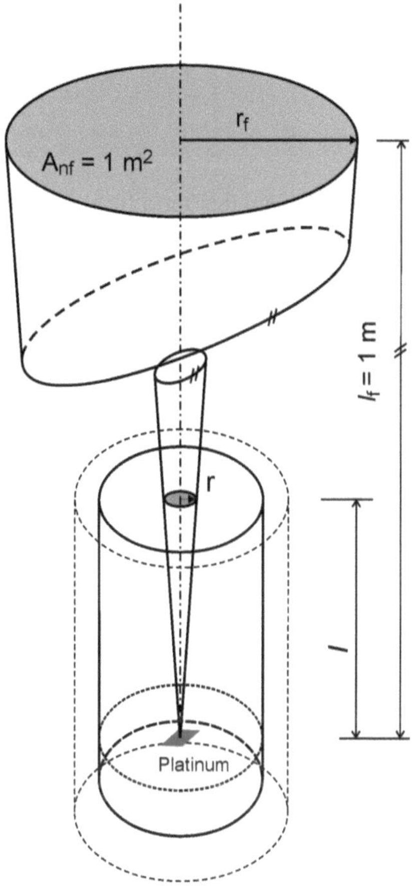

Fig. 3.1 The platinum source of photometric units

For the fictitious horizontal area illuminated from the platinum surface $A_{\mathrm{nf}} = 1\ \mathrm{m}^2$, the same solid angle has to be valid, i.e., $\omega = 1 = A_{\mathrm{nf}}/l_{\mathrm{f}}^2 = 1$; thus $l_{\mathrm{f}} = 1$ m, and $r_{\mathrm{f}} = \sqrt{1/\pi} = 0.5642$ m.

It has to be noted that:

- When the sphere radius $r = 1$, then the solid angle of the hemisphere is $\omega = 2\pi = 6.283185$ sr.
- For the whole sphere or all-around space, $\omega = 4\pi = 12.56637$ sr.
- If the solid angle is 1 sr, its cone angle is $\arctan(r/l) = 29°26'$.

In the case of a unity solid angle in both the pinhole aperture and the fictitious plane, formula (3.4) becomes

$$I_{\mathrm{vn}} = \Phi_{\mathrm{vn}}(\mathrm{cd}), \tag{3.5}$$

where Φ_{vn} is the luminous flux emitted within a solid angle by a uniform source element with a luminous intensity of 1 cd.

At the same time the luminous flux density on the upper fictitious circular plane A_{nf}, which is just 1 m^2, is equal to its normal illuminance E_{vn} as

$$E_{vn} = \frac{\Phi_{vn}}{A_n} \text{ (lx)}. \tag{3.6}$$

Note that another term for illuminance is the luminous exitance M_{vn}, defined as the luminous flux leaving a surface element per unit area, i.e., also in lumens per square meter.

When the plane A_n has an absolutely diffuse and perfectly white surface, its luminance is

$$L_{vn} = \frac{\Phi_{vn}}{A_n \omega} = \frac{I_{vn}}{A_n} = \frac{E_{vn}}{\omega} \text{ (cd/m}^2\text{)}. \tag{3.7}$$

Furthermore, if the solid angles are kept at unity value, then (3.7) is further simplified when the surface area $A_n = 1\ m^2$,

$$L_{vn} = \frac{\Phi_{vn}}{A_n} = \frac{I_{vn}}{A_n} = E_{vn} \text{ (cd/m}^2\text{)}, \tag{3.8}$$

which means that in SI units 1 cd/m^2 = 1 lm/(m^2 sr) = 1 lx/sr, and if two circular, square, or rectangular planes of unit area are placed normally to each other forming a unit solid angle, their luminance and illuminance are the same and directly proportional to their intensity and flux per unit area, respectively. Of course, such absolutely extreme simplifying conditions can be realized only in unit-calibrating facilities and when the unit system is standardized. In other words, to fully achieve the interrelationship of luminous terms, i.e., in the SI scheme based on the Meter Convention, the prerequisite is to accept the meter as a basic length unit which defines the area (m^2) as well as both angular radian and steradian units. Angular dimensions are also linked by trigonometric functions with length units, and the solid angle after (3.1) ($\omega = A_n/l^2$) is also expressed in area and length units; thus, the meter and solid angle play a substantial role in illuminance evaluations (see Appendix 3).

Of course, the interdependence of the luminous and radiant quantities has to be noted too; thus,

$$E_v = 683 \int_{\lambda=380}^{780} E_{e\lambda}\, V(\lambda)\, \mathrm{d}(\lambda) \text{ (lx)}. \tag{3.9}$$

Similar interrelations are valid for the luminous intensity and flux on respective radiant intensity and flux which can be used when the spectral distributions are available (e.g., as now in the case of spectral solar radiation).

The radiometry–photometry link is mainly dependent on the human spectral sensitivity now standardized by the $V(\lambda)$ function (CIE 1924, 1983) and on the

Planckian blackbody radiators at very high temperatures, which caused many experimental difficulties.

Because the realization and management of the primary luminous standard facility utilizing very hot platinum is rather difficult and because of the platinum surface the measurement of units has to be oriented vertically, usually secondary standard incandescent bulbs are calibrated for photometric bench measurements. Such incandescent lamps have a special small vertical wire element shining under a prescribed current voltage producing at 2,000–2,050 K the calibrating luminous flux or intensity. The advantage of the photometric bench is the ease of using horizontally directed light flux. Thus, in a typical dark (black) laboratory on a 4- or 6-m-long bench such a calibrated source is mounted in the exact normal position to the photometer sensor, with the adjustable distance or screen set at a defined solid angle enabling a check or recalibration of a particular instrument (e.g., the lux meter, luminance meter, or photocell). The advantage of a set of incandescent electric lamps is that one reaches the sufficiently high luminous intensities needed for daylight measurements, e.g., a 230-V and 500-W projector lamp can be calibrated for the range of approximately 650–950 cd, followed by a 1,000-W lamp in the range 1,250–1,950 cd, both in the vertical positions with an arrow mark indicating the side to face the photometer screen and placed over the bench zero meter mark. Thus, a perfect normal alignment of the standard source and calibrated cell as well as the distance of the calibrated photocell can be adjusted with the photometer bench distance measuring gauge.

In daylight theory, (3.8) is often used to define the sun and sky luminance, whereas illuminating engineers usually determine the luminance of artificial sources using the old, Bouguer definition as "the density of candles shining from the source surface or plane" after (3.3) ($L_{vn} = I_{vn}/A_n$). Therefore, if the designers of artificial lighting are asked to determine sky luminance, some embarrassment follows because the question is: How and where are there any candles in the sky and where is 1 m^2 on the sky or sun? So, the sun luminance can be calculated only from the luminous solar constant (LSC) when the solid angle of the sun is known.

3.2 Solar Constants and Extraterrestrial Luminous Parameters

In the history of humankind the sun has had an extraordinary priority of importance, as Leonardo da Vinci (1451–1519) expressed: "I do not know in the whole universe a bigger and mightier object, its light irradiates all celestial solids which are scattered in the whole universe." Its significance is especially appreciated:

- As the source of energy, radiation, and sunlight
- As the stimulator of life, growth, and evolution
- As the reference point of movement in the solar system
- As the basis of time measurements

- As the cause of dynamic day–night changes as well as periodic variations in yearly seasons
- As the principal element influencing climatic processes at the particular locations on Earth

The extraterrestrial solar radiation and sunlight are very difficult to define exactly because the source is so distant its parallel beams have to traverse the unknown universe and specific atmospheric layers with ever-changing content, qualities, transmittance, and scattering properties. Furthermore, the continual movement of the position of the sun with respect to any location complicates any predictions. Of course, many general influential facts are now known:

- The sun has a spherical shape with radius $r_s = 6.95508 \times 10^8$ m.
- Its distance from Earth changes year round owing to Earth's orbit within 147×10^6–152×10^6 km, with the mean distance expressed as the astronomical unit $AU = 1.4959787 \times 10^{11}$ m.
- Its virtual solid angle from ground level on Earth, influenced by the elliptic orbit of the sun, is also changeable: $\omega = 6.7905244 \times 10^{-5}$ sr $\pm$ 3.3%.
- The correlated temperature of the solar disk is in the range 5,500–6,300 K, which determines the solar spectrum as well as its luminance distribution (Kittler and Pulpitlova 1988; Darula et al. 2005).

Owing to the very long penetration of sunbeams through the quasi-vacuum of the universe, the so-called extraterrestrial density of the solar flux is determined at the outer border of the atmosphere and is expressed as a standard solar constant (SC) valid worldwide and generally present on the side of the globe exposed to sunshine. Only recently were measurements of the solar extraterrestrial spectral irradiance $E_{eo\lambda}$ (W/m^2 nm) as well as the SC in the total energy solar spectrum analyzed and published (Gueymard 2004). These made possible the derivation of the luminous solar constant (LSC). Both constants are determined for the mean Sun–Earth distance on the April 3 and October 5. The SC value of 1,366.1 W/m^2 is recommended. The spectral irradiance is shown in Fig. 3.2 together with the $V(\lambda)$ curve. With use of the integration and the four-point Lagrange interpolation scheme, the value of the LSC was calculated (Darula et al. 2005) after (3.9):

$$\mathrm{LSC} = 683 \int_{\lambda=380}^{780} E_{eo\lambda}\, V(\lambda)\, \mathrm{d}(\lambda)\ (\mathrm{lx}), \tag{3.10}$$

where the value of 683 lm/W is the maximal luminous efficacy of radiation in the 555–556-nm spectral range.

With use of $V(\lambda)$ and the modified $V_M(\lambda)$ function after CIE (1990), two LSC values were derived, i.e., 133,334 and 134,108 lx, respectively. Therefore, for engineering purposes an approximate average LSC = 133,800 lx was recommended.

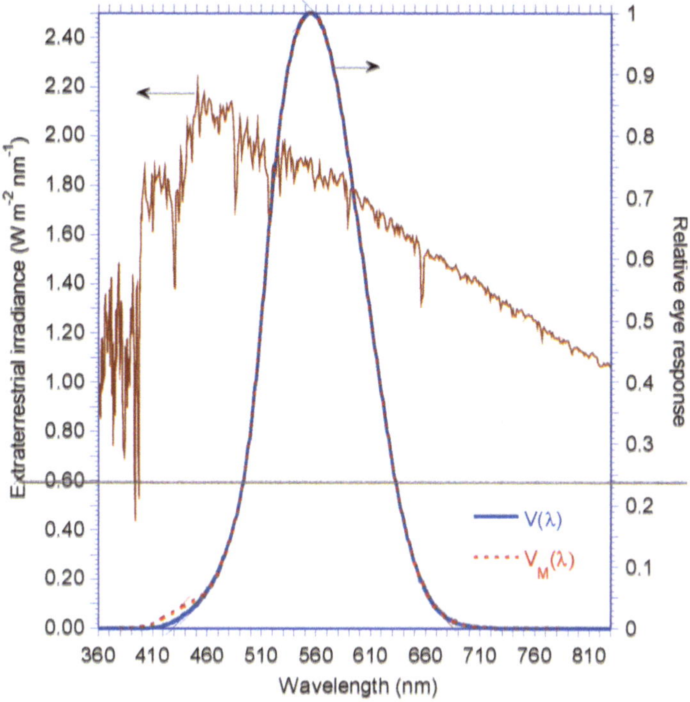

Fig. 3.2 Extraterrestrial solar irradiance in the visible spectrum comparable to International Commission on Illumination (CIE) standard photopic sensitivity $V(\lambda)$

On the basis of the LSC value, several other solar radiation and sunlight characteristics in the radiant–luminous twin system of solar terms, units, and parameters are determined (Kittler 1989). The parameters intensity, flux, radiance, and luminance are summarized in Tables 1 and 2 in the article by Darula et al. (2005) and thus are available for engineering and practical use.

In glare studies of the luminance of the solar disk, an extremely high luminance, as high as around 2×10^6 kcd/m^2, can be expected. In accordance with the new LSC value, the average solar extraterrestrial luminance is $L_{vs} = 1.9635 \times 10^9$ cd/m^2.

The LSC value is valid everywhere and is the same only during the day when the Earth–Sun distance is equal to the astronomical unit and if the parallel sunbeams are falling perpendicularly on the extraterrestrial fictitious plane, i.e., normal to the plane. In reality, however, this assumption is rarely met, so the LSC is usually applied as the daily corrected value E_{vo} due to the ellipticity of Earth's orbit $\in$ (Kittler and Pulpitlova 1988; Tregenza and Sharples 1993), e.g.,

$$E_{vo} = \text{LSC} \in = \text{LSC}\left[1 + 0.034\cos\frac{2\pi(J-2)}{365}\right] \text{(lx)}, \qquad (3.11)$$

where J is the day number (e.g., $J = 1$ on 1 January, $J = 365$ on 31 December).

The locally effective illuminance on the border of the local atmosphere is also influenced by geographical location and time during the day within that year. Therefore, the locally available sunlight is characterized best by the horizontal extraterrestrial illuminance E_v, i.e.,

$$E_v = E_{vo} \sin \gamma_s = E_{vo} \cos Z_s \text{ (lx)}, \tag{3.12}$$

where γ_s is the momentary sun elevation angle, called the solar altitude (the solar altitude is taken from the horizon) and Z_s, the zenith solar angle, is the angular distance of the sun from the zenith.

Owing to changes in the position of the sun during the year and day, the illuminance level E_v on the extraterrestrial fictitious horizontal plane represents the momentary maximum available sunlight in a particular locality in lux or kilolux at a certain time. In other words, it expresses the fact that a few hours after sunrise or before sunset there is locally less sunlight above the atmosphere than at noon and, at the same time, there are much higher illuminance levels in equatorial locations than in places far from the tropics. In addition, the solar altitude is also influenced by seasonal differences in the solar declination, as its well-known formula documents (see in Kittler and Mikler 1986 or Tregenza and Sharples 1993).

3.3 Momentary Sun Positions, Their Daily and Yearly Changes

Probably since hominids lived in the equatorial region, it has been known that the sun shadow thrown by a vertically placed stick indicates the angular sun position on the ground. The permanent changes of daytime and nighttime with the sunrise on the eastern side and sunset on the western side of the horizon indicate time as well as orientation. During the migration period, humans realized that their new settlements in the northern or southern territories were exposed to different sun paths on a sky vault and studied the consequences as well as the reason. The starting point was found in a special day, when the duration of the day and the duration of the night were exactly equal. This day was called the equinox day, which occurs twice a year everywhere, but with a different slope of the equinox sun path. In old Sumer and Egypt, owing to their capitals being located close to the geographical latitude of 30°N, this slope was traditionally in such a relation that a stick (later called gnomon) 5 units high casts at noon a 3-unit-long shadow (Vitruvius 1487; Kittler and Darula 2004, 2008). The oldest knowledge mentioned by Vitruvius in his manuscript finalized in 13 BC was probably found on the clay cuneiform tablets in 258–253 BC when the Chaldenian priest Berossos studied them in the old archives of the Esagila church in Babylon. In Roman times equinox ratios of gnomon to shadow units for Athens (4:3), Rome (9:8), Tarentum (11:9), and Rhodes (7:5) were determined.

Only after the development of trigonometric functions instead of these ratios could general formulae for the solar altitude and azimuth angular functions be developed.

The solar position is given by either the solar altitude taken from the horizon γ_s or the zenith solar angle Z_s as the angular distance from the zenith, whereas the orientation angle taken from north on the horizon is the solar azimuth. These angular solar coordinates are usually calculated after the following formulae either in degree form:

$$\sin \gamma_s = \cos Z_s = \sin \varphi \sin \delta - \cos \varphi \cos \delta \cos(15°H), \qquad (3.13a)$$

or in radiance form:

$$\sin \gamma_s = \cos Z_s = \sin \varphi \sin \delta - \cos \varphi \cos \delta \cos(\pi H/12), \qquad (3.13b)$$

where φ is the geographical latitude of the particular locality (positive north of equator and negative for the southern hemisphere) in degrees or radians, δ is the solar declination in degrees or radians, and H is the hour number in true solar time (TST) in the range 0–24.

The solar declination angle is the angle between the sun's rays and Earth's equatorial plane and can be more or less approximated using several formulae (see different formulae in Kittler and Mikler 1986 or Tregenza and Sharples 1993), e.g., one of the simpler is

$$\delta = 23.45 \sin\left[\frac{2\pi}{365(J-81)}\right] \text{(rad)}. \qquad (3.14)$$

The solar azimuth angle is also expressed via different hour angles from north or south, usually by two formulae one for morning and another for afternoon TST hours (Tregenza and Sharples 1993), e.g., taken from true north (Kittler and Mikler 1986):

- For $H \leq 12$,

$$A_s = \arccos[\cos \delta(\cos \varphi \tan \delta - \sin \varphi \cos(15°H))/\cos \gamma_s] \,(°). \qquad (3.15a)$$

- For $H > 12$,

$$A_s = 360° - \arccos[\cos \delta(\cos \varphi \tan \delta - \sin \varphi \cos(15°H))/\cos \gamma_s](°). \qquad (3.15b)$$

Of course, at TST the solar azimuth is 180° from the north cardinal point or 0° from the south. Today, computer algorithms are available to estimate the solar position (Reda and Afshin 2004) defining the solar zenith and azimuth angles with a high precision of ±0.0003°.

Computer calculations of sun positions can favor more precise algorithms as, e.g., those suggested by Angus and Muneer (1993), and recently it was shown by Parkin (2010) that the solar coordinates can be calculated advantageously also using a vector approach.

3.4 Propagation of Parallel Sunbeams Through the Atmosphere

Bouguer's (1726, 1760) basic contributions to daylight photometry can be summarized as follows:

- Respecting the spherical shape of the globe, he assumed the atmospheric envelope to be of the same shape, with a thickness of roughly 3,911 or 4,197 toises (equivalent to 7,623–8,180 m) in the zenith direction.
- According to the real solar altitude γ_s, he expected in the homogeneous zenith region the relative optical air mass to equal 1, which should increase roughly with the function $1/\sin\gamma_s$ to about 45°, but further the optical air mass was uncertain owing to the curvature of the atmosphere nearer the horizon. Thus, applying Marie's lucimeter, he used a creative method to measure the relative optical air mass during a clear night when the moon luminance could be matched with the feeble candle comparative source of the lucimeter. Furthermore, he chose the moon's altitude to simulate the summer solar altitude in his home town, Le Croisic, near Saint-Nazaire, in the range 0–62°. His tabulated results show a perfect correlation with current formulae for the dependence of relative air mass m on γ_s. With better instrumentation, roughly the same measurements of the relative air mass m were done and the values obtained were tabulated by Bemporad (1904) and afterward were simulated in formula form by Makhotkin (1960) under the assumption of a homogeneous atmospheric depth layer of 10,200 m. Today, a more precise formula by Kasten and

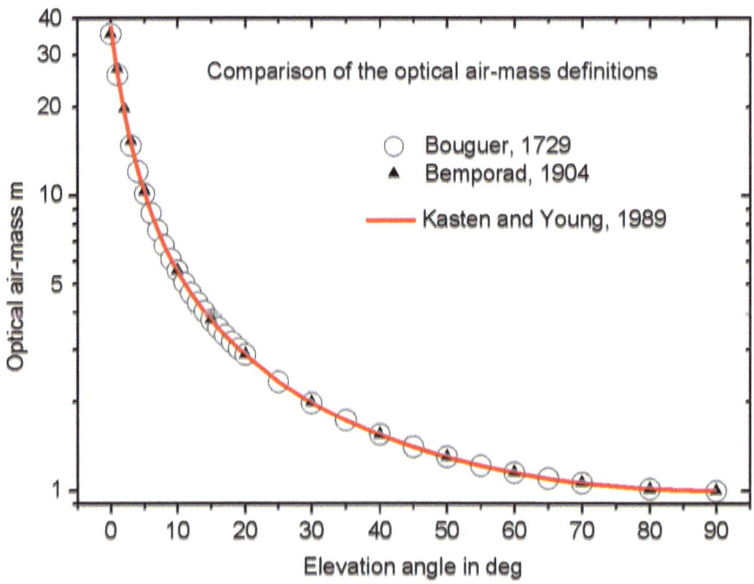

Fig. 3.3 Comparison of old Bouguer's data with those in Bemporade's tables and the current Kasten–Young formula

Young (1989) is applied in computer calculations. As evident from Fig. 3.3 all relative air mass values are very close to each other.

- Measuring the attenuation properties of air and water vapor layers when parallel sunbeams are transmitted through the atmosphere, Bouguer assumed that at ground level the extraterrestrial light flux is reduced by the optical air mass m and the total attenuation A. So he measured at sea level the ratio of fluxes at 66° sun height to that at 19° and found their ratio to be 3:2. Lambert (1760) mentioned these measurements and deduced $A = 0.089073$, which was according to his own measurements in Chur, Switzerland, very small as his A value was 0.229148. DiLaura as Lambert's translator (1760) commented that "Lambert was seriously misled when he used the results of the thermal experiment referenced here for luminous purposes ... a modern value of A is 0.0944, close to Bouguer's." However, in radiation calculations if $m = 1$, the extinction coefficient a_r is in the range 0.097–0.0995 (Kittler and Mikler 1986, p. 66), whereas for the visible spectrum the extinction coefficient of the absolutely clear and clean atmosphere is a_v=0.0957 (Clear 1982). If an ideal clear and clean atmosphere with turbidity $T_v = 1$ had been present during Bouguer's experiments, then his A should be approximately equal to a_v, but during Lambert's measurements the turbidity was probably higher. Thus, in modern calculations $A \ln 10 = 2.3026A$ can be replaced by $a_v T_v$, which suggests that in Bouguer's case T_v was roughly 2.14, whereas in Lambert's case, as is usual in Alpine mountainous countryside T_L at summertime noon was considerably higher.

Later it was suggested that the exponential function could be used instead of a logarithmic form to express the ratio of transmitted to incoming flux, which can be written in as

$$\frac{\Phi_{vg}}{\Phi_{vext}} = \exp(-\,a_v m T_v), \tag{3.16}$$

where Φ_{vg} is the luminous flux at ground level, Φ_{vext} is the simultaneous extraterrestrial luminous flux, a_v is the extinction coefficient of the absolutely clear and clean atmosphere, m is the relative optical air mass and is dependent on the solar altitude, and T_v is the luminous turbidity factor in the direction of the sunbeams.

Note that a_v, m, and T_v will be defined and explained later.

- Using several glass filters, Bouguer also tried to measure the relative luminance distribution on clear skies, but unfortunately he did not succeed in incorporating the results in his manuscript and nobody was able to include them in the posthumous book (Bouguer 1760).

At the same time as Bouguer's book was published in Paris, Lambert (1760) defined some basic principles of theoretical photometry, including those connected with sunlight and skylight:

- He introduced the simplifying assumption that light sources (e.g., the solar disk, the sky, or illuminated diffuse surfaces) have the same luminance within their extent.

- If further generalization and simplification is followed in mathematically defined calculations, then a uniform and unity luminance can be assumed constant all over the source (e.g., the sky vault), i.e., $L_v = 1$.
- Generally, the illuminance of any surface element is dependent on the luminance of the light source and its solid angle projected onto the illuminated plane.
- If the sky hemisphere can simulate the sky vault and its luminance is uniform and equal to unity and it has a radius also equal to unity, then the Lambertian "absolute" horizontal illuminance from an unobstructed sky outdoors within the half-space solid angle, i.e., 2π sr, produces illuminance on the horizontal plane equal to its projection on the hemisphere, i.e., 2π sr $= 2 \times 3.1415926$ sr.
- Planar light sources (e.g., the sky, windows, or other apertures) can be substituted, added, or subtracted only with relation to their solid angles, which is the rule of reciprocity and substitution.

Today, the extraterrestrial normal solar flux density as the LSC is known (Darula et al. 2005) after Bouguer's formula (3.16) can be applied in terms of illuminance values:

$$P_v = E_v \exp\left(-a_v\, m\, T_v\right)\,(\mathrm{lx}), \tag{3.17}$$

where E_v is the horizontal extraterrestrial illuminance after (3.12), a_v is the luminous extinction coefficient of the absolutely clear and clean atmosphere, which was defined by Clear (1982) and later published in Navvab et al. (1984),

$$a_v = \frac{1}{10.1 + 0.045\,m} \text{ or } a_v = \frac{1}{9.9 + 0.043\,m}, \tag{3.18}$$

m is the relative optical air mass and depends on the solar altitude, which was approximated by several authors (see in Kittler and Mikler 1986), but the Kasten and Young (1989) formula is recommended for solar altitude in degrees (in Fig. 3.3),

$$m = \frac{1}{\sin \gamma_s + 0.50572\,(\gamma_s + 6.07995^\circ)^{-1.6364}}, \tag{3.19}$$

and T_v is the luminous turbidity factor in the direction of the sunbeams and is determined as

$$T_v = \frac{\ln E_{vo} - \ln P_{vn}}{-a_v\, m}. \tag{3.20}$$

Originally, the turbidity factor in the whole solar spectrum was introduced in meteorology by Linke and Boda (1922), sometimes called the Linke turbidity factor T_L in the whole sun radiation spectrum. It is defined as a multiplying factor or ratio determining how many times more the measured normal sunbeam irradiance P_{en} is attenuated with respect to its simultaneous extraterrestrial value SC than would be the case for a clean and clear atmosphere with an extinction a_r. The luminous turbidity factor in the direction of sunbeams T_v also expresses the number of ideal clear atmospheres to be taken into account to reproduce the real measured turbidity or solar beam attenuation. In (3.20) is applied the normal direct sun illuminance $P_{vn} = P_v/\sin(\gamma_s)$.

It is evident that in reality not only horizontal planes exist; illuminance on sloped or vertical variously oriented planes also has to be calculated. All buildings have vertical walls, tilted or flat roofs, and various apertures within them such as windows, top lights, and hollow light guides. These are usually stable, permanently oriented, and cannot be moved with the changing sun position or the sky luminance patterns. Lambert realized that from the normal maximum directional effect of sunlight this effect vanishes on any sloped plane until the sunbeams are blocked altogether when the incidence angle is over 90° in any direction. The cosine function range from 1 to 0 simulates this effect exactly.

From the old gnomon tradition, the sun angular elevation from the horizon was taken as a basic sun position measure, now called solar altitude γ_s, so the sunbeam illuminance on an arbitrarily sloped and oriented plane $P_{v\beta}$ is reduced by the effect of the cosine value of the incidence angle and is expressed as

$$P_{v\beta} = P_{vn}\, cos\, i = \frac{P_v \cos i}{\sin \gamma_s}\ (\mathrm{lx}), \tag{3.21}$$

where P_{vn} is the normal illuminance when the sunbeams are reaching the illuminated plane in the perpendicular direction, i.e., in the direction of the plane normal, and thus

$$P_{vn} = \mathrm{LSC} \in \exp(-a_v\, m\, T_v)\ (\mathrm{lx}), \tag{3.22}$$

and P_v is the horizontal sunbeam illuminance:

$$P_v = P_{vn} \sin \gamma_s\ (\mathrm{lx}). \tag{3.23}$$

In the case of a sloped plane, cos i, i.e., the cosine of the incidence angle, is determined from the horizon as

$$\cos i = \cos \beta \sin \gamma_s + \sin \beta \cos \gamma_s \cos|A_s - A_n|, \tag{3.24}$$

where β, the slope angle of the illuminated plane taken from the horizon, is arbitrary, but is simplified for the horizontal plane position with any orientation (as $\cos \beta = 1$ and $\sin \beta = 0$) to $\cos i = \sin \gamma_s$, whereas in the case of vertical placement of the illuminated plane its orientation is taken into account as $\cos \beta = 0$ and $\sin \beta = 1$; thus, $\cos i = \cos \gamma_s \cos |A_s - A_n|$, i.e., the difference of the azimuth angles of the illuminated plane normal A_n from the sun meridian azimuth A_s with its absolute value has to be considered.

The space incidence angle i between the sunbeam direction and the direction of the normal of the sloped plane is generally also taken after spherical trigonometry, i.e., expressed by angles taken from the zenith; then,

$$\cos\, i = \cos\, Z_s \,\cos\, Z + \sin Z_s \,\sin Z \,\cos A_z, \tag{3.25}$$

where Z_s is the angular distance of the sun from the zenith and $\cos Z_s = \sin\gamma_s$, Z is the zenith angle of the illuminated plane normal, and A_z is the absolute difference between the solar azimuth and the azimuth of the normal plane.

This general formula is simplified again in the case of:

- Horizontal surfaces as $\cos Z = 1$ and $\sin Z = 0$; thus. $\cos\, i = \cos Z_s = \sin\gamma_s$.
- Vertical planes as $\cos Z = 0$ and $\sin Z = 1$; thus, $\cos\, i = \cos Z\, \cos A_Z$, which is simplified when the plane is oriented directly to the south and thus at noon TST is $\cos A_z = 1$.
- With sloped planes with a small tilt it has to be noted that in spite of $\cos A_z = 0$, $\cos\, i = \cos Z_s\, \cos Z$, thus a certain value indicates that some sunlight can reach this sloped plane.

3.5 Historical Basis of Daylight Photometry

Although some experiments with the rectilinear propagation of light were known at the beginning of the eighteenth century, the more concise experimentally and theoretically based mathematical expressions specifying and enabling daylight calculations or predictions were worked out by Bouguer and Lambert. Although Bouguer was primarily a good and thoughtful experimenter who was clever to use the first subjective luminance meter, Lambert concentrated more on fundamental principles to formulate them into a mathematical system. Both felt that the eye is a specific instrument of still unknown complex qualities and that light measurements depend on them. However, they assumed and expected that eyes should estimate quite precisely the equality of brightness or luminance between two adjacent objects or surfaces which can be seen simultaneously with the same pupil adjustment. In other words, the first photometry principle in comparing two different light qualities (intensity or luminance) was to vary one or both so that they would become equal. Their equality has to be controlled and judged by eyes only. The recognition of the visual stimuli or sensation automatically respected the spectral, i.e., $V(\lambda)$, function, the perfect contrast and sharpness sensitivity of human eyes, and the adaptation and accommodation dependence of visual appraisal.

Bouguer's contributions to daylight photometry can be summarized as follows:

- Respecting the spherical shape of the globe, he expected the atmosphere envelope to be of the same homogeneous thickness all over the globe, but owing to atmospheric pressure falling with height, regression to the vacuum state has to be expected.

- He assumed that the attenuation of sunbeams in the atmosphere under different solar altitude γ_s is dependent on the relative optical air mass, which decreases from the zenith ($m = 1$) first with the relative function $m = 1/\sin\gamma_s$, but suspected that closer to the horizon owing to curvature of the atmosphere this decrease is less rapid.
- Using Marie's lucimeter, he measured the relative optical mass of the atmosphere during the night of November 23, 1725, utilizing the brightness of the full moon during the clear night, which for him simulated the solar summer path, but in the brightness range that matched his candle-lighted meter (Bouguer 1726).
- He applied the same instrument with glass filters to measure sunbeam attenuation and the relative luminance pattern of the clear sky.
- He measured light attenuation properties of air and water layers simulating parallel beam losses in the atmosphere with the aim to determine their transparency by the determination of the ratio between the luminous flux density after the layer penetration and that of the incoming flux density; thus, for normal sunlight penetrating through the whole atmosphere the horizontal illuminance produced by parallel sunbeams that can be expected on the ground is P_{vn},

$$\frac{P_{vn}}{LSC} = \exp(-\tau), \tag{3.26}$$

and for any solar altitude the sunlight illumination on a horizontal plane is P_v,

$$\frac{P_v}{E_v} = \exp(-\tau m), \tag{3.27}$$

where $E_v = E_{vo} \sin\gamma_s$ is the extraterrestrial horizontal illuminance, τ is the total atmospheric extinction including pollution, i.e., transmittance of the total atmosphere for the sunbeam from the zenith, and m is the relative optical air mass.

Bouguer inherited much geometry and practical knowledge as well as his interest and use of measuring tools from his father, who taught navigation in the port town of Le Croisic near Saint-Nazaire, France. Besides his very precise definition of the relative air mass (Fig. 3.3), he studied the attenuation of sunbeams in the atmosphere by placing several glass plates on the lucimeter head and also found the probable transparency under the clear sky conditions, now expressed as the luminous extinction coefficient for an absolutely clean and clear sky a_v, roughly equal to 0.0900–0.0957. It seems that Bouguer's clear atmosphere of the assumed thickness of 7,469 toises had a very precise value of attenuation of $A = 0.089$.

Bouguer (1760) did not succeed in finalizing his manuscript before his death, so unfortunately his sky luminance pattern measurements are not completely documented. At the same time as Bouguer's book was posthumously published in Paris, his younger colleague Lambert already had in press his two books (Lambert 1759, 1760). The first book summarized his studies in geometry and perspective projection that helped him to imagine the importance of the sky image in the form of a hemisphere as well as the solid angle and its projections. In contrast to

Bouguer's interest in measurements and clear skies, Lambert was impressed by the winter foggy and overcast skies so frequent in Europe. These seemed to be very uniform, evenly homogenous and dull, thus producing minimal interior illuminance levels – dependent on window size. Nevertheless, Lambert's assumption of an absolutely unit and uniform sky luminance pattern $L_{\chi Z} = L_{vz} = 1$, i.e., currently 1 cd/m^2 with the sun absent, proved a valuable simplification especially in the mathematical expression, when in the general formula of interior horizontal illuminance E_{vi} is then only left the influence of the solid angle:

$$E_{vi} = L_{\chi Z} \omega_s \cos i = \omega_{ph} \, (\mathrm{lx}), \tag{3.28}$$

where the solid angle ω_s is defined as a surface area on a unit radius sphere (see the appendix in this chapter) which has to be projected onto the illuminated plane according to its normal incidence angle i. The projection after spherical trigonometry is defined by a formula similar to (3.25) but instead the angular distance of the sun from the zenith, the sky element has to be used (e.g., after (8.2)). For horizontal planes a simplification is expressed by $\cos i = \sin \varepsilon$,

$$\omega_{ph} = \omega_s \sin \varepsilon \, (\mathrm{rad}). \tag{3.29}$$

where ε is elevation angle of the sky element.

Thus, under the unobstructed sky hemisphere with unity radius and with unity luminance, the horizontal skylight illuminance from the whole sky vault is $E_h = D_v = \pi$, i.e., it is the projected area of the hemisphere onto the horizontal plane. However, this fictitious diffuse illuminance is very seldom used because in reality many conditions do not follow the historical photometric simplification.

3.6 Exterior Daylight Conditions Based on Regular Measurements at Ground Level

At ground level the global horizontal irradiance G_e, including sun and sky components, is the easiest measurable quantity which is regularly recorded by the meteorological network of worldwide observatories. Since the International Daylight Measurement Programme (IDMP) also global horizontal illuminance G_v is regularly measured, usually in 1-min steps during daytime at the general stations (IEA 1994, http://idmp.entpe.fr). Because the global sensors are stationary and unshaded from the sun position, these can occasionally record both sun and sky effects influenced by moving clouds from the whole unobstructed sky vault. Their stable fixed position means they need only periodic cleaning, recorder checks, and evaluation of daily illuminance recordings. The latter task will provide the overall illuminance level changes under momentary turbidity and cloudiness conditions. To specify these influences, it is necessary to measure simultaneously either the

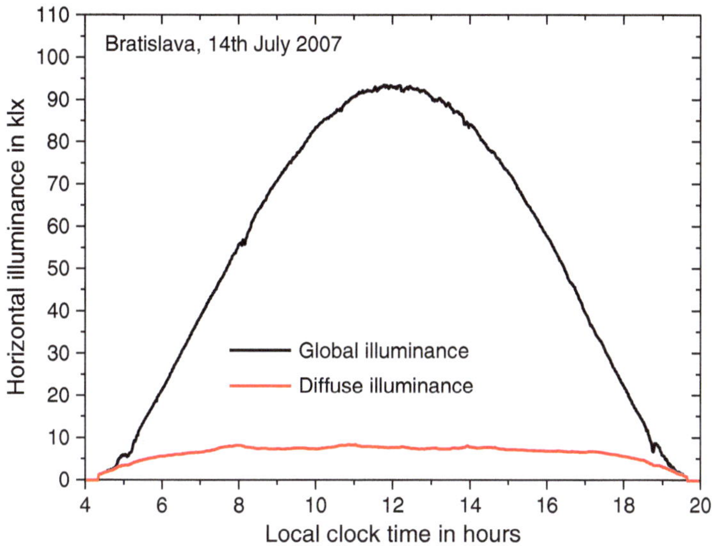

Fig. 3.4 Daily variations in global and diffuse illuminance during a typical clear day after 1-min measurements

diffuse skylight illuminance D_v (with a sun-shading disk or ring) or the parallel sunbeam illuminance P_v (using a sun tracker).

As the global horizontal illuminance has only these two components, either the skylight D_v or the parallel sunbeam P_v component can be calculated from the two measured levels, i.e.,

$$D_v = G_v - P_v \quad \text{or} \quad P_v = G_v - D_v \text{ (lx)} \tag{3.30}$$

or in ratio form normalized by extraterrestrial horizontal illuminance E_v

$$\frac{D_v}{E_v} = \frac{G_v}{E_v} - \frac{P_v}{E_v} \quad \text{and} \quad \frac{P_v}{E_v} = \frac{G_v}{E_v} - \frac{D_v}{E_v}. \tag{3.31}$$

All these quantities except E_v are measured at ground level on a horizontal plane and therefore include local atmospheric attenuation including light absorption and diffusion within the atmosphere.

At the IDMP general stations, normal recording of horizontal illuminances D_v and G_v and irradiance D_e and G_e levels, respectively, and sometimes measurements of simultaneous zenith luminance L_{vZ} in 1-min steps during daytime are made.

The ratios G_v/E_v, D_v/E_v, and P_v/E_v represent the atmospheric transmittance of global, diffuse, and beam illuminance in the real local instantaneous case and indicate also the presence of sunbeams and turbidity and cloudiness conditions.

The frequent representation of these recordings is shown in the graphical daily course in Figs. 3.4 and 3.5.

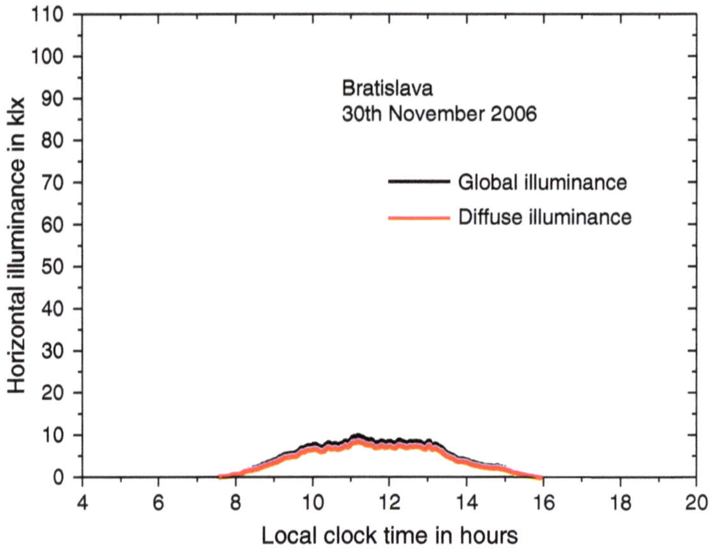

Fig. 3.5 Daily variations in global and diffuse illuminance during a typical overcast day after 1-min measurements

If homogeneity of the atmospheric layer is assumed, e.g., under clear and uniformly overcast skies, then two extremes can occur:

1. The ideal clear sky is characterized by the prevailing P_v or P_v/E_v levels and the penetration of sunbeams is detectable only when normal P_{vn} or horizontal $P_v = P_{vn} \sin\gamma_s$ are measured or can be calculated after (3.23), because only then can the turbidity in the direction of the sunbeam be estimated after

$$\frac{P_v}{E_v} = \frac{P_{vn}}{E_{vo}} = \exp\left(-a_v m T_v\right), \tag{3.32}$$

where the only unknown is the luminous turbidity factor T_v, thus

$$T_v = \frac{-\ln P_v/E_v}{a_v\, m}. \tag{3.33}$$

It is evident that the minute changes of solar altitude γ_s, relative optical air mass m, and luminous extinction a_v have to be calculated, because all measurements are registered in local clock time (LCT), whereas all necessary parameters have to take into account TST. Therefore, to evaluate regular measurements it is economical to apply a computer program for the recalculation of LCT to TST because of the differences due to the time zone, summer time shifts, as well as the equation of time expressing the eccentricity of Earth's orbit. So TST in H hours is

$$\mathrm{TST} = \mathrm{LCT} + \left(\frac{\lambda_z{}^\circ - \lambda_s{}^\circ}{15^\circ}\right) + \mathrm{ET} - \mathrm{ST}\,(\mathrm{h}), \tag{3.34}$$

where $\lambda_z°$ is the geographical longitude of the time zone in degrees, $\lambda_s°$ is the geographical longitude of the site (positive west of Greenwich) in degrees, ET is the equation of time in hours, and ST is summer time shift usually 1 h.

The equation of time depends also on the leap year cycle and thus different approximations are available (see in Kittler and Mikler 1986), but the one by Pierpoint (1982) was recommended and is accurate to within 40 s:

$$\mathrm{ET} = 0.17 \sin\left[\frac{4\pi\,(J-80)}{273}\right] - \sin\left[\frac{2\pi\,(J-8)}{355}\right] \ (\mathrm{h}). \tag{3.35}$$

It is only in the tropics at equinox noontime, when the sun is directly at the zenith and the atmosphere is absolutely clean without any pollution, that $m = 1$ and $T_v = 1$; thus $P_{vn}/E_{vo} = \exp(-a_v\, m\, T_v) = \exp(-a_v)$, which is approximately $\exp(-0.1) = 0.9048$, i.e., the maximum possible transmittance of parallel sunbeams through the Rayleigh atmosphere. It is interesting to note that the normal transmittance of standard window glass is roughly the same and therefore Bouguer was quite right to use glass pieces, up to 80, as neutral filters to measure sunbeam penetration.

2. In the other extreme of a densely overcast sky without any sunlight is P_v or $P_v/E_v = 0$ then $D_v = G_v$ usually in the range 0.05–0.40, most often 0.1 during the dull November and December days (Darula and Kittler 2004a, b).

Only a few IDMP research stations can afford the expensive sky scanner needed to measure the luminance distribution on the sky vault. Today, scanning luminance meters are used to record sky luminance in small solid angles in their acceptance angle (usually around 10°) gradually and quickly rotating in azimuth and altitude angles under the control of a microcomputer. Although several probes were tried with smaller acceptance angles to record small areas of sky variation and fluctuations, it is important to cover the whole sky vault by sky patches of identical shape and area. The scanning routine should be symmetrical about the zenith to suit the solar meridian and should contain a high proportion of the total sky vault with a simple control of the probe movements. Usually all designs of sky scanners follow the horizontal bands parallel to the horizon with a step rotation of 10–12°. Tregenza (1987) and Lynes (1988) proposed a scanning pattern of 145 or 151 circular patches with the angular width of zones between 12.47° or 1.13° respectively roughly covering 70% of the hemisphere, whereas Kelton and Murdoch (1986) suggested 137 sky patches to achieve a 4-min recording time for the whole scan. Later a sky-scanning luminance meter with fiber-optic detectors by Hayman (1989) followed the idea of reducing the complexity of the drive mechanism with possible improvement in accuracy and a reduction in cost.

All luminance meters as well as the sky scanners are calibrated in accordance with their acceptance angle, i.e., the solid angle in which they measure the luminance of any source or surface. Modern objective luminance meters have smaller acceptance angles (e.g., 1°) or the possibility to switch to several acceptance angles but sky scanners usually follow with the acceptance angle 10–12° the sky vault division by circular patches.

Regularly collected measurement data at any local IDMP station have to be quality-controlled, stored, and evaluated to characterize the local daylight climate (Kittler et al. 1992; Darula and Kittler 2005). It was evident from the start of regular daylight measurements that every day there are quite different situations owing to cloudiness and turbidity changes and including the sunrise beginning of daytime and its end with sunset. For the analysis of regular data it is therefore necessary to reproduce the daily courses in the monthly sequences to find typical days with quasi-homogeneous conditions or periods characterizing typical sky patterns influenced by different turbidities and cloudiness changes.

3.7 Evaluation of the Exterior Daylight Measurements

The most frequently used and simplest evaluation method is the representation of data recorded during a particular or typical day in the graphical courses of G_v and D_v as shown in Figs. 3.4 and 3.5 with dependence on the clock time of the measurements (Dumortier et al. 1995) or the momentary solar altitude. However, when a review of everyday data is documented graphically, a better classification of stable and homogeneous or cloudy and dynamic half days or whole days can be done. E.g. daily global and diffuse courses measured in 1-minute steps during a month are shown by Darula and Kittler (2011).

A more sophisticated method tries to identify daily, weekly, or monthly courses in G_v/E_v and D_v/E_v representation with respect to the changes of solar altitude either during randomly chosen days or under typical overcast or clear conditions (Fig. 3.6) or the combination of both usually in approximately half days as documented in Fig. 3.7. Quite often the weather changes occur during nighttime, but sometimes also noontime is critical probably owing to the evaporation of water resulting in partly cloudy or dynamic changes. These are documented in Fig. 3.7 when flowing cloud patches are causing frequent shading of direct sunlight. Clear days include quite high sunlight illuminance levels G_v/E_v and P_v/E_v which determine also momentary turbidity conditions when T_v is calculated after (3.33). For the same day as in Fig. 3.7, an example of the turbidity changes is shown in Fig. 3.8.

The different turbidity during clear sky days is often homogeneous and roughly follows that measured in the direction of sunbeams; thus, T_v is an important additional and complementary indicator of the proportion of the P_v/E_v and D_v/E_v ratio interrelation participating in the global transmittance of the momentary atmospheric state, i.e., on the total illuminance level G_v/E_v. In other words, the higher the turbidity factor T_v, the greater the reduction of sunlight, but owing to light scattering within the atmosphere the D_v/E_v ratio is elevated.

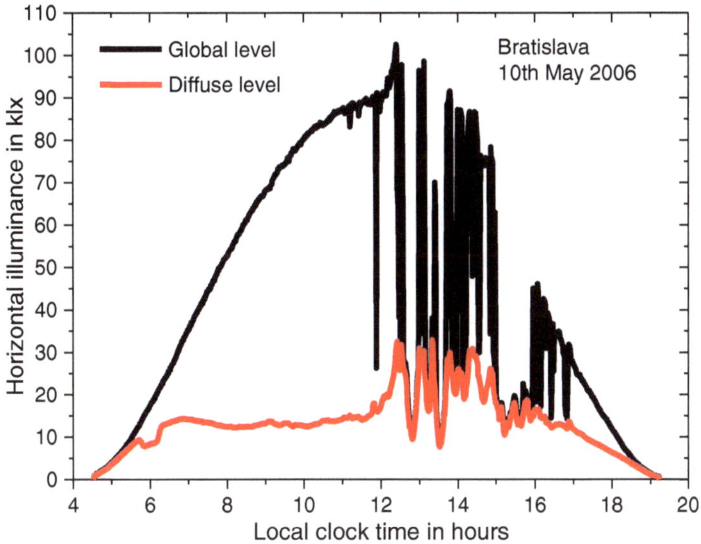

Fig. 3.6 The morning clear period is changed by in-moving clouds to the afternoon dynamic situation

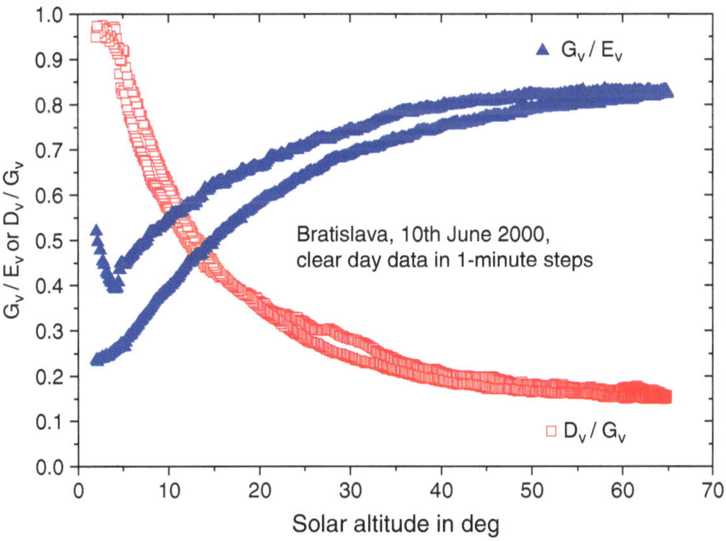

Fig. 3.7 Global and diffuse ratios during a clear day

The monthly distribution of exterior illuminance is sometimes characterized in triangular diagrams with all P_v/E_v, G_v/E_v, and D_v/E_v parameters, e.g., in Figs. 3.9 and 3.10. Such PGD diagrams also help to identify the seasonally characteristic features, e.g., either the prevailing sunless overcast days or the frequency of occurrence of sunny weather (Kittler 1995).

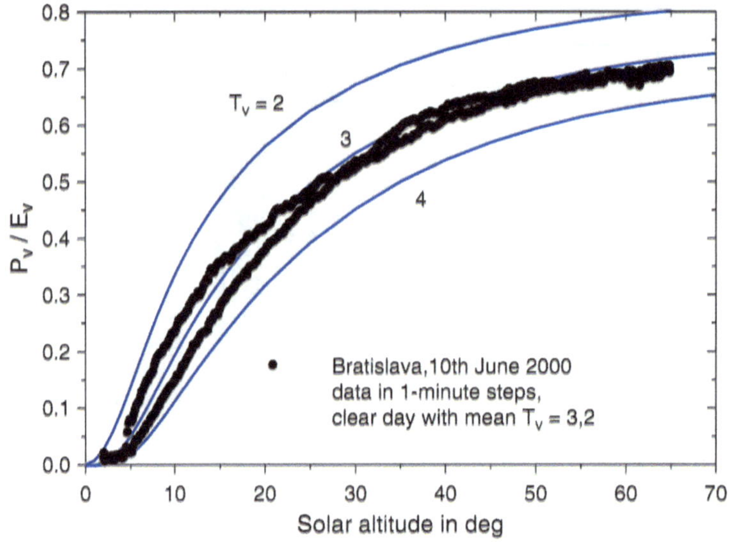

Fig. 3.8 Dependence of turbidity changes during a clear day on the sunbeam index

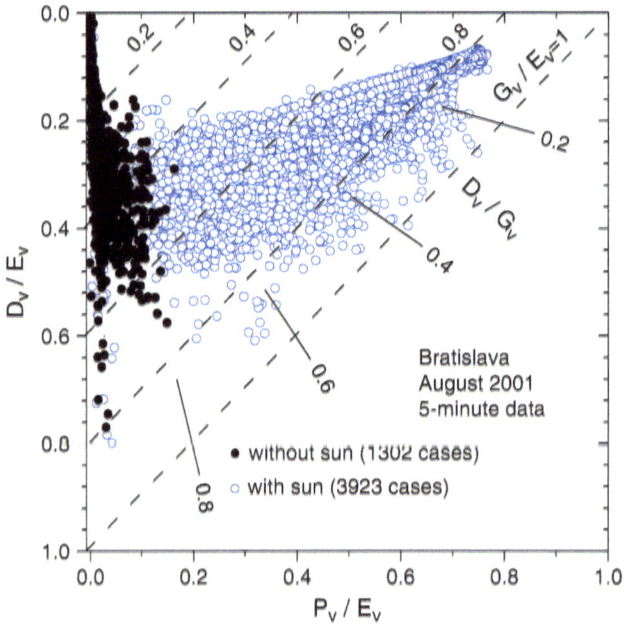

Fig. 3.9 Triangular PGD diagram with ratios characterizing clear sky conditions

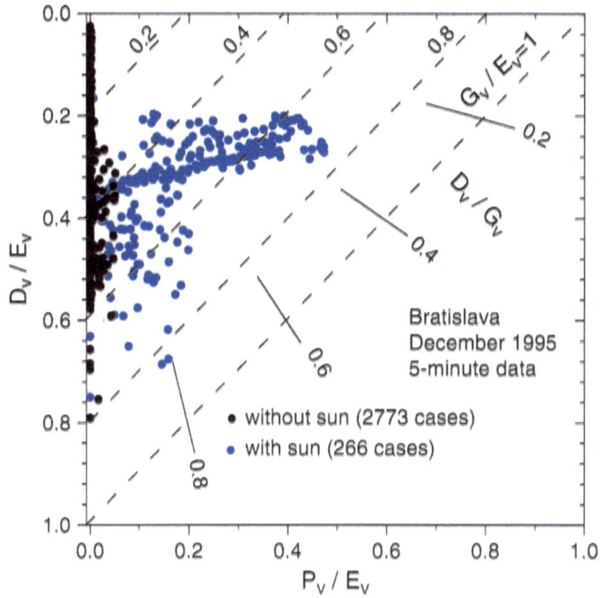

Fig. 3.10 Triangular PGD diagram with ratios characterizing monthly quasi-overcast sky conditions

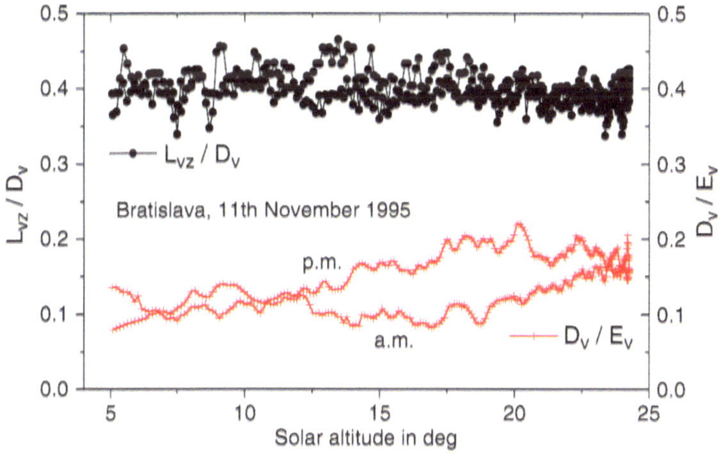

Fig. 3.11 A typical densely overcast day characterized by L_{vz}/D_v and D_v/E_v ratios using instantaneous 1-min measurements

The most interesting method is the evaluation of instantaneously measured zenith luminance and diffuse illuminance data. Their ratio under overcast sky conditions is very stable during the day as documented in Fig. 3.11, with the tendency to stay approximately around the 0.4 mean and simultaneously the D_v/E_v parameter is quite

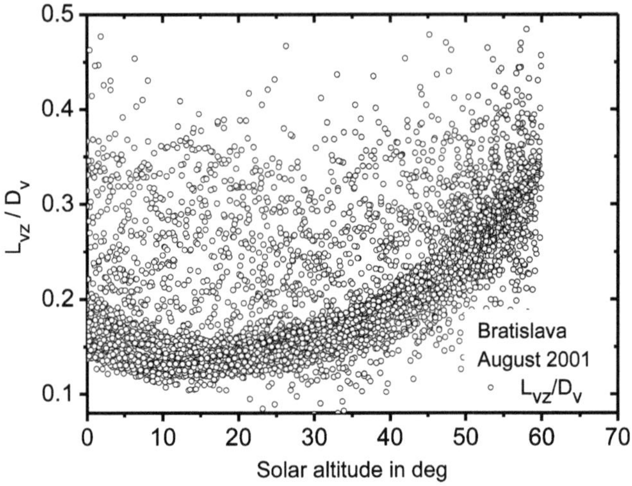

Fig. 3.12 Monthly L_{vz}/D_v ratios evaluated from 5-min averages

low, roughly from 0.1 to 0.2. In contrast, during clear days the ratios of L_{vz}/D_v change rapidly with rising solar altitude. From the monthly data in Fig. 3.12, the 5-min averages were used to avoid the densely covered area of points. Thus, the prevailing clear sky conditions during summer and also overcast and cloudy periods are represented.

So it seems that when simultaneous zenith luminance is measured, then momentary sky type character can be identified by either L_{vz}/D_v or D_v/L_{vz} parameters and compared with the ISO/International Commission on Illumination (CIE) standard general sky under the same momentary solar altitude. Both these ratios roughly classify the sky type, but it has to be realized that such ratios are quite precise only under close-to-ideal homogeneous conditions, i.e.:

- For the CIE overcast sky with gradation 1:3, L_{vz}/D_v is derived from the integration of the sky luminance pattern close to the value of 0.4, i.e., $D_v/L_{vz} = 2.5$, with absolutely no dependence on the solar altitude.
- For the uniform sky standard with no gradation at all, L_{vz}/D_v is equal to $1/\pi = 0.3183$, i.e., $D_v/L_{vz} = \pi = 3.1415$, also without any dependence on the solar altitude (Darula and Kittler 2004a, b), but the virtual Lambertian unity luminance sky is an extraordinary case because when $L_{vz} = 1$, $D_v = \pi = 3.1415$. In reality such a sky with a uniform luminance of 1 cd/m^2 is very seldom in the presunrise moments or under dusk conditions, but the level of 1 kcd/m^2 with $D_v = 3.1415$ klx in dense fog could happen during daytime.
- Partially cloudy skies are usually nonhomogeneous and thus their L_{vz}/D_v or D_v/L_{vz} ratios are spread in a wide range between overcast and clear skies.
- Homogeneous clear skies or those with different turbidity content which are usually quasi-homogeneous can be characterized by a fluent rise of L_{vz}/D_v ratios with increasing solar altitude as evident in Fig. 3.12. For clear skies, a fluent

change of the L_{vz}/D_v ratio from 0.14 to 0.50 at 60° solar altitude was found (Darula and Kittler 2005). Similar L_{vz}/D_v or D_v/L_{vz} ratio courses can be found for all clear sky ISO/CIE (2003, 2004) standards.

Of course in real measured cases the atmospheric homogeneity is often distorted by various irregularities caused by different turbidity, cloudiness type, cloud cover, and cloud placement on the sky vault. Therefore, when evaluating such relations over a longer period, e.g., 1 month, one finds the spread of L_{vz}/D_v ratios is quite irregular although certain prevailing sky types can be identified as in Fig. 3.12, where the densely covered area of points indicates the dominant prevailing clear sky conditions during summer.

The IDMP research stations also own sky scanners and thus can regularly measure sky luminance distributions, which enable luminance patterns or solids to be drawn (Kittler et al. 1992, 1998). Furthermore, from luminance sky scans also scattering indicatrix and gradation functions can be evaluated, as will be explained in Chap. 5.

Long-term regular outdoor measurements enable analysis of the daylight availability not only in terms of illuminance levels but also in terms of the frequency of occurrence of sky types in different months, seasons, or years. Such information is needed to better define the local daylight climate and its long-term occurrence probability.

3.8 Luminous Efficacy Measured and Modeled

The luminous efficacy is defined as the ratio of luminous flux in lumens emitted by a light source over all visible wavelengths to the total radiant flux in watts. On the border of the atmosphere, the maximum sun radiation can be measured. The sun, with its surface temperature around 6,000 K, radiates like a blackbody in the continuous spectrum, with the peak output in the visible part of the spectrum (Fig. 3.2). The luminous efficacy of the extraterrestrial sunlight is equal to the ratio of the solar constants, i.e., LSC/SC = 133,334/1,366.1 = 98.1685 lm/W (Darula et al. 2005).

When the sun radiation penetrates the atmosphere, it is absorbed and scattered owing to molecules, water vapor, various gases, aerosols and particles. When illuminance and irradiance meters are precisely calibrated, simultaneously measured levels at Earth's surface can be compared and calculated as the global, diffuse, and direct luminous efficacies of radiation in lux per watt per square meter (equivalent to lumens per watt) and thus these can be estimated for any instance and sky condition, i.e., $\mathrm{Eff}_g = G_v/G_e$, $\mathrm{Eff}_d = D_v/D_e$, and $\mathrm{Eff}_b = P_v/P_e$ (Littlefair 1985; Perez et al. 1990).

The luminous efficacy is generally defined in two wavelength ranges:

$$\mathrm{Eff} = \frac{K_m \int_0^\infty \Phi_{e,\lambda} V(\lambda)\, \mathrm{d}(\lambda)}{\int_0^\infty \Phi_{e,\lambda}\, \mathrm{d}(\lambda)} \, (\mathrm{lm/W}), \quad (3.36)$$

$$\mathrm{Eff_v} = \frac{K_\mathrm{m} \int_{380}^{780} \Phi_{\mathrm{e},\lambda} V(\lambda)\, \mathrm{d}(\lambda)}{\int_{380}^{780} \Phi_{\mathrm{e},\lambda}\, \mathrm{d}(\lambda)} \, (\mathrm{lm/W}), \tag{3.37}$$

where $\Phi_{\mathrm{e},\lambda}$ is the spectral distribution of the radiant flux at wavelength λ, $K_\mathrm{m} = 683\,\mathrm{lm/W}$ is the maximal luminous efficacy for the normalized wavelength $\lambda_\mathrm{m} = 555$ nm, and $V(\lambda)$ is the relative spectral luminous efficacy defined by the CIE standard curve.

Formula (3.37) allows the luminous efficacy to be evaluated only in the range of the visible spectrum. As radiation of the total spectrum (from $\lambda = 0$ to infinity) is impossible to measure exactly, the luminous efficacy can be approximated for practical applications after (3.37) from the measurements with an illuminance meter and with a pyranometer as the global, diffuse, or direct luminous efficacy.

Owing to different states of the atmosphere, the spectral composition of global sunlight and skylight, or diffuse skylight alone, can vary substantially and affect luminous efficacies. Sunlight, especially, is partly scattered by the atmospheric media and attenuated by air mass; thus, global efficacy is rather unstable and dependent on the momentary solar altitude. The three main influences on the sunlight transmitted through the atmosphere are light scattering by air molecules, aerosols (tiny particles of dust and water droplets), and various gases, e.g., water vapor, ozone, and carbon dioxide (Littlefair 1985; Molineaux et al. 1995; Muneer and Kinghorn 1997).

In recent years several authors have proposed the calculation of daylight illuminance via luminous efficacy, using irradiance data measured by pyranometers at local meteorological stations or from satellite images. All of these illuminance levels are biased by errors caused by specific luminous efficacy differences and ranges in the measured solar and sky irradiance values. Such differences were analyzed and documented by Littlefair (1985) and Vartiainen (2000), who summarized several methods and calculated values of the luminous efficacy of beam/direct solar radiation, clear sky diffuse and global efficacy under clear skies, and diffuse efficacy under overcast skies. Littlefair (1988) also mentioned some possible uncertainties in experimental measurements of the luminous efficacy and documented them using Garston measurements gathered during 1 year from April 1984 to March 1985. Perez et al. (1987, 1990) published data validated from ten American and three European sites with a categorization of insolation conditions under clear and overcast skies depending on the solar zenith angle, and defined sky clearness and brightness as well as atmospheric precipitable water vapor content. Navvab et al. (1988), Olseth and Skartveit (1989), Wright et al. (1989), Chung (1992), Muneer and Angus (1995), Dumortier (1997), Robledo and Soler (2003), Tsikaloudaki (2005), and Souza et al. (2006) tried to formulate efficacy models for various conditions, e.g., clear sky, overcast sky, and cloudy sky, with local validation or models with various types of parameterization, e.g., applying clearness index, cloudiness ratio, or brightness index.

Several sky radiance/luminance distribution models, e.g., the all-weather model of Perez et al. (1993) and all sky models of Igawa et al. (2004), use luminous efficacy relations for the conversion of radiance to luminance.

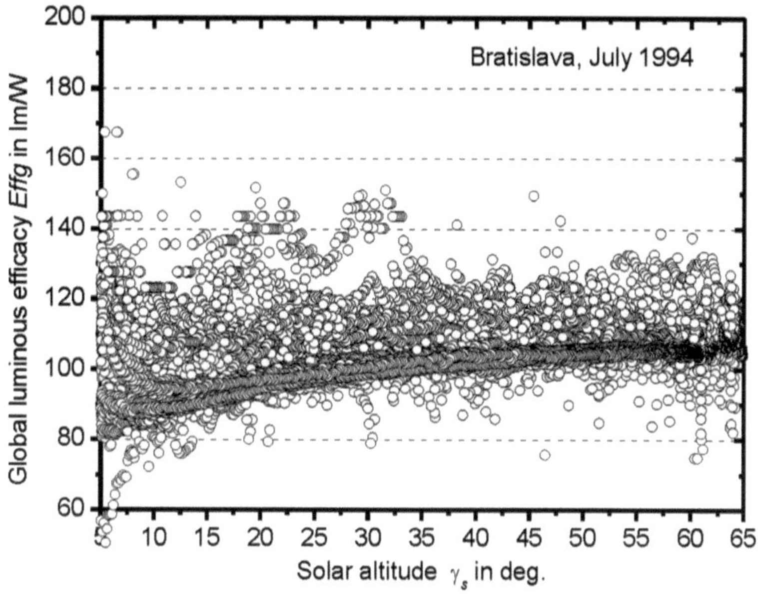

Fig. 3.13 Global luminous efficacy occurrence in July 1994 in Bratislava

Some models for predicting the year-round illuminance availability or its variability sometimes need to convert input data based on irradiance measurements. However, the CIE IDMP stations obtain calibrated illuminance and irradiance data only in a few places irregularly located around the world. The worldwide net of meteorological stations covers places in all climatic zones and provides internationally agreed meteorological data such as irradiance, sunshine duration, and cloud cover. In recent decades methods for application of satellite images from the visible channel to determination of irradiances and illuminances were published (Olseth and Skartveit 1989; Janjai et al. 2003, 2008) using the luminous efficacy.

Such models based on the luminous efficacy procedures are complicated and introduce several uncertainties (Darula and Kittler 2008). Owing to continual changes of atmospheric conditions, the illuminance and irradiance levels are also permanently changing. Instantaneous or averaged data can be given with regard to various types of parameterization. Figures 3.13 and 3.14 show occurrences of instantaneous global luminous efficacies $\mathrm{Eff_g}$ in relation to the solar altitudes after measurements in Bratislava during July and November 1994, respectively. Significant differences can be observed between these two figures. Figure 3.13 shows conditions for a typical month with higher sunshine duration. During the summer period when solar altitudes are rising to 64°, the November data are restricted only to 27°. Many values in July are concentrated around the curve starting at $\mathrm{Eff_g} = 80$ lm/W and culminating at $\mathrm{Eff_g} = 100$ lm/W, whereas the prevalent occurrence of data do not exceed $\mathrm{Eff_g} = 120$ lm/m.

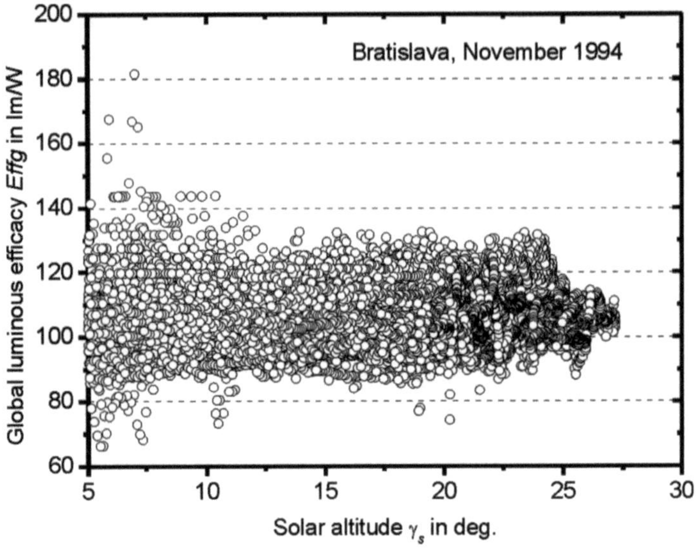

Fig. 3.14 Occurrence of global luminous efficacy changes in November 1994 in Bratislava

In winter, Fig. 3.14 shows that lower sunshine duration values of $\mathrm{Eff_g}$ are dispersed more uniformly around $\mathrm{Eff_g} = 110$ lm/W, which corresponds to data also in the review of Littlefair (1985). Owing to prevailing overcast and cloudy skies in November 1994, the light penetrating through the atmosphere is scattered, resulting in diffuse skylight. Also, independence of monthly data with respect to the solar altitude is caused by this effect. Monthly luminous efficacies inform more about sunny or cloudy situations but their simulations can lead to high errors in the conversion of irradiance data to illuminance data.

More valuable information can be obtained by studying measured instantaneous illuminance and irradiance data because the human eye responds to momentary influences and does not perceive any averaged or cumulative values commonly applied in energy calculations. The illuminance as well as irradiance levels permanently change during every day in accordance with changes of the sun's altitude, cloud cover, and turbidity of the atmosphere. Four characteristic daily illuminance and irradiance courses can be identified under clear, cloudless, overcast sunless, and cloudy with abrupt changes of sunny and sunless situations as well as a dynamic course pattern representing frequent changes of sunny and sunless periods (Darula and Kittler 2004c, 2011). Luminous efficacies are thus influenced by atmospheric conditions; therefore, typical states have to be considered.

During clear days owing to the absence of clouds, the sky luminance distribution is homogeneous and the luminous efficacy is fluently changing. The typical daily illuminance courses of such ideal clear days with the high and low noon solar altitudes are presented in Figs. 3.15 and 3.17, respectively. Simultaneous luminous efficacy depends on the solar altitude with differences that can be observed during

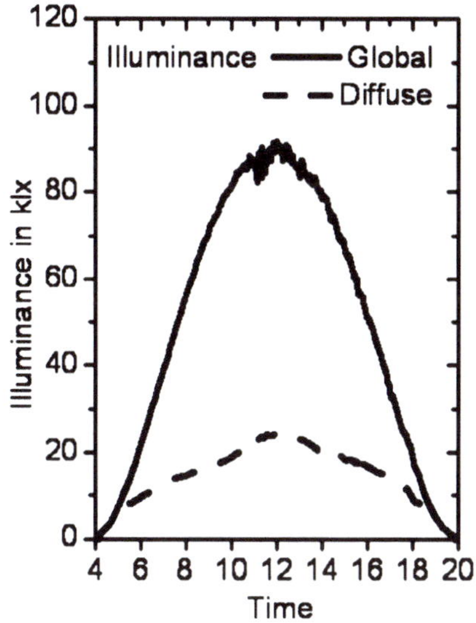

Fig. 3.15 Typical illuminance changes during a clear summer day on July 3, 1994

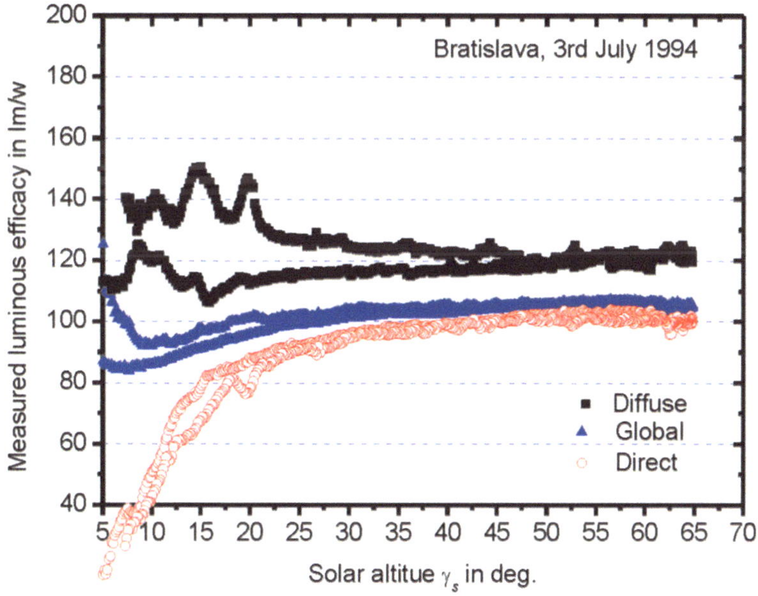

Fig. 3.16 Changes of luminous efficacies during a July day

both clear days (Figs. 3.16 and 3.18). The lower values of the direct luminous efficacy occur early in the morning and late in the evening. The comparison of direct luminous efficacy changes on July 3 with those on November 22, 1994 in the solar altitude interval 5–20° indicates influences of the various air mass differences on these days. The November direct efficacy curve starts at higher values because of

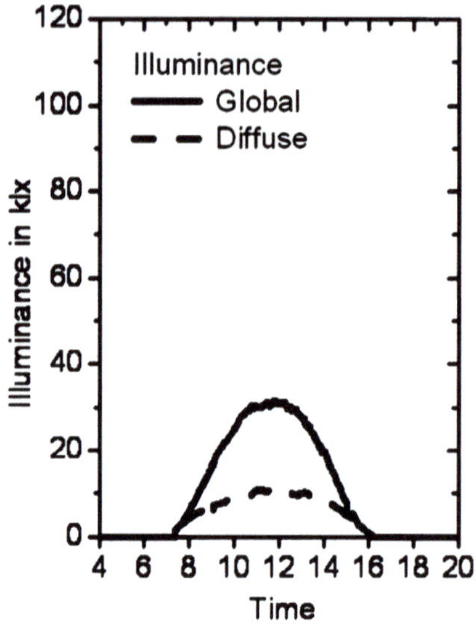

Fig. 3.17 Illuminance courses during a November day

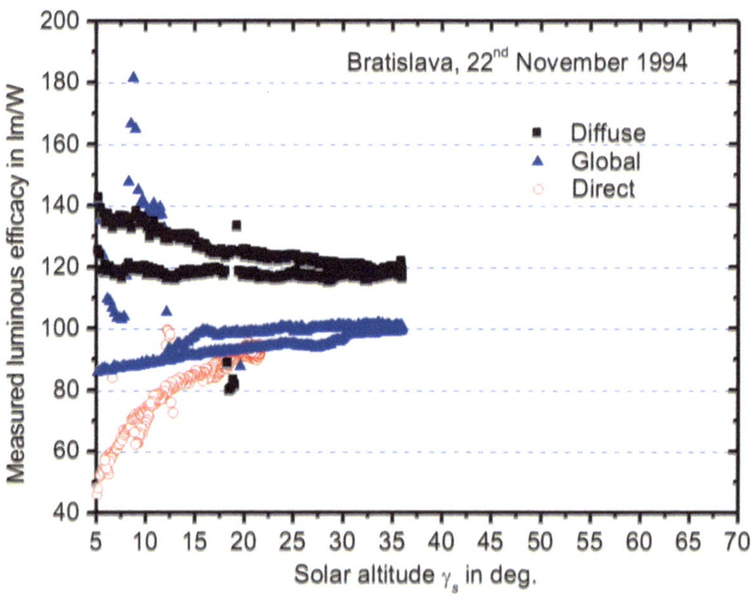

Fig. 3.18 Global, diffuse, and direct luminous efficacy changes during a clear November day

the higher water vapor content. Global and diffuse luminous efficacies show similar trends during a clear day. Although the global values are lower than the diffuse values, both are significantly different in the early morning and late in the evening.

At the other extreme, during overcast days only sky diffuse radiation or light is received on the ground. In this case the sky is completely covered by clouds of

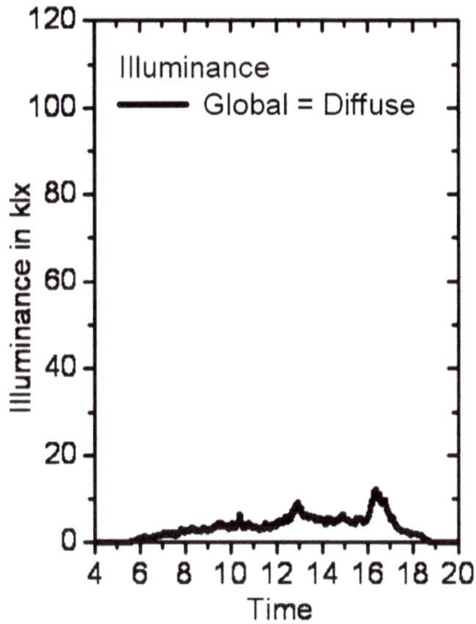

Fig. 3.19 Global and diffuse illuminance are equal on overcast August 26, 1994

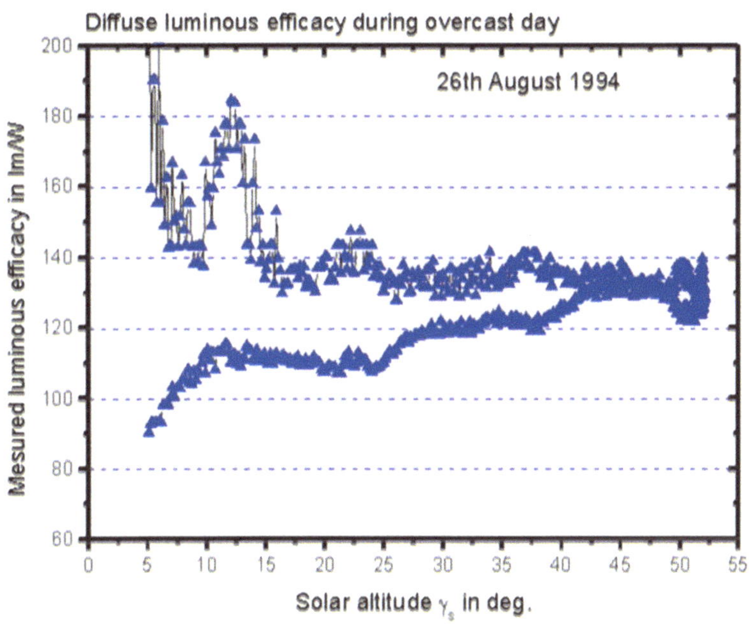

Fig. 3.20 Global luminous efficacy changes during a summer overcast day

various types and thicknesses. The illuminance levels are much lower, depending on the solar altitude as shown in Figs. 3.19 and 3.21. Although August is not typical for overcast skies, with higher solar altitudes quite low daily illuminance levels can occur and shift toward late afternoon owing to cloudiness conditions (Fig. 3.19 and 3.20).

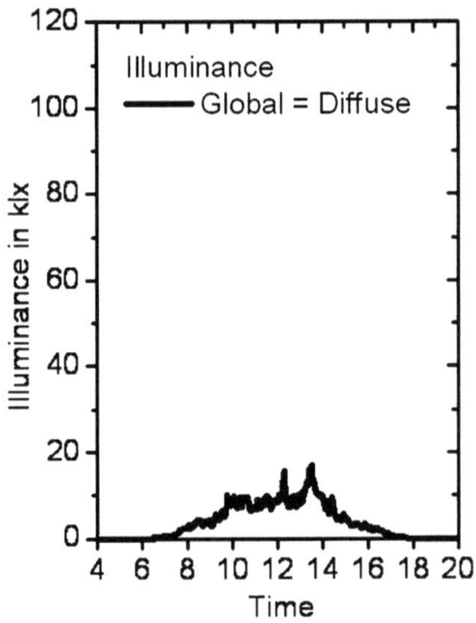

Fig. 3.21 Illuminance course during 9 March 1994

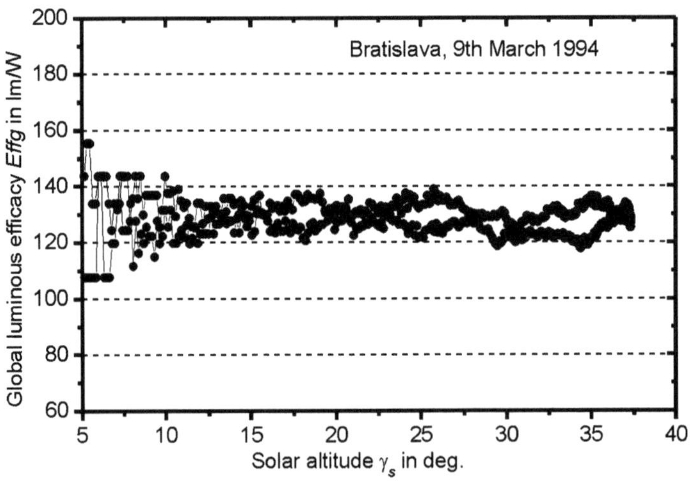

Fig. 3.22 Global luminous efficacy changes during an overcast spring day

The March day is similar to the overcast winter day with low solar altitudes. Small variations in luminous efficacy courses can also be observed during these selected days. Higher differences between morning and afternoon global luminous efficacy were measured on August 26, 1994 than on March 9, 1994. Whereas the highest values of global luminous efficacy on August 26, 1994 were recorded in the early morning, the lowest are in the late afternoon. A different trend of luminous efficacy course can be observed on March 9, 1994, when differences between morning and afternoon values are small and the daily course is relatively stable, which represents quite homogeneous atmospheric conditions in Fig. 3.22.

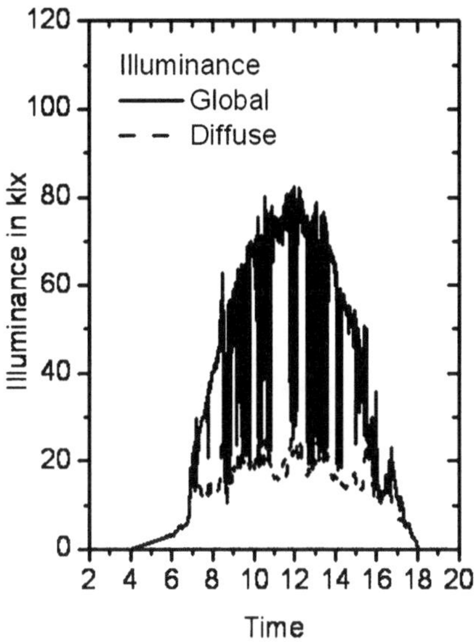

Fig. 3.23 Dynamic illuminance courses on September 6, 1994

During overcast days, typical quite high luminous efficacy levels are around $Eff_g = Eff_d = 130$ lm/W.

The most complicated daylight conditions can be expected during days with rapid changes in the global illuminance daily course. The diffuse illuminance course on a dynamic day is quite fluent. The direct illuminance randomly varies from zero to values equal to the difference between the global and the diffuse illuminance (Fig. 3.23). The global luminous efficacy in Fig. 3.24 is unstable, with random higher differences influenced by the presence or absence of sunlight. It seems that the basic efficacy is influenced mostly by the diffuse level of concentrated data approximately around $Eff_d = 105$ lm/W from the whole sky hemisphere without showing extremes of the sun contribution. This is quite logical because the diffuse illuminance sensor is shaded against sunlight and permanently measures only the influence of the sky. The separation of the diffuse efficacy course in Fig. 3.25 and the direct efficacy in Fig. 3.26 represents the extreme instantaneous variations of the direct solar luminous efficacy in the range roughly between 40 and 400 lm/W. Furthermore, the direct luminous efficacy varies in occurrence and level very frequently but randomly during dynamic days. However, the basic concentration of the direct efficacy level resembles the usual trend found under clear sky conditions (Figs. 3.16 and 3.18). Very high occasionally occurring direct solar efficacy values are probably caused by light transmittance through bright clouds while at the same time radiation absorption by dense cloud layers

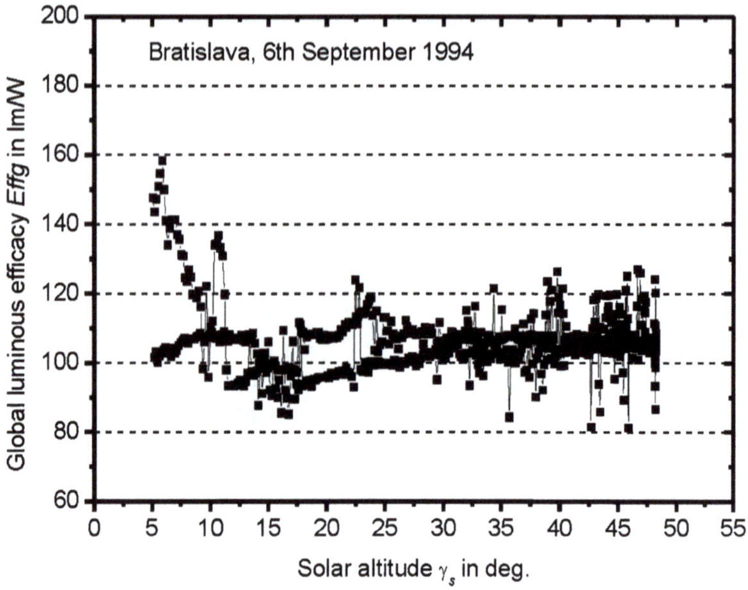

Fig. 3.24 Global luminous efficacy on a day with dynamic illuminance changes

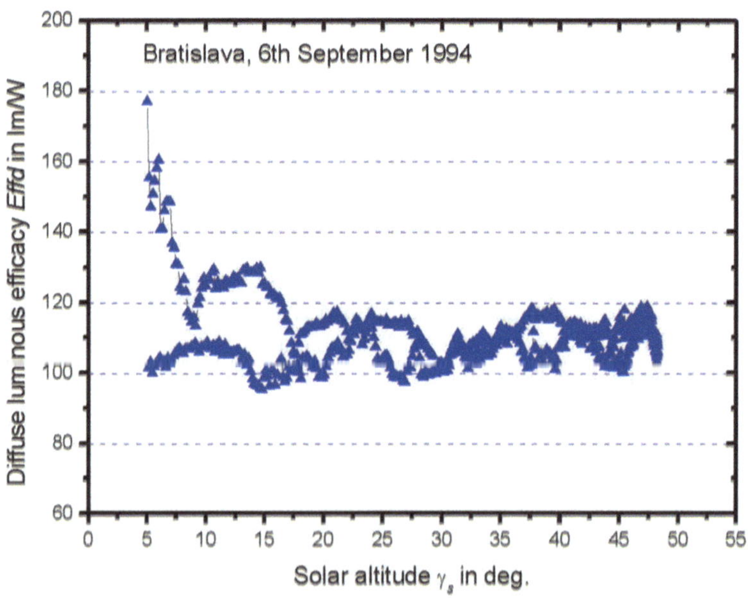

Fig. 3.25 Diffuse luminous efficacy in relation to the solar altitude

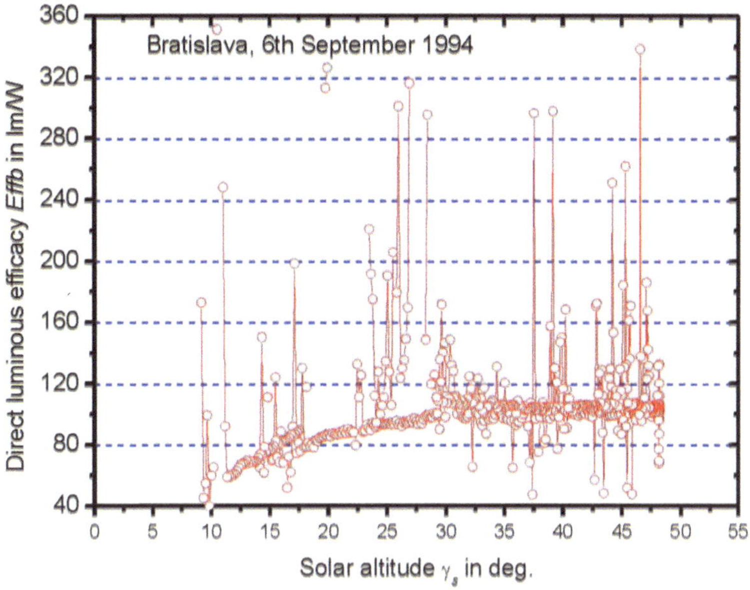

Fig. 3.26 Direct solar luminous efficacy changes during a dynamic illuminance day

substantially reduces the radiant flux. As a consequence, the luminous efficacy value is extremely high.

In some practical calculations, constant values of luminous efficacy are applied to convert irradiance data to luminous data. In the more advanced methods, authors have tried to empirically predetermine the dependence of luminous efficacy on the solar altitude, clearness index, cloud ratio, etc. for clear, overcast, or medium sky conditions.

Sometimes even the influence of water vapor, gases, or aerosols on the luminous efficacy is considered (Perez et al. 1987; Olseth and Skartveit 1989; Molineaux et al. 1995). Littlefair (1985), Olseth and Skartveit (1989), and Molineaux et al. (1995) noticed also that atmospheric turbidity described by Linke and Boda (1922) has an impact on the luminous efficacy. The study of the turbidity factor T_L in relation to the solar altitude, based on the data recorded during cloudless days in 1994 at the Bratislava CIE IDMP station, found values in the range $T_L = 1.5 - 4.5$, thus giving the direct luminous efficacy dependence shown in Fig. 3.27a, c. More frequent turbidities in the range 2.5–3.5 concentrate the efficacy to a narrow strip. The T_L influence is roughly the same until its value is over 4.5, when measured efficacy data are more separated and shifted toward lower values (Fig. 3.27d).

A slightly different trend of diffuse luminous efficacy can be observed during cloudless situations. Owing to the clear atmosphere, decreasing efficacy values in the range $T_L = 2.5 - 3.5$ are influenced by the solar altitude (Fig. 3.28a). Higher

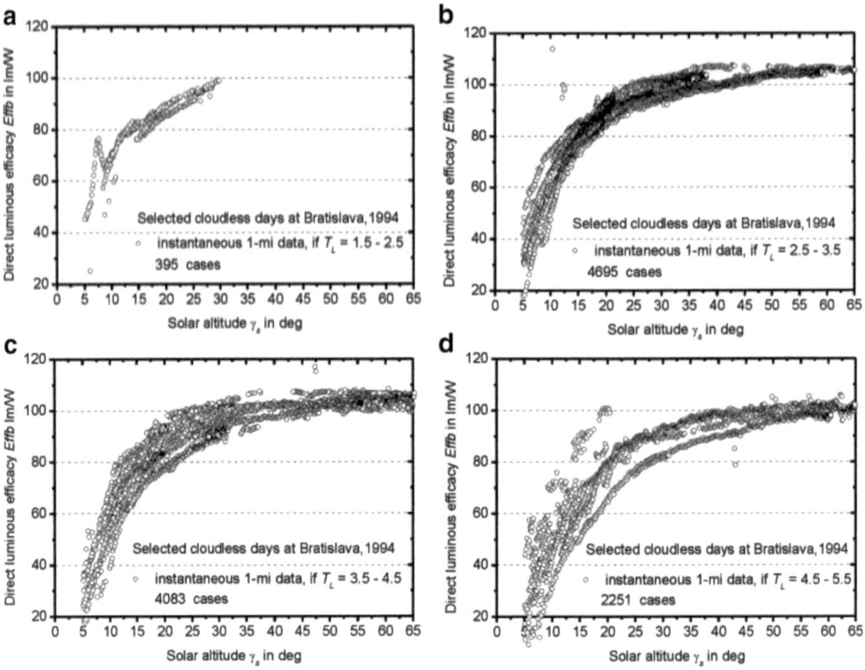

Fig. 3.27 (**a**) Direct luminous efficacy under lowest turbidity. (**b**) Yearly influence of low turbidity. (**c**) Efficacy range at medium turbidity. (**d**) Luminous efficacy under very high turbidity

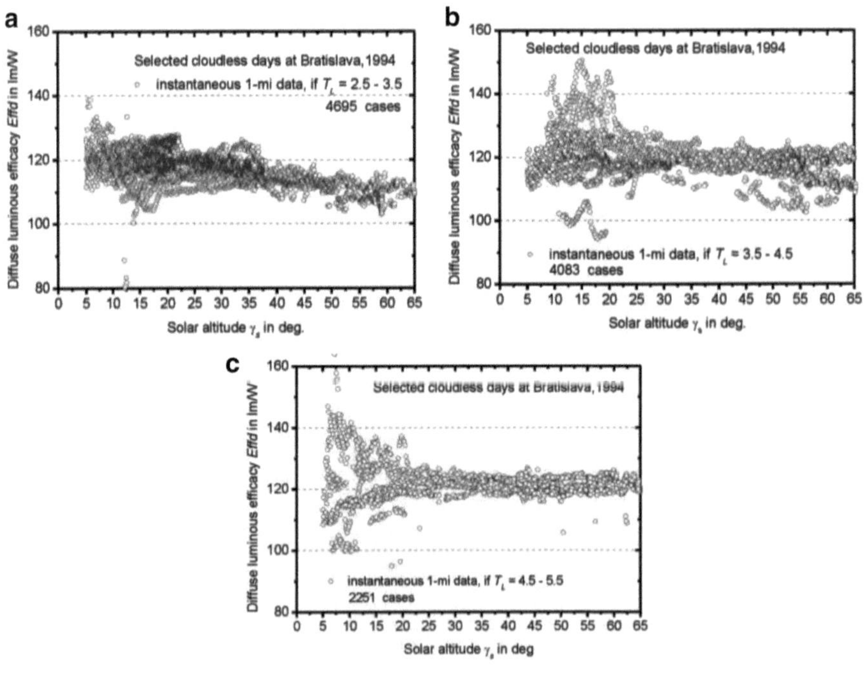

Fig. 3.28 (**a**) Diffuse efficacy range on clear days. (**b**) Medium turbidity influences. (**c**) Dependence of diffuse luminous efficacy occurrence during cloudless days on higher atmospheric turbidity

content of water vapor and aerosols in the atmosphere results in higher turbidities and more intensive light scattering. Therefore, occurrences of diffuse luminous efficacy are independent of the solar altitude from $\gamma_s > 20°$ when $T_L = 3.5 - 4.5$ in Fig. 3.28b or when $T_L = 4.5 - 5.5$ in Fig. 3.28c.

As described above, the luminous efficacy varies over the whole year, from day to day, and also in daily sequences linked to illuminance and irradiance daily occurrence courses. Each of these sequences generates unique daylighting conditions.

In illuminating engineering practice, it is important in methods based on luminous efficacy to check which procedures or formulae are applied and their range of validity. Application procedures based on constant luminous efficacy or methods considering only clear, overcast, intermediate, average daylight, or yearly weather simulations can lead to misleading results. Therefore, uncertainties and possible errors have to be identified and documented in such cases (Darula and Kittler 2008).

3.9 Partial Conclusions

The effectiveness of sunlight or solar energy depends on several facts:

- The momentary extraterrestrial density of the luminous flux reaching the outer border of the atmosphere, which is also the extraterrestrial illuminance on a fictitious plane placed perpendicularly to sunbeams called the "luminous solar constant" (LSC).
- The momentary horizontal extraterrestrial illuminance $E_v = E_{vo} \sin \gamma_s$, which is comparable to illuminance on the horizontal ground surface produced by parallel sunbeams P_v and the ratio P_v/E_v characterizes the transmittance of sunbeams through the atmosphere and the local availability of sunlight at ground level.
- The atmospheric transmittance, which is dependent on the primary extinction coefficient a_v, the momentary relative optical air mass m, and the luminous turbidity factor T_v.
- The sun position defined by the angular solar altitude and azimuth.
- The daily, monthly, or yearly sun path.
- The momentary or longer-term cloudless or shaded sun position defined by relative sunshine duration in a half day, day, month, or year.
- The inclination angle i of a particular plane illuminated by solar beams defined by the plane normal and its cosine function, i.e., $\cos i$, which is determined by its angular distance from the sun position.

The specific sunlight and skylight conditions in an arbitrary location and at a particular time can be determined only by regular measurements. The results of such investigations can indicate typical daily, monthly, and seasonal changes in the local daylight climate. It is important to analyze such changes using a parameterization with photometric units taking into account also the difference between the radiation and luminous measurements.

The luminous efficacy expressed either by the global and diffuse daylight or the direct sunlight efficacy at ground level is influenced by the spectral composition of the visible and infrared flux penetrating different atmospheric layers. The content of the atmosphere, owing to pollution particles with their concentration in different cloud types and covers, determines the broadband and spectral absorption, reflectance, and final transmittance. Visible and infrared fluxes are preferentially transmitted through a specific atmosphere and luminous efficacy at ground level is also influenced by momentary variations in air mass (Henderson 1970). As all of these influences must be considered together with the uncertainties already mentioned, the practice of converting radiation data to luminous data is suspect and possible errors have to be identified and documented in such cases (Darula and Kittler 2008).

To learn more about the local daylight climate it is necessary to establish an IDMP station, to obtain regular and long-term measured data that can be analyzed and evaluated (e.g., after Kittler et al. 1992; Kittler and Darula 2002). More stations in the equatorial, tropical, and subtropical regions are needed to characterize daylight climate and specific conditions there. Information about the IDMP network with the general and research stations is summarized at http://idmp.entpe.fr and some data are available for general study and possible application. Unfortunately, even researchers collecting IDMP data seldom evaluate their measurements to determine all specific daylight conditions in detail to learn more about the local daylight climate. Owing to energy-saving policy, it is important to determine the year-round availability of sunlight and skylight. Long-term measurements (at least 5–10-year of regularly recorded data) seem to be sufficient to determine seasonal, e.g., monthly or yearly, changes and the frequency of sequential overcast, cloudy, or clear half days or periods with dynamic weather. Because the effective utilization of daylight is expected to reduce electricity consumption in interiors, it has to be realized that window size, orientation, and placement have to take into account the sky luminance within the solid angle with respect to exterior obstructions and shading. Thus, luminance sky patterns are a prerequisite to determine the energy-saving contribution of daylight. In this task no virtual Lambert sky can replace the real local daylight situations.

Appendix 3

Comparison of Solid Angle Calculation and Stereographic Representations

Lambert's knowledge of geometry and photometry (1760) and principles of perspective images (1759) as well as the importance of solid angles and their projections, i.e. cosine transformations, were the basis of skylight calculations. The understanding that the illumination of extensive space or planar light sources

Fig. A3.1 Solid angle for the whole space and half space

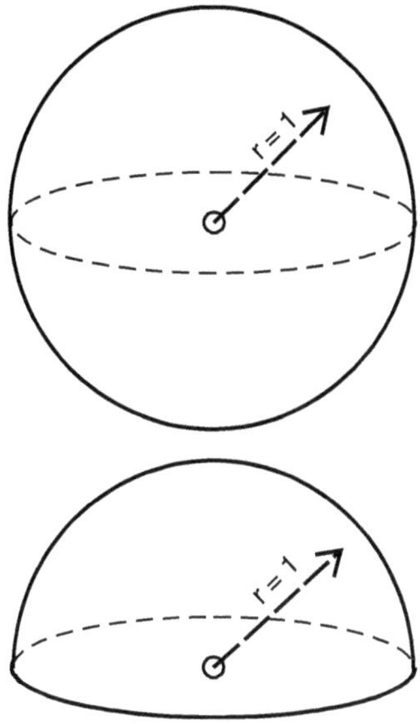

depends on their abstraction in the three-dimensional space, represented in a sphere or hemisphere, is fundamental to the science of photometry. Without their transformation by conical or pyramidal sections created by circular or rectangular light apertures the solid angle calculation methods could not exist. The introduction of spherical trigonometry and the definitions of steradian and radian units have to be understood.

The solid angle ω is defined by the surface area on a sphere with unity radius (basis of the unit-sphere methods). This spherical surface area is defined by the conical angle drawn from the sphere center, so:

- In the case of the whole space encircling the center point, it contains the whole sphere surface area (Fig. A3.1)

$$A_{O}=\omega_{O} = 4\pi = 12.5664\,(\mathrm{sr}). \tag{A3.1}$$

- In the case of a hemisphere in any arbitrary position,

$$A_{\cap} = \omega_{\cap} = 2\pi = 6.28315\,(\mathrm{sr}). \tag{A3.2}$$

- In the case of a conical cut of the sphere with the spherical area subtending the rotational cone with an angular extent θ (Fig. A3.2),

$$A_{\supset} = \omega_{\supset} = 2\pi(1 - \cos\,\theta)(\mathrm{sr}). \tag{A3.3}$$

- So, for example, when the human binocular visual field (Fig. 2.5) extends roughly $\theta = 60^{\circ}$, its solid angle is $\omega_{\supset} = \pi(1 - \cos\,60^{\circ}) = 3.14159\,(\mathrm{sr})$,

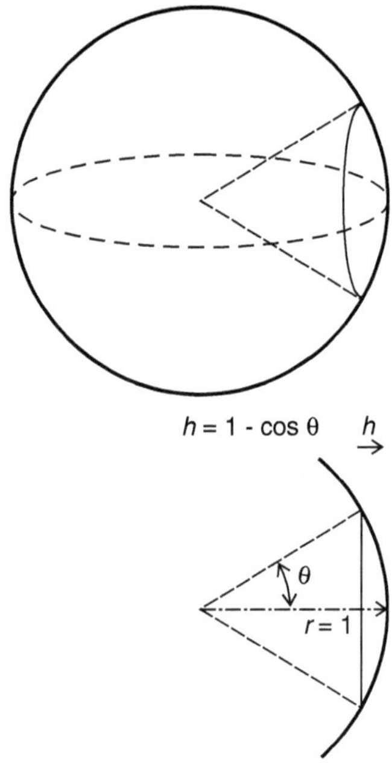

Fig. A3.2 Solid angle formed by a cone

whereas the inner area of sharp vision with $\theta = 30°$ has a solid angle of only 0.84179 sr.
In the case of a distant round object with disk radius r_d and surface area S_d at distance r_S (i.e., the radius of a fictitious big sphere r_S from its center), the calculation of the solid angle has to follow after

$$\omega_C = \frac{S_d}{r_S^2}\,(\mathrm{sr}). \tag{A3.4}$$

Further, the solar disk has the solid angle measured from ground level on the equinox day: $r_S = 1.4959787 \times 10^{11}$ m and $r_d = 6.9626 \times 10^8$ m (Darula et al. 2005). Then,

$$\omega_C = \frac{S_d}{r_S^2} = \frac{\pi r_d^2}{r_S^2} = 6.7905 \times 10^{-5}\,(\mathrm{sr}). \tag{A3.5}$$

A space object, such as a tiny aerosol in the atmosphere, faces the whole space around it and can be abstracted as a point with a solid angle of 4π. Any planar surface element owing to its position (slope and orientation) is exposed only to a half space of $\pi/2$ sr in all directions from the plane normal. Thus, a virtual hemisphere can be imagined from the center of the surface plane subtending a solid angle equal to the half space, i.e., $\pi/2$ sr.

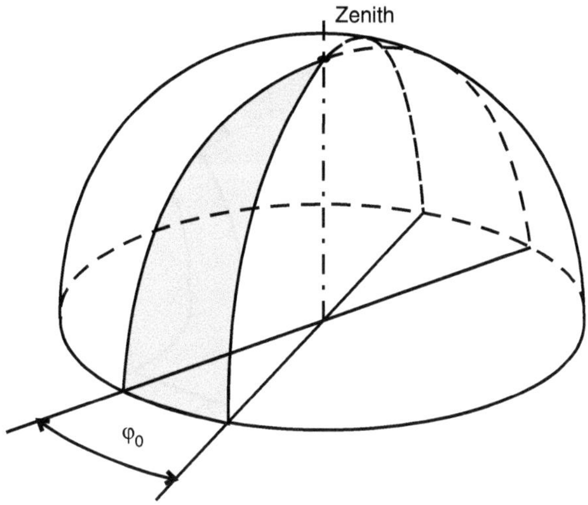

Fig. A3.3 Vertical half lune

In daylight theory, a specific sky hemisphere on a horizontal plane has a special significance, because if its horizon is unobstructed it represents the maximum available diffuse skylight outdoors D_v that can be utilized at ground level under any turbidity or cloudiness conditions. This horizontal illuminance is sometimes also called the scalar quantity. The scalar illuminance is equal to the average illuminance over the whole surface and is the weighted illuminance received from the upper sky hemisphere. The skylight scalar illuminance transmits the light flow indoors to the position, orientation, and solid angle of any aperture (Lynes et al. 1966).

When solid angles from vertical endless apertures or windows have to be calculated, then it is convenient to use the image of a spherical vertical lune on a unit sphere (Fig. A3.3), e.g., the half lune (on a unit-radius sphere) with the bottom edge on the horizon, or an endlessly high window with its sill at the level of the illuminated horizontal plane has a surface area or solid angle

$$\omega_{\wedge} = \varphi_0 = \pi \frac{\varphi_0^{\circ}}{90^{\circ}} \,(\mathrm{sr}). \tag{A3.6}$$

Thus, when the lune width is extended to a half hemisphere, i.e.,
$\varphi_0 = 180^{\circ}$, then

$$\omega_{\wedge} = \omega_{\cap} = \pi,$$

or for the full hemisphere, $\omega_{\wedge} = \omega_{O} = 2\pi$.

If the lune height is restricted by a circular upper edge with width angle φ_{01} (Fig. A3.4), then it is related to the plane φ_0 angle as $\varphi_{01} = \varphi_0 \cos \varepsilon_0$; thus,

$$\omega_{\wedge} = \varphi_0 \,(1 - \cos \varepsilon_0)(\mathrm{sr}). \tag{A3.7}$$

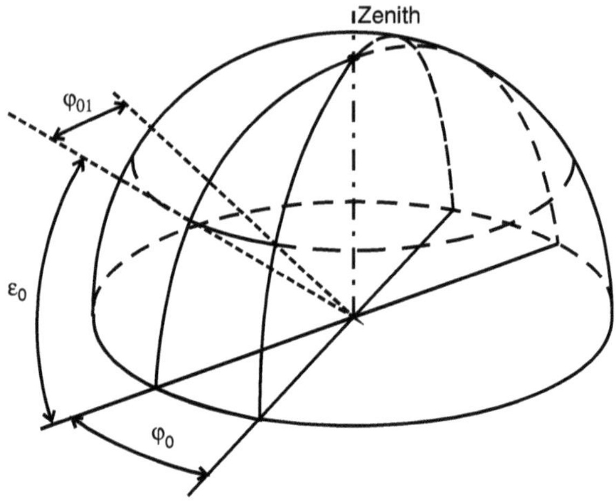

Fig. A3.4 Solid angle of a vertical lune cut by an elevation angle

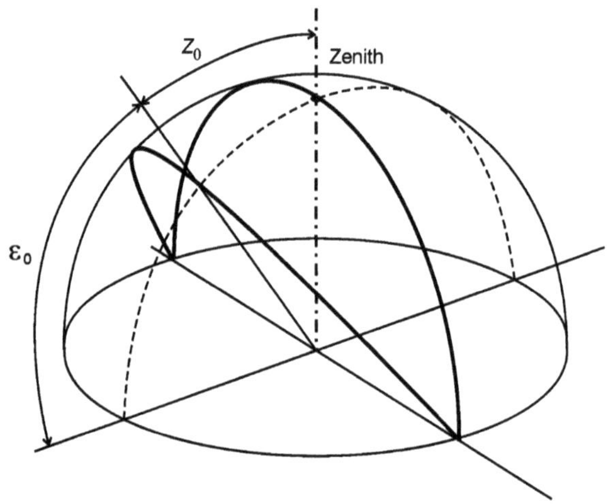

Fig. A3.5 Horizontal lune

The division of the sphere into regular lunes is also applied in cartographic representation of the globe to determine time zones, but a double system is used to determine any location on the surface of the globe after geographical longitude (lunes) and geographical latitude (rings). The latter is given by horizontal circles parallel to the equator circle in angular steps.

When solid angles from two-sided endless horizontal apertures or windows have to be calculated, then it is convenient to use the image of a spherical horizontal lune on a unit sphere (Fig. A3.5), which is determined after

$$\omega_{\subset\supset} = 2Z_0(\text{rad, sr}), \tag{A3.8}$$

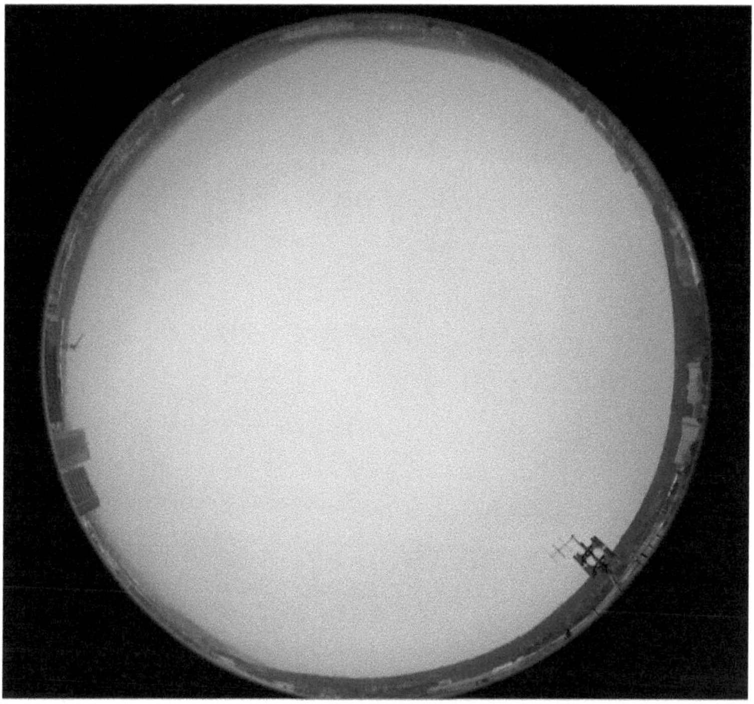

Fig. A3.6 Fish-eye image of an overcast sky

and similar to the case of a vertical lune, according to the restriction of the width angle or the extended zenith angle, the particular solid angles for streets, squares, or roof lights can be estimated.

As indicated above, in a fixed view of any environment the luminance patterns in the eye's central field of vision (30° around the fovea axis) are most relevant, and the solid angle is determined as a circular cut of a sphere projected by the eye lens on the near-spherical retinal surface. Such an image can also be projected onto a sensitive film inside a camera accepting that image from any viewing position and orientation. So, also the luminance distribution on the whole sky vault can be taken by a fish-eye lens reproducing the whole upper-hemispherical solid angle (e.g., Fig. A3.6 shows an overcast sky and Fig. A3.7 shows a clear overcast sky luminance pattern). If there are local obstructions of the skyline close to the horizon, e.g., hills, buildings, or trees, the sky solid angle could be restricted. Outdoor architectural spaces such as squares, streets, or courtyards sometimes have complex obstructions of different shape, height, and orientation which, from any point, can be captured by a fish-eye photograph (Fig. A3.8). It is interesting to note that owing to the usual vertical and horizontal edges of buildings, often the rectangular structure of architectural spaces containing house fronts, walls, windows, etc. can obstruct solid angles following the vertical and horizontal lunes.

Fig. A3.7 Fish-eye image of a clear sky

Additional to the obstructions on the horizon, a 180° photograph documents the momentary cloud presence, cloudiness cover of the vault, and cloud distribution and type in a particular plan projection of the whole sky half space. The cloud camera of Hill (1924) was probably the first with a lens adjustment for such a purpose, and later solutions using concave mirrors, the "heliopanorama camera" of Pers (1933) and the globoscope of Pleijel (1952, 1954), have now been replaced by fish-eye lenses. For the evaluation of solid angle shading the projection system is important. The globoscope image because of the paraboloidal mirror shows a stereographic projection of the surroundings, whereas the fish-eye lens produced by Nikon in the twentieth century had an equidistant projection. For example the newer Nikon digital camera with a Sigma fish-eye lens has an equisolid angle projection (which was used for Figs. A3.6–A3.8). The projection definitions and a solid angle diagram with some other advantages of full-field cameras were described by Longmore (1964).

Of course, when illuminance of surface elements on horizontal, sloped, or vertical planes has to be calculated, then the solid angle projected on such a plane has to be multiplied by the cosine of the incidence angle, as in case of a horizontal element (see the appendix in Chap. 8).

Fig. A3.8 Fish-eye image of an obstructed horizon and partly cloudy sky

However, from the same "daylighting" aperture solid angles different and transient sky luminance patterns will be seen. In the case of a uniform sky luminance or under overcast skies with vertically graded luminance patterns and the absence of sunlight, the aperture orientation has no influence. Conversely, with variable sky luminance conditions, it is not difficult to imagine that a horizontal planar element can see from a certain interior position also sky patches close to or even directly the position of the sun. Thus, scalar illuminance does not characterize accurately an extremely directional sky luminance pattern. Therefore, the vector/scalar ratio is only a convenient measure of the asymmetry of the illumination solid and, together with the vector direction, relates to the perceived directional strength of the light flux entering the interior (Cuttle 1978).

The sky and sun luminance distributions form a solid body which can be characterized by its sections and their cover curves called indicatrices and can also be used with the scalar and vector quantities as the basic properties of light field or photic field (Gershun 1928, 1936, 1958; Moon and Spencer 1981).

References

Angus, R.C. and Muneer, T.: Sun position for daylight models: Precise algorithms for determination. Lighting Res. Technol., **25**, 2, 81–83 (1993)

Bemporad, A.: Zur Theorie der Extinktion des Lichtes in der Erdatmosphäre. Mitteilungen der Grossh. Sternwarte zu Heidelberg, **4**, 1–78 (1904)

Blondel, A.: Rapport sur les unités photométriques. Congrés International des Electriciens. Geneva (1896)

BIPM – Bureau International des Poids et Mesures: Compt. Rendu 16 Conf., Gén. Poids et Mesures, 100 (1979)

Blewin, W.R. and Steiner, B.: Redefinition of the Candela and Lumen. Metrologia, **11**, 97–104 (1975)

Bouguer, P.: Comparaison de la force de la lumiere du soleil, de la lune et des plusieurs chandelles. Histoire de l'Acad. Roy. des Sc., Paris (1726)

Bouguer, P.: Traité d'Optique sur la gradation de la lumiére. Paris (1760), Latin translation by Richtenburg, J.: Bougeri Optice de Diversis Luminis Gradibus Dimetiendis. Vienna-Prague (1762), Russian translation by Tolstoy, N.A. and Feofilova, P.P.: Opticheskiy traktat o gradaciyi sveta, edited and commented by A.A. Gershun, Izd. Acad. Sci. USSR, Moscow (1950), English translation by Knowles Middleton, W.E.: Pierre Bouguer's optical treatise on the gradation of light. University of Toronto Press, Toronto (1961)

Chung, T.M.: A study of luminous efficacy of daylight in Hong Kong. Energy and Buildings **19**, 1, 45–50 (1992)

Clear, R.: Calculation of turbidity and direct sun illuminance. Memo to Daylight Goup. LBL, Berkeley (1982)

CIE – Commission Internationale de l'Eclairage: Compte Rendu, p.67 or The basis of physical photometry. Publ. CIE 18.2 (1924, 1983)

CIE – Commission Internationale de l'Éclairage: CIE 1998 2° spectral luminous efficiency function for photopic vision. Publ. CIE 86–1990, CB CIE Vienna (1990)

CIE – Commission Internationale de l'Éclairage: Spatial distribution of daylight – CIE Standard General Sky. CIE Standard S 011/E:2003, CB CIE Vienna (2003)

Darula, S. and Kittler, R.: New trends in daylight theory based on the new ISO/CIE Sky Standard 1. Zenith luminance on overcast skies. Build. Res. Journ., **52**, 3, 181–197 (2004a)

Darula, S. and Kittler, R.: New trends in daylight theory based on the new ISO/CIE Sky Standard 2. Typology of cloudy skies and their zenith luminance. Build. Res. Journ., **52**, 4, 245–255 (2004b)

Darula, S. and Kittler, R.: Sunshine duration and daily courses of illuminances in Bratislava. International Journal of Climatology, **24**, 14, 1777–1783 (2004c)

Darula, S. and Kittler, R.: New trends in daylight theory based on the new ISO/CIE Sky Standard 3. Zenith luminance formula verified by measurement data under cloudless skies. Build. Res. Journ., **53**, 1, 9–31 (2005)

Darula, S., Kittler, R., Gueymard, C.: Reference luminous solar constant and solar luminance for illuminance calculations. Solar Energy, **79**, 5, 559–565 (2005)

Darula, S. and Kittler, R.: Uncertainties of luminous efficacy under various skies. Proc. Balkan Light 2008, 315–322 (2008)

Darula, S. and Kittler, R.: Sunlight and skylight availability. Advances in Energy Research. Vol.1, Chapter 4, 133–182, Nova Publ., NY (2011)

Dumortier, D., Avouac, P., Fontoynont, M.: Proposal for the processing of recorded daylight data. Proc. 23rd Session CIE, Vol.1, 208–211 (1995)

Dumortier, D.: Evaluation of luminous efficacy models according to sky types and atmospheric conditions. Proc. LuxEuropa Conf., Amsterdam, 1068–1080 (1997)

Euler, L.: Réflections sur les divers degrees de lumiére du soleil et des autres corps célestes. Histoire de l'Acad. Roy. des Science et belles Lettres, 280–310, Berlin (1750)

Gueymard, C.: The sun's total and spectral irradiance for solar energy applications and solar radiation models. Solar Energy, **76**, 4, 423–453 (2004)

Gershun, A.A.: Svetovoye pole ot poverkhnostnykh izluchateley. (In Russian: The light field from surface radiators). Trudy GOI, 38 (1928)

Gershun, A.A.: Svetovoye pole. Gostechizdat, Moskva (1936), English translation by Moon, P. and Timoshenko, G. in Journ. of Mathem. and Physics, **18**, 51–151, MIT, (1939)

Gershun, A.A.: Izbranye trudy po fotometriyi i svetotecknike. (In Russian. Selected works in photometry and illuminating engineering.) State Publ. of Physic and Mathem. Literature, Moscow (1958)

Hayman, S.: Sky-scanning luminance meter: Proposal using fibre-optic photocells. Light. Res. and Technol., **21**, 4, 195–196 (1989)

Henderson, S.T.: Daylight and its spectrum. Hilger Ltd., London (1970)

Hovila, J., Toivanen, P., Ikonen, E., Ohno, Y.: International comparison of the illuminance responsivity scales and units of luminous flux maintained at the HUT (Finland) and the NIST (USA). Metrologia, **39**, 219–223 (2002)

IEA – International Energy Agency: Broad-band visible radiation data acquisition and analysis. A technical report of task 17: Measuring and modelling spectral radiation affecting solar systems and buildings. Vol. 2: 180 6 Simulation of Seasonal Variations in the Local Daylight Climate Satellite Report No. 1 World Network of Daylight Measuring Stations (IDMP), ASRC Albany, 569 NY (1994)

Igawa, N., Koga, Y., Matsuzawa, T., Nakamura, H.: Models of sky radiance distribution and sky luminance distribution. Solar Energy, **77**, 137–157 (2004)

ISO – International Standardisation Organisation: Spatial distribution of daylight – CIE Standard General Sky. ISO Standard 15409:2004 (2004)

Janjai, S., Jantarach, T., Laksanaboonsong, J.: A model for calculating global illuminance from satellite data. Renewable Energy, **28**, 15, 2355–2365 (2003)

Janjai, S., Tohsing, K., Nunez, K., Laksanaboonsong, J.: A technique for mapping global illuminance from satellite data. Solar Energy, **82**, 6, 543–555 (2008)

Kasten, F. and Young, A.T.: Revised optical air mass tables and approximation formula. Applied Optics, **28**, 22. 4735–4738 (1989)

Kelton, W. and Murdoch, J.: Sky luminance measurements and analysis. Proc. Intern. Daylighting Conf., Long Beach, 9–13 (1986)

Kepler, J.: Ad vitellionem paralipomena quibus astronomiae pars optica traditur. Frankfurt (1604)

Kittler, R. and Mikler, J.: Základy využívania slnečného žiarenia. (In Slovak. Basis of the utilization of solar radiation.) Veda Publ., Bratislava (1986)

Kittler, R. and Pulpitlova, J.: Základy využívania prírodného svetla. (In Slovak. Basis of the utilization of daylight.) Veda Publ., Bratislava (1988)

Kittler, R.: A new concept for standardizing the daylight climate state. Proc. Lux Europa Congress, Budapest, 303–312 (1989)

Kittler, R., Hayman, S., Ruck, N., Julian, W.: Daylight measurement data: Methods of evaluation and representation. Light. Res. and Technol., **24**, 4, 173–187 (1992)

Kittler, R.: The general relationship between sky luminance patterns and exterior illuminance levels. Proc. 23rd Session CIE, Vol.1, 217–218 (1995)

Kittler, R., Darula, S., Perez, R.: A set of standard skies characterising daylight conditions for computer and energy conscious design. Res. Report of the American-Slovak grant project US-SK 92 052, 240 pp. (1998)

Kittler, R. and Darula, S.: Parametric definition of the daylight climate. Renewable Energy, **26**, 177–187 (2002)

Kittler, R. and Darula, S.: Analemma, the ancient sketch of fictitious sunpath geometry – Sun, time and history of mathematics. Architectural Science Review, **47**, 2, 141–144 (2004)

Kittler, R. and Darula, S.: Applying solar geometry to understand the foundation rituals of "Old Kingdom" Egyptian pyramids. Architectural Science Review, **51**, 4, 407–412 (2008)

Lambert, J.H.: Photometria sive de mensura et gradibus luminis, colorum et umbrae. Augsburg (1760), German translation by Anding, E.: Lamberts Fotometrie. Klett Publ., Leipzig (1892), English translation and introduction notes by DiLaura, D.L.: Photometry, or on the measure and gradation of light, colors and shade. Publ. IESNA, N.Y. (2001)

Linke, F. and Boda, K.: Vorschläge zur Berechnung des Trübungsgrades der Atmosphäre aus den Messungen der Intensität der Sonnenstrahlung. Meteorol. Zeitschr. **39**, 6, 161–166 (1922)

Littlefair, P.J.: The luminous efficacy of daylight: a review. Light. Res. Technol., **17**, 4, 162–177 (1985)

Littlefair, P.J.: Measurements of the luminous efficacy of daylight. Light. Res. Technol., **20**, 4, 177–188 (1988)

Lynes, J.: Alternative subdivision of the sky vault. Lighting Res. Technol., **20**, 1, 33–37 (1988)

Maurolyco, F.: Photismi de lumine. Venezia (1575), English translation: The Photismi de Lumine of Maurolycus. A Chapter in Late Medieval Optics. New York (1940).

Makhotkin, L.G.: Ekvivalent massy Bemporada (In Russian. Equivalent of the air mass by Bemporad). Trudy GGO, 100, 15–16 (1960)

Meshkov, V.V.: Osnovy svetotekhniki. (In Russian. Basis of illuminating engineering.) State Energy Publ., Moscow (1957)

Molineaux, B., Ineichen, P., Delaunay, J.J.: Direct luminous efficacy and atmospheric turbidity improving model performance. Solar Energy, **55**, 2, 125–137 (1995)

Moon, P. and Spencer, D.E.: The Photoc Field. MIT Press, Boston (1981)

Muneer, T. and Angus, R.C.: Luminous efficacy: evaluation of models for the United Kingdom, Lighting Res. Tech., **27**, 71–77 (1995)

Muneer, T. and Kinghorn, D.: Luminous efficacy of solar irradiance: Improved models. Light. Res. Technol., **29**, 4, 185–191 (1997)

Navvab, M., Karayel, M., Ne'eman, E., Selkowitz, S.: Analysis of atmospheric turbidity for daylight calculations. Energy and Buildings, **6**, 2–4, 293–303 (1984)

Navvab, M., Karayel, M., Ne'eman, E., Selkowitz, S.: Luminous efficacy of daylight. Proc. CIBSE Nat. Light. Conf., 409 (1988)

Ohno, Y.: Improved Photometric Standards and Calibration Procedures at NIST. Proceedings of NCSL 1996 Workshop and Symposium, Vol.1, 343–353 (1996)

Olseth, J.A. and Skartveit, A.: Observed and modelled hourly luminous efficacies under arbitrary cloudiness. Solar Energy, **42**, 221–233 (1989)

Parkin, R.E.: Solar angles revisited using a general vector approach. Solar Energy, **84**, 6, 912–916 (2010)

Perez, R., Webster, K., Stewart, R., Barron, J.: Variations of the luminous efficacy of global and diffuse radiation and zenith luminance with weather conditions. Solar Energy, **38**, 33–44 (1987)

Perez, R., Ineichen, P., Seals, R., Michalsky, J., Steward, R.: Modeling daylight availability and irradiance components from direct and global irradiance. Solar Energy, **44**, 5, 271–289 (1990)

Perez, R., Michalsky, J., Seals, R.: An all-weather model for sky and luminance distribution. Solar Energy, **50**, 3, 235–245 (1993)

Pierpoint, W.: Recommended practice for the calculation of daylight availability. Journ. IESNA, **13**, 4, 381–392 (1982)

Reda, I. and Afshin A.: Solar position algorithm for solar radiation applications. Solar Energy, **76**, 577–589 (2004)

Robledo, L. and Soler, A.: Modeling the luminous efficacy of diffuse solar radiation on inclined surfaces for all sky conditions. Energy Conversion and Management, **44**, 1, 177–189 (2003)

Souza, R.G., Robledo, L., Pereira, F.O.R., Soler, A.: Evaluation of global luminous efficacy models for Florianópolis, Brazil. Building and Environment, **41**, 10, 1364–1371 (2006)

Toivanen, P., Kärhä, P., Manoochehri, F., Ikonen, E.: Realization of the unit of luminous intensity at the HUT. Metrologia, **37**, 131–140 (2000)

Tregenza, P.: Subdivision of the sky hemisphere for luminance measurements. Light. Res. and Technol., **19**, 1, 13–14 (1987)

Tregenza, P. and Sharples, S.: Daylight algoritms. Report ETSU, University of Sheffield (1993)

Tsikaloudaki, K.: A study on luminous efficacy of global radiation under clear sky conditions in Athens, Greece. Renewable Energy, **30**, 4, 551–563 (2005)
Vartiainen, E.: A comparison of luminous efficacy models with illuminance and irradiance measurements. Renewable Energy, **20**, 265–277 (2000)
Vitruvius, M.P.: De architectura libri X. Rome (13 BC, 1487), English translation by Morgan, M.H.: The ten books on architecture. Harvard Univ. Press, Harvard (1914) and by Granger, F.: Of Architecture, The ten books. Heinemann, London (1970)
Wright, J., Perez, R., Michalsky, J.J.: Luminous efficacy of direct irradiance: Variations with insolation and moisture conditions. Solar Energy, **42**, 387–394 (1989)

References to Appendix

Cuttle, C.: The new photometry. In: Developments in lighting-1. Edited by J.A. Lynes, 1–24. Applied Science Publ., London (1978)
Darula, S., Kittler, R., Gueymard C.: Reference luminous solar constant and solar luminance for illuminance calculations. Solar Energy, 79, 559–565 (2005)
Gershun, A.A.: Svetovoye pole ot poverkhnostnykh izluchateley. (In Russian: The light field from surface radiators). Trudy GOI, 38 (1928)
Gershun, A.A.: Svetovoye pole. (Translated by P. Moon and G. Timoshenko in Journ. of Mathem. and Physics18, 51–151, MIT, 1939.) Gostechizdat, Moscow (1936)
Gershun, A.A.: Izbranye trudy po fotometriyi i svetotecknike. (In Russian. Selected works in photometry and illuminating engineering.) State Publ. of Physic and Mathem. Literature, Moscow (1958)
Hill, R.: A lens for the whole sky photographs. Roy. Met. Soc. Journ., **50**, 211, 227–235 (1924)
Lambert, J.H.: Die freye Perspektive, oder Anweisung, Jeden Perspektivischen Aufriß von freyen Stücken und ohne Grundriß zu verfertigen. Heidegger und Compagnie, Zurich (1759)
Lambert, J.H.: Photometria sive de mensura et gradibus luminis, colorum et umbrae. Augsburg (1760), German translation by E. Anding: Lamberts Fotometrie. Klett Publ., Leipzig (1892), English translation by D.L. DiLaura: Photometry, or On the measure and gradation of light, colors and shade. Publ. IES Amer., New York (2001)
Longmore, J.: The full-field camera photometric technique in lighting research. The Photographic Journ., 104, 5, 133–141 (1964)
Lynes, J.A., Jackson, G.K., Burt, W., Cuttle, C.: The flow of light into buildings. Transac. Illum. Eng. Soc., **31**, 65–91 (1966)
Moon, P., Spencer, D.E.: The Photic Field. MIT Press, Boston (1981)
Pers, R.: Inluence des obstacles, montagnes ou batiments sur la durée d'insolation Appareil pour la photographie des nuages. Assoc. Fran. Pour l'Advancement des Science, 212–218 (1933)
Pleijel, G.: Globoskopet: Ett instrument för solljus- och dagsljusanalyser. (In Swedish), Byggmästaren, **31**, 10, 219–224 (1952)
Pleijel, G.: The computation of natural radiation in architecture and town planning. Statens nämnd för byggnadsforskning, Stockholm (1954)

Chapter 4
Propagation of Light in the Atmospheric Environment

4.1 Scattering and Absorption Phenomena in a Turbid Environment

Any daytime photometric measurements or observations outdoors reflect the momentary conditions of sunlight and skylight illuminance at ground level. The natural light sources determine the photic field adjusted by the atmospheric environment. Every constituent of the atmosphere, air molecules, aerosol and dust particles, water vapor droplets, and other gaseous compounds, forms the actual state of the photic field. Basically, the light propagated through the atmosphere undergoes scattering and absorption phenomena with different efficiencies across the visible spectrum. To characterize the daylight behavior it is necessary to understand the underlying physics, especially the radiative transfer of monochromatic radiation.

Systematic observations under cloud-free conditions have shown that the blue color of the sky is more intense at elevated altitudes, i.e., in mountain regions with low pollution levels. Lord Rayleigh (John William Strutt) explained this effect theoretically in his excellent study (1871). He originally considered the scattering by nonabsorbing targets that are much smaller than the wavelength of incident radiation. This condition is perfectly satisfied for air molecules. However, Kerker et al. (1978) showed that the accuracy of Rayleigh theory decreases as the absolute value of the complex refractive index of scatterers increases. In particular, it can be important for the smallest aerosol particles. Generally, the presence of aerosol particles in the scattering medium makes it turbid. Here, the turbidity is defined as the ratio of total (aerosol plus molecular) attenuation to the reference (molecular) attenuation. Although atmospheric gases may also absorb the incident radiation, their absorption bands are quite narrow and thus have negligible influence on human visual perception. This is the reason why gaseous absorption is traditionally disregarded in daylight simulations, although this may result in some errors.

Rayleigh theory dictates that the intensity of scattered radiation is inversely proportional to the fourth power of the wavelength λ, i.e., the red light is scattered

R. Kittler et al., *Daylight Science and Daylighting Technology*,
DOI 10.1007/978-1-4419-8816-4_4,

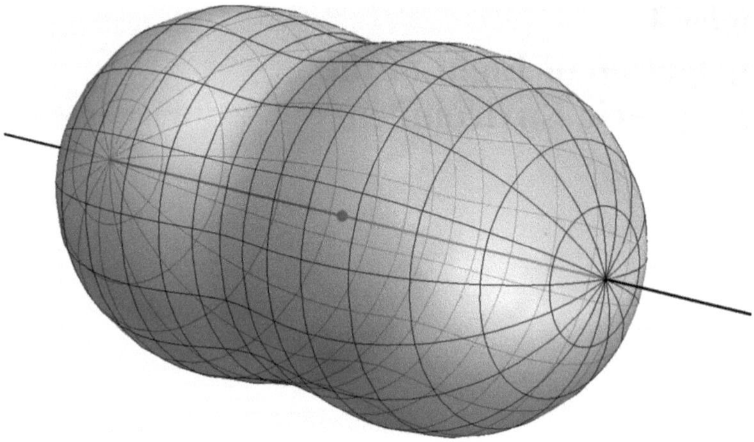

Fig. 4.1 Three-dimensional model of the scattering phase function in the Rayleigh approximation. A scattering particle is situated at the central point. The beam of light propagates from the *left* to the *right*

less efficiently than the blue light. In nonabsorbing media, the electromagnetic energy that is removed from an incident beam of light represents the portion that is scattered and forms skylight. As losses due to scattering dominate at the blue edge of the spectrum, the diffuse light of a cloudless sky also appears blue. Another facet of this effect is the red color of the solar disk during sunshine and sunset. The red light is scattered inefficiently; thus, the energetic losses at the red end of the visible spectrum are smaller than those at shorter wavelengths. Therefore, the solar disk appears red at low sun altitudes.

In principle, the scattering of electromagnetic radiation by a random medium is a complex process that can be mathematically formulated as a dedicated solution of Maxwell's equations subject to boundary conditions. The air molecules are small enough so they respond to the incident electromagnetic wave as elementary dipoles showing the symmetry of the scattering pattern along the forward to backward directions of beam incidence. In Rayleigh theory the forward to side scatter intensities are in the ratio 2:1 (see Fig. 4.1) and the complete dimensionless scattering pattern can be expressed as follows;

$$p_{\mathrm{R}}(\theta) = \frac{3}{4}(1 + \cos^2\theta), \tag{4.1}$$

where θ is the scattering angle, i.e., the angle that is contained by directions of incident and scattered beams. The relative scattering pattern $p_{\mathrm{R}}(\theta)$ is also sometimes called the normalized Rayleigh scattering phase function (Lenoble 1985).

The normalization condition requires that the integration of the scattering phase function over the whole solid angle must be 4π, thus satisfying the law of energy conservation. Since the Rayleigh formula (4.1) is independent of α, the angle measured in the plane perpendicular to the direction of beam incidence, the integration is

$$\int_{4\pi} p_{\mathrm{R}}(\theta)\mathrm{d}\Omega = \int_{0}^{\pi} p_{\mathrm{R}}(\theta)\,\sin\theta\,\mathrm{d}\theta \int_{0}^{2\pi} \mathrm{d}\alpha = \frac{3\pi}{2}\int_{0}^{\pi}(1+\cos^2\theta)\sin\theta\,\mathrm{d}\theta = 4\pi, \tag{4.2}$$

where $\mathrm{d}\Omega = \sin\theta\,\mathrm{d}\theta\,\mathrm{d}\alpha$ is the elementary solid angle.

Unfortunately, the atmosphere can scarcely be considered as an absolutely clean environment, even if observations are made under cloudless conditions at elevated altitudes. Normally, the atmosphere is constantly polluted by aerosol particles, showing a large spatial and temporal heterogeneity (Cadle 1966). Although the particles are concentrated mainly near the ground, some fraction of the aerosol population can be easily transported long distances or to elevated altitudes (e.g., owing to turbulent mixing, or by wind). This distribution strongly depends on the particle sizes. Typically, the airborne particles are of different origin, implying different microphysical properties, i.e., different size and shape distributions, chemistry, and also internal topologies. For instance, the particles emitted by vehicles are quite small, with sizes between 50 and 200 nm, and their residence time in the atmosphere is about 1 week. These particles may significantly contribute to the pollution in large cities, such as Vienna (Horvath et al. 1988). In contrast to on-road particles, soil particles are completely different in chemistry and their sizes may exceed 1 μm. Traditionally, atmospheric particulate matter is characterized by three main modes: nucleation, accumulation, and coarse (Berner and Lürzer 1980). Each of these modes shows specific scattering features. The nucleation-mode particles with radius a much smaller than the wavelength of incident radiation scatter the radiation proportionally to a^6 (see Sect. 6.3 in Van de Hulst 1957), whereas the scattering cross section of large particles (coarse mode) is a function of a^2. The scattering properties of ideally spherical and homogeneous particles of any size can be simulated by the conventional Mie theory (Mie 1908). The optical response of any particle to the incident electromagnetic radiation is usually expressed in terms of the effective cross section of the particle. That is not surprising when one thinks in simple geometrical relations. If a spherical particle is embedded in a nonabsorbing medium and illuminated by a plane wave, that particle behaves as an opaque target (circular disk in the case of spherical shape) and thus the net rate at which electromagnetic energy crosses the particle is proportional to its cross section. Thus, in a very rough approximation, the amount of monochromatic energy $F_{0\lambda}\pi a^2$ is removed from a beam with irradiance $F_{0\lambda}$. Following this concept, the scattering cross section C_{sca} in square meters is introduced into the light-scattering theory to characterize the efficiency of scattering by a single particle (Bohren and Huffman 1998). Basically, C_{sca} (m^2) is the ratio of the power scattered by the particle to the irradiance of an electromagnetic beam illuminating the particle. In a similar way, the absorption cross section C_{abs} is defined as the ratio of the power absorbed by the particle to the irradiance. Conveniently, the normalized efficiency factors for scattering and absorption, $Q_{\mathrm{sca}} = C_{\mathrm{sca}}/(\pi a^2)$ and $Q_{\mathrm{abs}} = C_{\mathrm{abs}}/(\pi a^2)$, are used rather than cross sections.

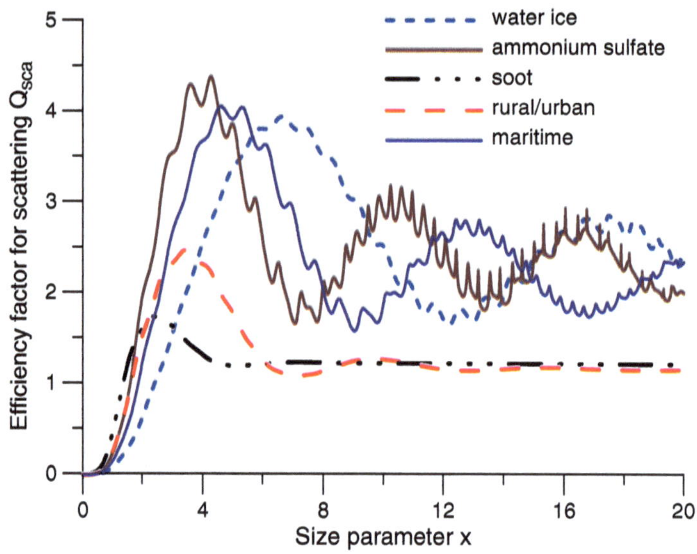

Fig. 4.2 Efficiency factors for scattering. The higher the value of Q_{sca}, the more efficiently the particle scatters. The size parameter x is equal to $2\pi a/\lambda$

Whereas dimensionless Q_{sca} and Q_{abs} typically range from 0 to 5, C_{sca} and C_{abs} continuously increase with increasing particle radius. It is well known that efficiency factors are oscillatory functions of size parameter x, which is defined as $2\pi a/\lambda$ (Figs. 4.2 and 4.3). Q_{sca} and Q_{abs} play a pivotal role in evaluation of scattering and absorption properties of polydisperse aerosol systems that coexist in Earth's atmosphere.

As evident from Fig. 4.2, both the slightly absorbing maritime aerosols and ammonium sulfate particles scatter twice as efficiently as water ice if the size parameter is about 4. In contrast, the opposite behavior is observed for $x \approx 8$. Such a shift is predominately due to different values of the real part n_r of complex refractive indices $n = n_r + \mathrm{i}\, n_i$ of these aerosol constituents. However, it is important to emphasize that the imaginary part n_i of the aerosol refractive index is responsible for absorption (Born and Wolf 1997). This can be easily shown for a plane wave propagating in an absorbing medium. With use of the concept of complex functions, the electric **E** and magnetic **H** fields of the electromagnetic beam of light can be expressed as $\mathbf{E}_0 \exp(\mathrm{i}\mathbf{k}\cdot\mathbf{x} - \mathrm{i}\omega t)$ and $\mathbf{H}_0 \exp(\mathrm{i}\mathbf{k}\cdot\mathbf{x} - \mathrm{i}\omega t)$, respectively. These formulae are compatible with Maxwell's equations. Here **k** is the wave vector and **x** is any vector characterizing the direction of beam propagation. In addition, ω is the angular frequency and t is the time. Because $\mathbf{k} = \omega c^{-1}(n_r + \mathrm{i}n_i)\hat{\mathbf{e}}$, the electric vector is $\mathbf{E}_0 \exp\{-(\hat{\mathbf{e}}\cdot\mathbf{x})\omega n_i/c\}\exp\{\mathrm{i}(\hat{\mathbf{e}}\cdot\mathbf{x})\omega n_r/c - \mathrm{i}\omega t\}$, where c is the speed of light and $\hat{\mathbf{e}}$ is a unit vector parallel to **k**.

It is evident that the amplitude of the electric field exponentially decreases as n_i increases. The corresponding formula can also be found for the magnetic field.

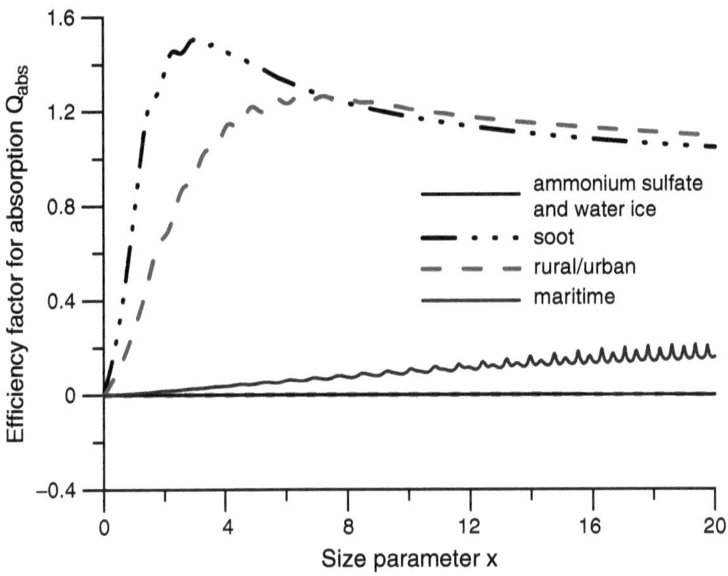

Fig. 4.3 Efficiency factors for absorption. The higher the value of Q_{abs}, the more efficiently the particle absorbs. The size parameter is equal to $2\pi a/\lambda$

In the case of combustion waste particles the value of n_i is quite high, thus implying an enhanced absorption efficiency (which is consistent with Fig. 4.3). One of the interesting features presented in Fig. 4.2 is a ripple structure with scattering efficiencies. This is not an artifact due to the finite accuracy of numerical computations, but it is a real effect of the optical resonances on spherical nonabsorbing (or slightly absorbing) particles. In principle, the distance between resonances can be related to the microphysical properties of the scattering particle as is apparent from the analysis by Chýlek (1990).

The particles in the atmosphere are grouped into large populations. More commonly, the aerosol systems are classified as continental, marine, and maritime. Each of these physical models can be characterized by a set of physical parameters, especially the size distribution function $f(a)$ in reciprocal meters to the fourth power and the mean effective refractive index n (Zuev and Krekov 1986). The function $f(a)$ can be understood as the total number of particles N contained in a unit volume of the air with radii in the interval $\langle a, a + \Delta a \rangle$. Mathematically, $f(a) = \mathrm{d}N/\mathrm{d}a$. If S_{11} is the scattering phase function for a single particle (Bohren and Huffman 1998), then the bulk scattering phase function of an ensemble of aerosol particles with size distribution $f(a)$ can be given by the formula

$$p_{\mathrm{A}}(\theta) = \frac{\lambda^2}{\pi k_{\mathrm{sca,A}}} \int_0^\infty S_{11}(x, \theta) f(a)\, \mathrm{d}a, \tag{4.3}$$

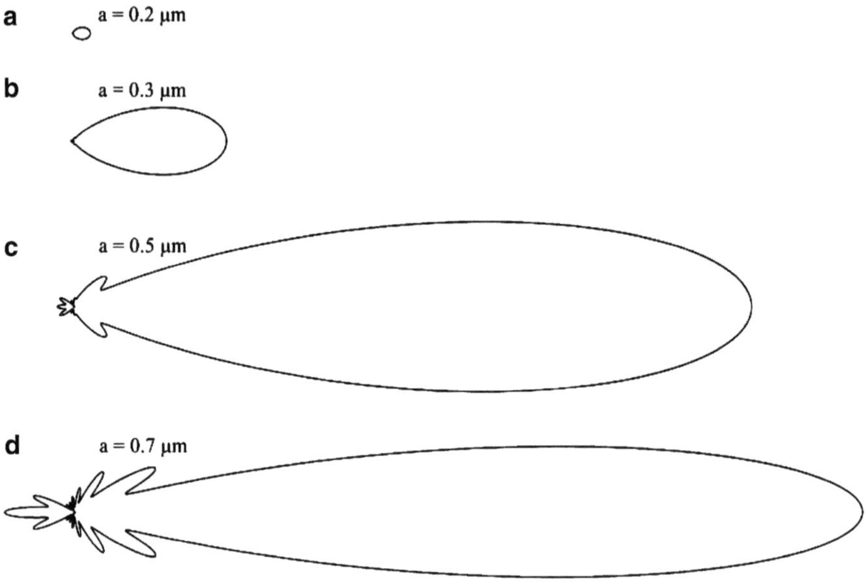

Fig. 4.4 Scattering phase functions S_{11} for single particles of different radii a. The wavelength of incident radiation is $\lambda = 550\,\mathrm{nm}$. The refractive index of aerosol particles is assumed to be $n = 1.5 + 0i$. The beam of light propagates from the *left* to the *right*

where $k_{\mathrm{sca,A}}\,(\mathrm{m}^{-1})$ is the volume scattering coefficient (Deirmendjian 1969; McCartney 1977):

$$k_{\mathrm{sca,A}} = \pi \int_0^{\infty} a^2 f(a) Q_{\mathrm{sca}}(a)\,\mathrm{d}a\,(\mathrm{m}^{-1}). \tag{4.4}$$

Exemplary patterns of the phase function S_{11} for single spherical particles of different radii are presented in Fig. 4.4.

It is evident from Fig. 4.4 that the larger the particle, the more light is scattered in its forward hemisphere. At the same time, the total amount of scattered energy increases with particle size – which is documented by a proportional increase of the intensities in all directions.

Formulae (4.3) and (4.4) are applicable to spherical particles. But usually the physical processes in the atmospheric environment do not support production of ideally spherical particles; therefore, S_{11} needs to be averaged over all particle orientations, thus giving $\langle S_{11} \rangle$. Similarly, Q_{sca} must be replaced by $\langle Q_{\mathrm{sca}} \rangle$ in (4.4) when irregularly shaped particles are taken into account. For nonspherical particles, a is defined as the radius of a volume-equivalent sphere.

There is no doubt that real particles are neither spheres nor symmetric homogeneous objects. As the nonspherical particles are randomly oriented in space, the orientational averaging becomes important. Assuming a monodisperse system of equally shaped particles, the orientational averaging translates into numerical computation of the triple integral

$$\langle S_{11}\rangle(x,\theta) = \frac{1}{8\pi^2}\int_0^{2\pi} \mathrm{d}\phi_{\mathrm{E}} \int_0^{\pi} \sin\Theta_{\mathrm{E}}\,\mathrm{d}\Theta_{\mathrm{E}} \int_0^{2\pi} S_{11}(x,\theta,\phi_{\mathrm{E}},\Theta_{\mathrm{E}},\Psi_{\mathrm{E}})\mathrm{d}\Psi_{\mathrm{E}}, \tag{4.5}$$

where the concept of Euler angles $\phi_{\mathrm{E}}, \Theta_{\mathrm{E}}, \Psi_{\mathrm{E}}$ is used to characterize the actual orientation of the particle with respect to the direction of beam incidence (Draine and Flatau 2007).

4.2 The Factors Influencing Light-Beam Propagation in the Atmosphere

Sunbeams entering the top of the atmosphere lose their energy continuously owing to absorption and scattering effects. The sum of these effects is called extinction. If the direction of beam propagation is characterized by the solar zenith angle Z_{S}, the spectral flux density of solar radiation $F(\lambda, h, Z_{\mathrm{S}})$ is reduced to $F - \mathrm{d}F$ after interaction with an elementary volume of the atmosphere $\mathrm{d}V$ at the altitude h occurs (Fig. 4.5). Then the amount of electromagnetic radiation removed from the incident flux density is

$$\mathrm{d}F(\lambda, h, z_{\mathrm{S}}) = -k_{\mathrm{ext}}(\lambda, h)F(\lambda, h, z_{\mathrm{S}})\mathrm{d}s\,(\mathrm{W\,m^{-2}\,nm^{-1}}), \tag{4.6}$$

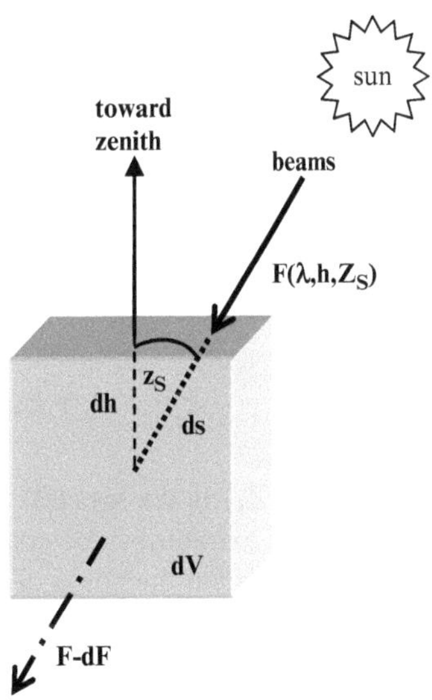

Fig. 4.5 Extinction of sunbeams after interaction with elementary volume $\mathrm{d}V$ situated at altitude h

where $k_{ext}(\lambda, h)$ is the volume extinction coefficient of an elementary volume of the atmosphere and ds is the path length of sunbeams as evident from Fig. 4.5. The extinction coefficient $k_{ext}(\lambda, h)$ has the dimension of reciprocal meters and comprises all attenuating constituents in dV. A usual target of lighting research is to simulate the daylight behavior, i.e., the bulk effects of the electromagnetic radiation in the visible spectral range. Therefore, the main focus is on the atmospheric constituents with broader attenuation spectra.

Aerosols and air molecules form the continua and attenuate the radiation at all wavelengths, specifically in the visible range. Analogous with (4.4), the aerosol extinction coefficient can be expressed as

$$k_{ext,A} = \pi \int_0^\infty a^2 f(a) Q_{ext}(a) \mathrm{d}a \ (\mathrm{m}^{-1}), \tag{4.7}$$

where $Q_{ext}(a) = Q_{sca}(a) + Q_{abs}(a)$ is the efficiency factor for extinction. The extinction coefficient of air molecules is due to Rayleigh scatter:

$$k_{ext,R}(\lambda) = k_{sca,R}(\lambda) = \frac{8\pi^3 (n_R^2 - 1)^2}{3\lambda^4 N_R} \ (\mathrm{m}^{-1}), \tag{4.8}$$

where N_R and n_R are the concentration and refractive index of air molecules.

Integrating (4.6) over the optical path, the flux density becomes

$$F(\lambda, h, Z_S) = F_0(\lambda, h, Z_S) \exp\left\{ - \int_s k_{ext}(\lambda, h) \, \mathrm{d}s \right\} \ (\mathrm{W\,m^{-2}\,nm^{-1}}). \tag{4.9}$$

The total transmittance along the beam trajectory from the top of the atmosphere to the ground can be obtained as

$$T(\lambda, Z_S) = \exp\left\{ - \int_0^\infty \frac{k_{ext}(\lambda, h)}{\cos Z_S} \mathrm{d}h \right\}. \tag{4.10}$$

Frequently, (4.9) and (4.10) are written in a shortened form:

$$F(\lambda, Z_S) = F_0(\lambda) \exp\left\{ - \frac{\tau(\lambda)}{\cos Z_S} \right\} \ (\mathrm{W\,m^{-2}\,nm^{-1}}). \tag{4.11}$$

where $F(\lambda, Z_S)$ is the spectral flux density of direct solar radiation at the ground, $F_0(\lambda)$ is its extraterrestrial component assuming normal sunbeam incidence, and $\tau(\lambda)$ is the optical thickness of the atmosphere.

In addition to gaseous components that typically absorb in narrow spectral bands, ozone or water vapor may influence the solar radiation across a wider wavelength interval, thus possibly affecting the daylight simulations. Unlike the precipitable water that evolves quite quickly, the total ozone amount contained in

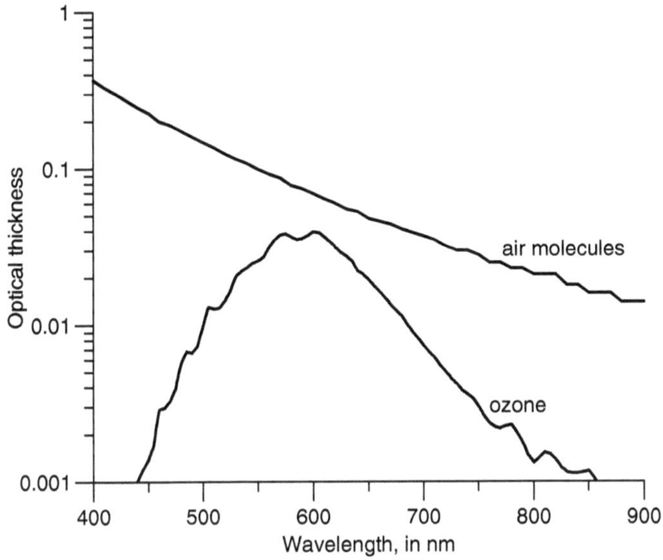

Fig. 4.6 Spectral profiles of the Rayleigh (molecular) and ozone optical thicknesses of the atmosphere, respectively. The total amount of ozone is 0.3 atmospheric centimeters

the atmospheric column fluctuates slowly. Assuming an average value of the total ozone amount of 300 Dobson units, the optical thickness of ozone may be half of the Rayleigh optical thickness at $\lambda \approx 600\,\mathrm{nm}$ (Fig. 4.6).

The molecular optical thickness is an integral of $k_{\mathrm{ext,R}}(\lambda)$ over the whole atmosphere. Although not indicated in (4.8), both N_{R} and n_{R} depend on the altitude h. According to Zuev and Krekov (1986) or Teillet (1990), the Rayleigh optical thickness can be written in the simpler form

$$\tau_{\mathrm{R}}(\lambda) = \frac{0.00879}{\lambda^{4.09}}. \tag{4.12}$$

The optical thickness of aerosol particles,

$$\tau_{\mathrm{A}}(\lambda) = \int_0^\infty k_{\mathrm{ext,A}}(\lambda, h)\,\mathrm{d}h, \tag{4.13}$$

is a function of their microphysical characteristics, especially the size distribution function. Figure 4.7 documents the possible differences between spectral profiles of $\tau_{\mathrm{A}}(\lambda)$ for three model size distributions.

The factor $1/\cos Z_{\mathrm{S}}$ in (4.10) can be removed from the integral and replaced by the well-known optical air mass m (Kasten and Young 1989). The function m was introduced in meteorology in a tabulated form much earlier, e.g., by Bemporad (1904). However, such a substitution results in two errors. The first one relates to

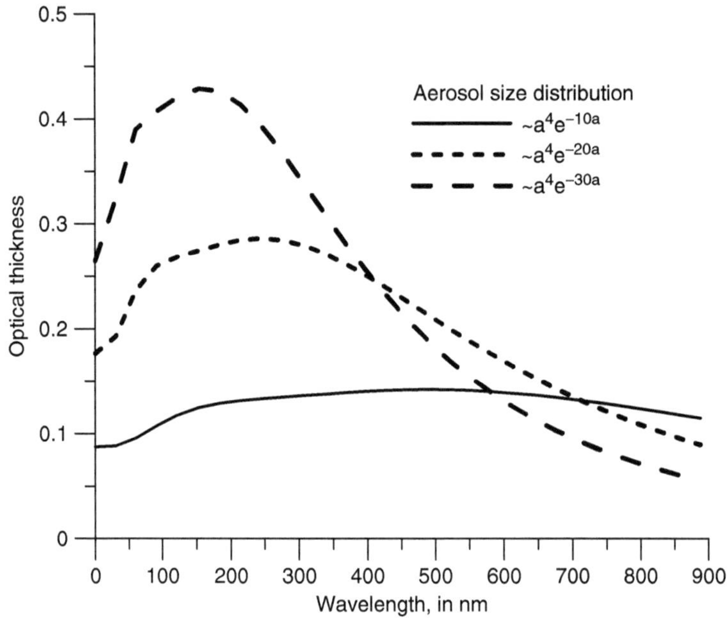

Fig. 4.7 Spectral profiles of the aerosol optical thickness assuming three different size distributions. The nonabsorbing particles with refractive index $n = 1.5 + 0.0\mathrm{i}$ are taken into consideration

the fact that the molecular atmosphere is vertically stratified. As a consequence, the zenith angle changes as the beam propagates through the atmospheric environment. The physics behind this phenomenon is the well-known Snell's law of refraction (Knight 2004). In respect to Snell's law, (4.10) should be written in a more correct form:

$$T(\lambda, Z_S) = \exp\left\{ - \int_0^\infty \frac{k_{\mathrm{ext}}(\lambda, h)}{\cos \tilde{z}(\lambda, Z_S)} \mathrm{d}h \right\}, \tag{4.14}$$

where $\tilde{z}$ is the zenith angle of sunbeams at altitude h resulting in Z_S being observed at the ground. The second potential error originates because $k_{\mathrm{ext}}(\lambda, h)$ is the attenuation coefficient for all atmospheric constituents that coexist at altitude h. However, these substances are not identically distributed along the vertical and therefore the optical masses of water vapor, aerosols, ozone, and other atmospheric elements cannot be easily replaced by a single function. In practical computations, (4.14) is used in the form

$$T(\lambda, Z_S) = \prod_{i=1 \ldots I} T_i(\lambda, Z_S), \tag{4.15}$$

where $T_i(\lambda) = \exp\{m_i(Z_s)\tau_i(\lambda)\}$ is the transmission function for a beam propagated along an inclined trajectory and computed for the ith atmospheric constituent (e.g., for aerosols, air molecules, ozone, water vapor, etc.).

The atmospheric constituents are usually well mixed and thus constantly influence the sunbeams through absorption and scattering. The relative contribution of the scattering to the bulk extinction can be described by single-scattering albedo $\tilde{\omega}$, which, in general, can be expressed as the ratio

$$\tilde{\omega} = \frac{k_{\text{sca}}}{k_{\text{sca}} + k_{\text{abs}}}. \tag{4.16}$$

In the case of air molecules, the attenuation is due to Rayleigh scatter, so $\tilde{\omega}_{\text{R}} \approx 1$. Single-scattering albedo of aerosol particles varies with their sizes and chemistry, but also with their morphologies. Many particles scatter efficiently in the visible spectral range, thus suggesting the elevated values of $\tilde{\omega}_{\text{A}}$. For instance, desert dust particles or sulfates (Torres et al. 2002) are almost nonabsorbing, so their single-scattering albedo is large. But carbonaceous species and black carbon absorb too much, resulting in $\tilde{\omega}_{\text{A}}$ smaller than 0.5 (strongly absorbing black carbon particles can have $\tilde{\omega}_{\text{A}} \approx 0.2$) (Horvath et al. 2002). The portion of the attenuated radiation that is scattered in all directions undergoes the same processes as the original sunbeam. This means the scattered radiation is absorbed and secondarily scattered, and consequently it contributes to the diffuse component of ground-reaching radiation. This process is called multiple scattering and can be theoretically solved depending on the atmospheric optical thickness $\tau(\lambda)$. If $\tau(\lambda)$ is smaller than approximately 0.5, the higher scattering orders can be evaluated using the method of successive orders (MSO) (van de Hulst 1980; Wendisch and von Hoyningen-Huene 1991).

4.3 Singly and Multiply Scattered Diffuse Light

The electromagnetic beams entering an elementary volume d$V$of the atmosphere (Fig. 4.5) are subject to scattering. The angular behavior of scattered radiation is ruled by many factors, especially the microphysical properties of aerosol particles as well as the relative concentrations of the atmospheric constituents. In an aerosol–molecular medium, the scattered signal at wavelength λ is given as the sum of scattering phase functions ((4.1), (4.3)) weighted by the corresponding volume scattering coefficients ((4.8), (4.4)), i.e.,

$$k_{\text{sca}}(\theta) = \frac{p_{\text{R}}(\theta)}{4\pi} k_{\text{sca,R}} + \frac{p_{\text{A}}(\theta)}{4\pi} k_{\text{sca,A}} \, (\text{m}^{-1}\text{sr}^{-1}). \tag{4.17}$$

Since the scattered beams propagate in all directions, the phase functions are divided by the whole solid angle, which is 4π. Any elementary volume acts as a source of electromagnetic radiation that irradiates the surrounding atmosphere and this process results in further scattering events (see Fig. 4.8).

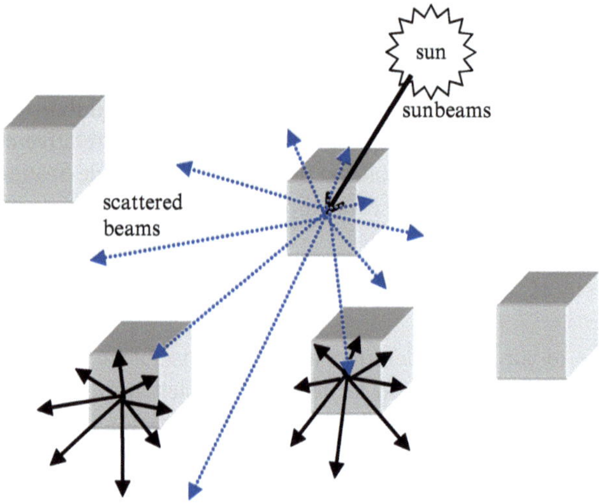

Fig. 4.8 The principle of multiple scattering in Earth's atmosphere. An elementary volume irradiated by the electromagnetic wave scatters the radiation in all directions. The directionality of the scattered radiation is ruled by (4.17)

The radiance observed at the ground can be obtained as a superposition of all independent signals from all elementary volumes of the atmosphere. However, the complexity of the problem increases with the number of scattering events, even in the Rayleigh scattering regime (Dave 1964; Kattawar et al. 1976). Even if the polarization effects are disregarded, the geometry of beam propagation makes the solution difficult (Kocifaj 1992). When the atmosphere is optically thin with $\tau(\lambda)$ below 0.5, the MSO is usually applied to determine the total radiance $L_{eT}(\lambda, Z, A)$, where Z and A are the zenith angle and azimuth of the observed sky element. A basic assumption in this method is that the efficiency of the scattering rapidly decreases with increasing number of scattering events. Then, $L_{eT}(\lambda, Z, A)$ can be formulated as a finite series:

$$L_{eT}(\lambda, Z, A) = \sum_i L_{ei}(\lambda, Z, A)\,(\mathrm{W\,m^{-2}\,sr^{-1}}), \tag{4.18}$$

where it is usually sufficient to confine the series to two or three terms. The index i is used for numbering the scattering events (orders). In the first-scattering-order approximation $L_{eT}(\lambda, Z, A) = L_{e1}(\lambda, Z, A)$. The expressions for ground-reaching radiance in the first-scattering-order approximation can be formulated in terms of the MSO easily. This is demonstrated in the following derivations. Let the sunbeams undergo only one scattering event at altitude h. Then in accordance with (4.9) the flux density of electromagnetic radiation incident on the elementary volume situated at altitude h is

$$F(\lambda, h, Z_S) = F_0(\lambda, h, Z_S)T(\lambda, h, Z_S)\,(\mathrm{W\,m^{-2}\,nm^{-1}}), \tag{4.19}$$

where, analogous with (4.10), the altitude-dependent transmission function can be expressed as follows:

$$T(\lambda, h, Z_S) = \exp\left\{ -\int_h^\infty \frac{k_{ext}(\lambda, h')}{\cos Z_S}\, dh' \right\}. \tag{4.20}$$

The intensity of scattered radiation depends on the direction of sunbeams incident on the elementary volume as well as on the direction of the scattered beam. Translated to spherical geometry, the scattering angle θ is

$$\cos\theta = \cos Z_S \cos Z + \sin Z_S \sin Z \cos(A_S - A), \tag{4.21}$$

where A_S and A are azimuth angles of the sun and the observed sky element, respectively. The scattered signal is proportional to the product of $k_{sca}(\theta)$ (4.17) and $F(\lambda, h, Z_S)$ (4.19). Since the path length of a beam in dV is $dh/\cos Z$, the first-scattering-order radiance observed at the ground is

$$L_{e1}(\lambda, Z, A) = \int_0^\infty F(\lambda, h, Z_S) k_{sca}(\theta, h) \frac{T(\lambda, 0, Z)}{T(\lambda, h, Z)} \frac{dh}{\cos Z} (\mathrm{W\,m^{-2}\,sr^{-1}}). \tag{4.22}$$

Here $k_{sca}(\theta, h)$ is a decreasing function of altitude h and θ is defined by (4.21). The ratio of transmission functions $T(\lambda, 0, Z)/T(\lambda, h, Z)$ characterizes the attenuation of the scattered beam along the trajectory from the elementary volume to the ground. The refraction of light beams in a stratified atmosphere is a function of altitude, so the term $\cos^{-1} Z$ is generally replaced by the optical air mass as a function of Z, i.e., $m(Z)$. Equation (4.22) then reads

$$L_{e1}(\lambda, Z, A) = m(Z) \int_0^\infty F(\lambda, h, Z_S) k_{sca}(\theta, h) \frac{T(\lambda, 0, Z)}{T(\lambda, h, Z)}\, dh \,(\mathrm{W\,m^{-2}\,sr^{-1}}). \tag{4.23}$$

The singly scattered beams propagate to the upper as well as the lower hemispheres, thus generating the upward and downward radiances. Part of the electromagnetic radiation propagating upward is definitely lost to space and can be observed by satellites orbiting Earth. However, some fraction of upward-emitted radiation is scattered downward and consequently detected at the ground as diffuse, multiply scattered radiation. Also singly scattered radiation originally directed downward is added to this multiply scattered component, and both are characterized as a contribution of higher scattering orders. The MSO enables determination of the contribution of the ith scattering order on the basis of the known $(i - 1)$th scattering order. In other words, the contribution of higher scattering orders can be computed recursively using a general concept. Kocifaj and Lukáč (1998) expressed the ith radiance as a triple integral of the $(i - 1)$th radiance, accepting also the Lambertian reflectance of Earth's surface.

In a more general case, the integrodifferential radiative transfer equation has to be solved (Minin 1988). Transforming the altitude-based coordinate system to the oppositely oriented optical-thickness-based coordinate system, the radiative transfer equation results in the following solution for the total radiance:

$$L_{eT}(\tau_h, Z, A) = \int_{\tau_h}^{\tau} J(\tau', Z, A) \exp\left\{\frac{\tau_h - \tau'}{\cos Z}\right\} \frac{d\tau'}{\cos Z} H\left(\frac{\pi}{2} - Z\right)$$
$$- \int_{0}^{\tau_h} J(\tau', Z, A) \exp\left\{\frac{\tau_h - \tau'}{\cos Z}\right\} \frac{d\tau'}{\cos Z} H\left(Z - \frac{\pi}{2}\right) \; (\mathrm{W\,m^{-2}\,sr^{-1}}), \quad (4.24)$$

where $\tau = \tau_R + \tau_A$ is the total optical thickness of the atmosphere computed using (4.12) and (4.13), τ_h is the corresponding optical thickness at altitude h, and $H(x)$ is the Heaviside function, which is a unity if $x>0$ and zero for $x<0$. All the functions in (4.24) except for Z and A depend on wavelength, but this is not indicated for the sake of brevity. The function J is called a source function and is used to generate the radiance. After many nontrivial manipulations, J can be written in the form

$$J(\tau_h, Z, A) = \frac{\tilde{\omega}}{4\pi} F_0 p(\tau_h, \theta) \exp\left\{-\frac{\tau_h}{\cos Z_S}\right\} + \tilde{\omega} \int_0^{2\pi} dA' \int_0^1 \frac{p(\theta)}{4\pi} \frac{d\mu'}{\mu'}$$
$$\int_{\tau_h}^{\tau} J(\tau', \arccos \mu', A') \exp\left\{\frac{\tau_h - \tau'}{\mu'}\right\} d\tau' - \tilde{\omega} \int_0^{2\pi} dA' \int_{-1}^{0} \frac{p(\theta)}{4\pi} \frac{d\mu'}{\mu'}$$
$$\int_0^{\tau_h} J(\tau', \arccos \mu', A') \exp\left\{\frac{\tau_h - \tau'}{\mu'}\right\} d\tau' \; (\mathrm{W\,m^{-2}\,sr^{-1}}), \quad (4.25)$$

where $\tilde{\omega}$ (4.16) is the single-scattering albedo of the atmosphere at altitude h, and the pair $[A', \mu' = \cos Z']$ characterize the azimuth and cosine of the zenith angle measured in the local coordinate system at a given optical depth τ'. The scattering phase function $p(\tau_h, \theta)$ of the molecular–aerosol medium satisfies the same normalization condition as used for Rayleigh scatter (4.2), i.e., $\int p(\tau_h, \theta) d\Omega = 4\pi$. The quantity F_0 on the right-hand side of (4.25) is the extraterrestrial flux density of direct solar radiation and was introduced in (4.11). At first glance it is evident that (4.25) can be solved iteratively, starting with some well-chosen zero approximation. Among many possibilities, the zero approximation of $J(\tau_h, Z, A)$ can be calculated from the known flux density of direct solar radiation:

$$J^0(\tau_h, Z, A) = \frac{\tilde{\omega}}{4\pi} F_0 p(\tau_h, \theta) \exp\left\{-\frac{\tau_h}{\cos Z_S}\right\} \; (\mathrm{W\,m^{-2}\,sr^{-1}}), \quad (4.26)$$

where the diffuse component of the radiation field has been disregarded. In the second step, $J^0(\tau', Z', A')$ replaces $J(\tau', \arccos\mu', A')$ on the right-hand side of (4.25). As a result, the first iteration of source function $J^1(\tau_h, Z, A)$ is obtained. The entire procedure is continuously repeated until the deviation of $|J^n(\tau_h, Z, A) - J^{n-1}(\tau_h, Z, A)|$ is smaller than a predefined error margin. Afterward, the final solution is placed into (4.24) and the spectral radiance in a turbid atmosphere with multiple scattering effects is determined this way. The MSO is efficient if the number of scattering orders is 1–2, but it becomes inefficient if higher scattering orders need to be simulated. This is because superimposed triple integrals lead to extremely increased computing time. The integrodifferential equation is a more convenient approach, but its numerical implementation is nontrivial.

4.4 Relation Between Scattering Phase Function and Indicatrix

The diffuse skylight observed at the ground can be theoretically modeled as the integral of spectral radiance:

$$L_{\mathrm{VT}}(\tau, Z, A) = \int_0^\infty V(\lambda) L_{eT}(\lambda, \tau, Z, A)\, \mathrm{d}\lambda\, (\mathrm{cd/m^2}), \tag{4.27}$$

where $V(\lambda)$ is the spectral luminous efficiency for an individual observer, otherwise called the photopic luminosity function (Stroebel and Zakia 1993). This function peaks at 555 nm and drops to zero at the boundaries of the visible spectrum. Although the derivations shown in the previous section suggest that an analytical solution for the light field at the ground cannot be found without heavy simplifications, $L_{\mathrm{VT}}(\tau, Z, A)$ is traditionally approximated by the product of two independent broadband functions, i.e., gradation function and indicatrix (Kittler et al. 1997; Li et al. 2005). This simplification follows the radiative transfer solved in a single-scattering approximation (Tousey and Hulburt 1947).

Gradation is assumed to be an explicit function of the observational zenith angle,

$$\varphi(Z) = 1 + a \exp\left[\frac{b}{\cos Z}\right], \tag{4.28}$$

whereas the indicatrix,

$$f(\chi) = 1 + c\left\{\exp(\chi d) - \exp\left(\pi \frac{d}{2}\right)\right\} + e\cos^2\chi, \tag{4.29}$$

varies only with the scattering angle χ. $a, b, c, d,$ and e are constants serving as scaling parameters for modeling the sky luminance patterns. Unlike the convention introduced in this chapter, Kittler et al. (1998) used the Greek symbol χ as the scattering angle, defined as the angular distance from the sun position and θ was the scattering angle in an elementary atmospheric volume. The indicatrix is a function that characterizes the angular behavior of scattered light in relative units.

Following the simplifications given above, (4.27) reduces to

$$L_{\mathrm{VT}}(\tau, Z, A) = L_Z \frac{\varphi(Z) f(\chi)}{\varphi(0^\circ) f(Z)}\ (\mathrm{cd/m^2}). \tag{4.30}$$

Here L_Z is the absolute zenith luminance and the product of $\varphi(0)$ and $f(Z)$ is luminance in relative units.

Nevertheless, the derivations previously presented clearly document that the relationship between the scattering phase function $p(\tau_h, \theta)$ and indicatrix $f(\chi)$ is not straightforward; $p(\tau_h, \theta)$ is an altitude-dependent function and even if the columnar phase function

$$p(\theta) = \frac{1}{\tau_h} \int_0^{\tau} p(\tau_h, \theta)\, \mathrm{d}\tau_h \tag{4.31}$$

is taken into consideration, there is still a problem with the separation of variables. Namely, although (4.30) assumes that indicatrix $f(\chi)$ and gradation function $\varphi(Z)$ can be isolated, it is not consistent with the more general solution ((4.24), (4.25)). Note, that the physical meaning of (4.31) is the averaging of $p(\tau_h, \theta)$ over the whole atmospheric column. However, using the single-scattering approximation in terms of the MSO (4.23), one can obtain the following formula after some mathematical manipulations:

$$\begin{aligned} L_{\mathrm{VT}}(\tau_V, Z, A) = & F_{0V} \frac{m(Z)}{m(Z_S) - m(Z)} \left[\mathrm{e}^{-m(Z_S)\tau_V} - \mathrm{e}^{-m(Z_S)\tau_V} \right] \cdot \frac{1}{4\pi} \\ & \left(\frac{p_{\mathrm{RV}}(\theta)\tau_{\mathrm{RV}} + p_{\mathrm{AV}}(\theta)\omega_{\mathrm{AV}}\tau_{\mathrm{AV}}}{\tau_V} \right) (\mathrm{cd/m^2}). \end{aligned} \tag{4.32}$$

Here, the wavelength-dependent characteristics have been smoothed and averaged over the visible spectral range. As a result, the spectral characteristics have been mapped to their "visual" broadband equivalents (V-indexed quantities in (4.32)). The parameter ω_{A} in (4.32) is the single-scattering albedo of aerosol ensembles. Now, the expression for $L_{\mathrm{VT}}(\tau, Z, A)$ is the product of two functions. The first one varies with Z, whereas the second one is a complex function of scattering angle θ. Such a result coincides well with the model (4.28–4.29).

4.5 Partial Conclusions

Light scattering in a turbid atmosphere is a complex phenomenon that requires special theoretical treatment to determine the spectral radiance and luminance measured at the terrestrial surface. Even if the homogeneous atmosphere is considered, the single-scattering approach might become inaccurate if the optical thickness of the atmosphere is large enough, say, above 0.5. Assuming the monochromatic radiation is propagated within the aerosol–molecular environment, one can define the spectral turbidity as the ratio of the total optical thickness to its Rayleigh component, i.e., $T(\lambda) = [\tau_R(\lambda) + \tau_A(\lambda)]/\tau_R(\lambda)$ (Zakey et al. 2004). This parameter was used for the first time by Linke (1922) in his excellent study. Analogically with Linke's turbidity parameter, the corresponding luminous equivalent T_V is also well accepted in illuminating engineering applications. The value of T_V can be determined experimentally and can be approximated by $T_V \cong 10\,\tau_V$, where τ_V is the spectrally averaged optical thickness of the atmosphere weighted by the photopic luminosity function $V(\lambda)$. If $\tau_V \approx 0.5$, the MSO can succeed in determining the spectral radiances (and integral luminance) when the turbidity of the cloudless atmosphere T_V is below 5. It was shown previously that gradation and indicatrix functions can be isolated also in the single-scattering approximation (4.32), but such an approach is correct only for T_V smaller than 5. Unfortunately, under real conditions the atmosphere is neither optically thin nor cloudless. This is the reason for the wide spread of the experimentally determined parameters $a, b, c, d,$ and e ((4.28), (4.29)) that should be otherwise constant for a given sky type. The characterization of real skies by the twin set of gradation function $\varphi(Z)$ and indicatrix $f(\chi)$ can be partly improved if multiple scattering is incorporated into the empirical models at least in a form similar to that presented in Kocifaj (2009). The problem with multiple scattering is especially important for clouds (Kokhanovsky and Macke 1999; Platnick 2001), which can have optical thicknesses 10 times or more larger than the typical optical thickness of a cloudless atmosphere. Multiple scattering can smooth the sky luminance patterns significantly, especially under overcast conditions. The optical behavior of a cloud depends on its spectral optical thickness, spectral reflectance, water content, altitude, position in the sky, size, shape, etc. In the new generation of sky luminance modeling, comprehensive consideration of these characteristics still does not exist. Thus, there is great scientific incentive for further daylight research to finally produce a more universal sky luminance/radiance model for turbid and cloudy atmospheres.

Appendix 4

Comparison of Trials to Measure and Model the Whole Range of the Indicatrix and Gradation Functions

Owing to the main influences of the momentary sunbeam direction and optical air mass, a homogeneous atmosphere can be characterized by two symmetrical functions determining the scattering and diffusion of skylight:

1. Under the clear sky conditions, the momentary air molecules and water vapor as well as aerosol particle content influence the scattering, and transmission

properties of the atmospheric layers. Thus, extraterrestrial sunbeams are spread in all directions into space and the resulting luminance distribution can be characterized by quasi-symmetrical luminance solids that are usually determined by their cover curves called indicatrices. Principally three basic types of indicatrices according to the presence of turbidity and cloudiness can occur:

- Under very clear and clean atmospheres the forward and backward scattering is roughly the same and close to the Rayleigh indicatrix.
- Under higher turbidities and partly cloudy conditions the forward scattering is prolonged and backward scattering is reduced, forming only a small tail.
- When dense cloudiness and/or fog scatter the sunbeams, so that the position of their original light source, i.e., the sun's position in the sky is no longer detectable, the luminance solid has the shape of a sphere, and the indicatrices are uniform and the relative indicatrix function is equal to 1.

2. Owing to the thickness of the atmospheric cover of the globe, the second basic influence on the sky luminance distribution is caused by the atmospheric optical air mass through which the sunbeams have to pass. Of course, the smallest thickness is in the direction of the zenith, then gradually rising toward the horizon, but the vertical gradation of sky luminance characterized by the gradation function has several tendencies:

   - Under clear and clean atmospheres when the sky is blue, the denser atmospheric layers close to the horizon owing to their turbidity have usually higher luminance and cause the gradation function increase toward the horizon.
   - In special cases when the dense fog or diffuse cloudiness produces totally uniform sky luminance (Lambert sky), the gradation function becomes unity.
   - Dull overcast skies are extremely dense in the horizontal directions and so the zenith luminance is twice or 3 times as high than at the horizon. Thus, the gradation function has to show this vertical luminance drop prevailing and characteristic for overcast sky conditions. The drop is usually associated with a unity indicatrix.

Because in arbitrary atmospheres both these indicatrix and gradation functions have multiplying influences, their true measurements can be done only when one function is stable so the other can be separated.

The possibility to measure the indicatrix function on the solar almucantar, i.e., the horizontal circle containing the sun position, or using the sky almucantar at different altitude where scan data are available was described by Kittler (1969, 1993) in connection with the CIE trend to standardize the clear sky luminance distribution (Kittler 1967) which was adopted by CIE (1973).

Sky luminance measurements were manually conducted only seldom and even less frequently analyzed, (e.g., clear sky patterns were measured at Bratislava on selected clear mornings in 1961 by Kittler 1962 to find the scattering effects). In 1990–1991 much quicker special indicatrix scans were performed also manually with a portable luminance meter on a tripod enabling the rotation around the solar almucantar.

However, neither simultaneous sky illuminances D_v nor zenith luminances L_{vZ}were available; thus, the appropriate sky type after L_{vZ}/D_v could not be identified or checked. A more sophisticated study deemed wise before sky luminance scans could be evaluated. The first American sky luminance measurements were also made manually, by Kimball and Hand (1921a, b), in Washington and in Chicago, and later measurements were made by Karayel et al. (1984) in San Francisco. From June 1985 to December 1986, data were recorded by a sky luminance scanner produced by Pacific Northwest Laboratories and used by an LBL team and are known as Berkeley scans. From roughly 16,000 all-sky scans, 88 typical sky conditions were selected for indicatrix and gradation analysis by Perez for US-SK project (Kittler et al. 1998). This early American scanner had a special scanning net with 10° vertical steps which was quite well suited for indicatrix and gradation analysis. Australian scans were recorded by a German PRC Krochmann scanner in Sydney (Hayman 1992) and Japanese scans from Tokyo (Igawa 1992) used an EKO scanner with the scanning net proposed by Tregenza (1987) with 12° vertical steps and various azimuth increments. Computer programs were provided for the American and European sky subdivisions and so during the research project Kittler et al. (1998) could analyze not only indicatrix functions, but also gradation functions from sky scans measured in Berkeley, Sydney, and Tokyo.

To derive scattering indicatrix courses from either luminance measurements along the sun or any sky almucantar or from a regular sky scan it is important to determine first the position of the normalizing sky element and its normalizing luminance $L_{90°}$. For different scan systems the scan step angle in appropriate computer programs has to be respected in order to calculate interpolated data of $f(90°)$ placement on the solar or sky almucantar, which is given generally after (A4.1) or (A4.2),

$$\chi = \arccos(\sin \varepsilon \sin \gamma_s + \cos \varepsilon \cos \gamma_s \cos A)\ (°), \tag{A4.1}$$

or when zenith angles are used,

$$\chi = \arccos(\cos Z_S \cos Z + \sin Z_S \sin Z \cos A)\ (°). \tag{A4.2}$$

Thus, for the normalizing function $f(90°)$ on the solar almucantar where $\varepsilon = \gamma_s$ or $Z_S = Z$ with$\chi = 90°$, i.e., $\cos \chi = 0$, the azimuth of this sky element taken from the sun meridian A (Fig. A4.1) is

$$A = \arccos(-\tan^2 \gamma_s) = \arccos(-\mathrm{cotg}^2 Z_s)\ (°). \tag{A4.3a}$$

Because $\cos A$ has a minus value, the A angle is in the second quadrant; therefore, e.g., the actual $A = 180° - \arccos(-\tan^2\gamma_s)$, i.e., larger than 90°.

For any sky almucantar with zenith angle Z and scattering angle equal to 90°:

$$A = \arccos\left[\frac{-\cos Z_s}{\sin Z_s}\frac{\cos Z}{\sin Z}\right] = \arccos(-\mathrm{cotg}\, Z_s\, \mathrm{cotg}\, Z)\ (°). \tag{A4.3b}$$

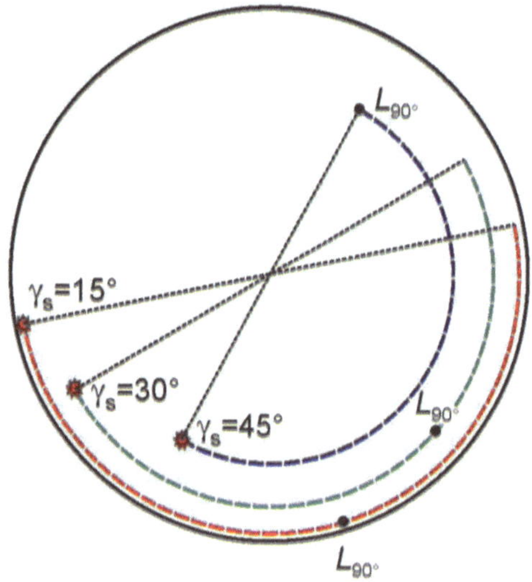

Fig. A4.1 Sun almucantars at different solar altitudes with $L_{90°}$ and $f(90°)$ positions

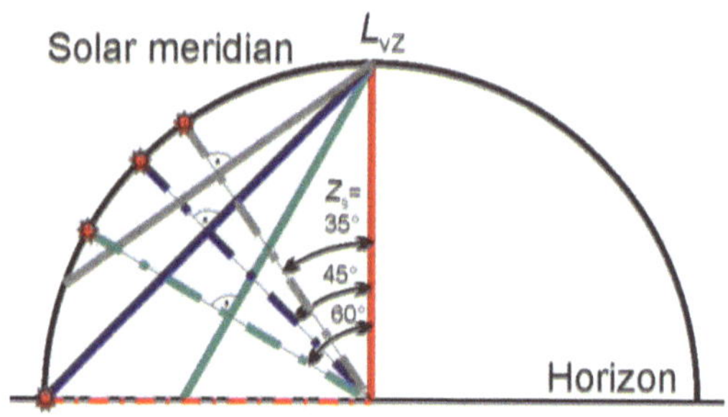

Fig. A4.2 Section circles for defining gradation courses

When sky scans are used for the indicatrix analyses, the regular scan step will usually not suit the needed A azimuth distance from the solar meridian; thus, interpolation between nearest luminance readings has to be applied. After the normalizing luminance $L_{90°}$ has been obtained, all measured luminances along the sun/sky almucantars can also be normalized as $L_Z/L_{90°}$ to describe the indicatrix course.

Interpolation is needed also for the gradation function $\varphi(Z)$ as a more spatial distribution of luminances has to be taken into account. The assumption to determine the gradation function has to avoid the influence of indicatrices, i.e., the measured luminance on a section circle of the sky hemisphere which passes the zenith and has the same angular extent from the sun position has to be found. Such section circles according to different sun positions are indicated in Fig. A4.2, which assumes the section is placed through the actual solar meridian. To find the full span of the gradation function, the sky scans with relatively low sun heights have to be selected to satisfy the assumption of a constant value of $f(\chi)$ corresponding to $\chi = Z_S$, and to obtain also the normalizing $\varphi(0°)$function at the zenith

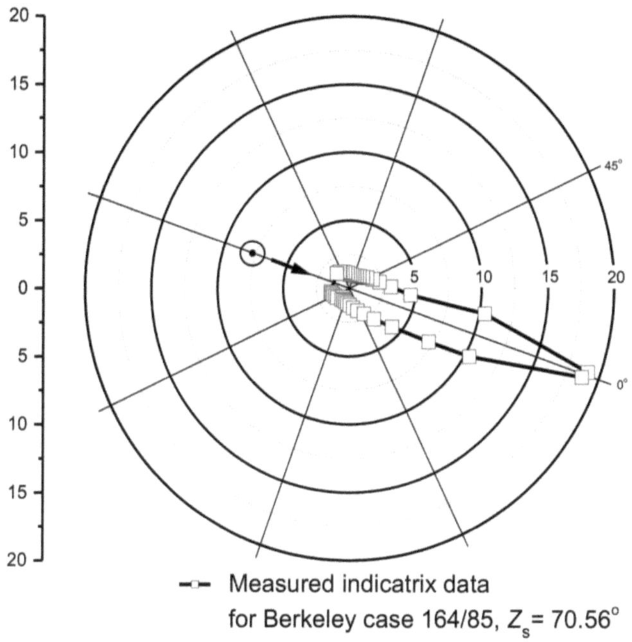

Fig. A4.3 Clear sky indicatrix in a polar plot with a linear scale

corresponding to the normalizing zenith luminance L_{vZ}. Owing to the assumption of the constant scattering influence on the section circle $\varphi(Z_s) = \varphi(0°)$ and thus $\cos Z_s = \cos Z_S \cos Z + \sin Z_S \sin Z \cos A$; thus,

$$A = \arccos\left[\frac{\cos Z_s(1 - \cos Z)}{\sin Z_s \sin Z}\right] (°). \quad \text{(A4.4)}$$

So for direct measurements and analysis of the indicatrix and gradation there is a disadvantage – for the high sun positions both functions are reduced and incomplete. But if such analysis has to be done using sky scans some error might also be caused by interpolation of luminances within the recorded scan net. To avoid these discrepancies, an ideal scanning step system could follow the solar almucantar and the "gradation circle" through the zenith and the same χ angle from the sun would be needed. The computer-regulated tracking along these circles indicated in Figs. A4.1 (in plan projection) and A4.2 (in section) could help the analysis precision.

To draw the indicatrix or gradation course either a polar plot (Fig. A4.3) resembling the luminance body or a semilogarithmic plot (Fig. A4.4), which seems to be more illustrative, can be used.

As an example, in the latter plot sky almucantar data were used to document the Berkeley clear sky case measured on June 13, 1985 at 5:41 p.m. archived as 164/85

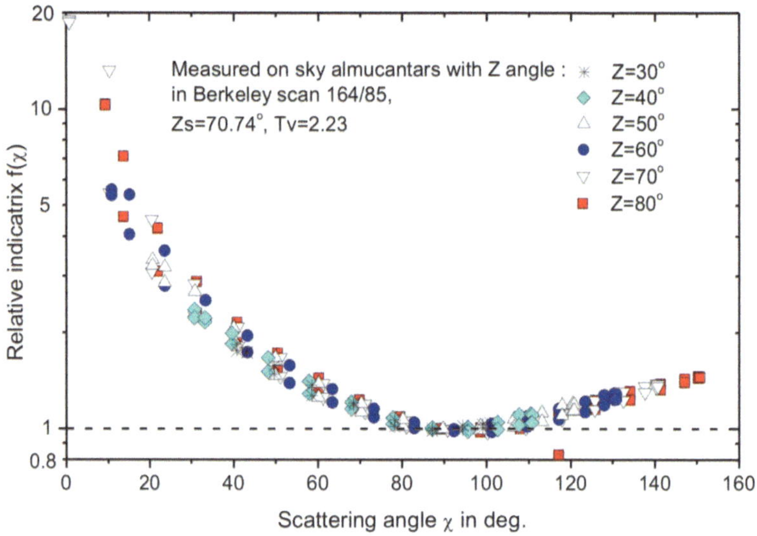

Fig. A4.4 Clear sky indicatrix in a semilogarithmic plot after measured luminance values

(the isoline luminance map made from scan data is shown in Fig. A4.5). The indicatrix analysis shown in Fig. A4.4 was done for six sky almucantars with zenith angles $Z = 30° - 80°$ in $10°$ steps. It is evident that under homogeneous conditions all indicatrix formations are in a condensed spread area with points close to each other, whether lower- or higher-placed sky almucantars are analyzed. However, the lower the sky almucantar and closer to sun position, the longer the forward and backward indicatrix course as is seen from the comparison of the red and cyan marks.

For the same scan also relative gradation was analyzed, showing almost a linear course in the polar plot in Fig. A4.6 but the real rather steep increase of luminance toward the horizon is evident in the semilogarithmic diagram in Fig. A4.7.

Further examples comparing measured indicatrix and gradation courses in typical cases of other clear and overcast skies derived from Berkeley scans were shown by Kittler et al. (1996). Similar analysis within a US–Slovak project (Kittler et al. 1998) were made from further selected sky scan measurements recorded in Berkeley as well as in Tokyo and Sydney. Further comparisons with the standard indicatrix and gradation functions can be found in the PhD thesis of Markou (2006), who analyzed sky scans measured in Watford, England, whereas Kobav (2006, 2009) utilized his own sky scanning data gathered in Lyon, France.

In addition to several historic measurements and approximations defining the gradation decrease on the overcast sky, e.g., by Schramm (1901) and (Kähler 1908), or Moon and Spencer (1940), few tried to determine the changes of the gradation function in the whole range, i.e., from the overcast to the clear homogeneous sky. A trial was published by Igawa et al. (1997) with the approximation following a sine function (4.18) or cosine function (4.19), i.e.,

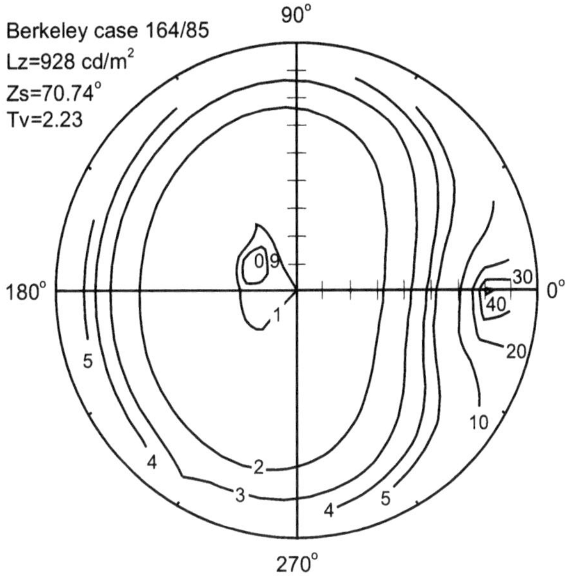

Fig. A4.5 Example of an isoline interpretation of a luminance scan

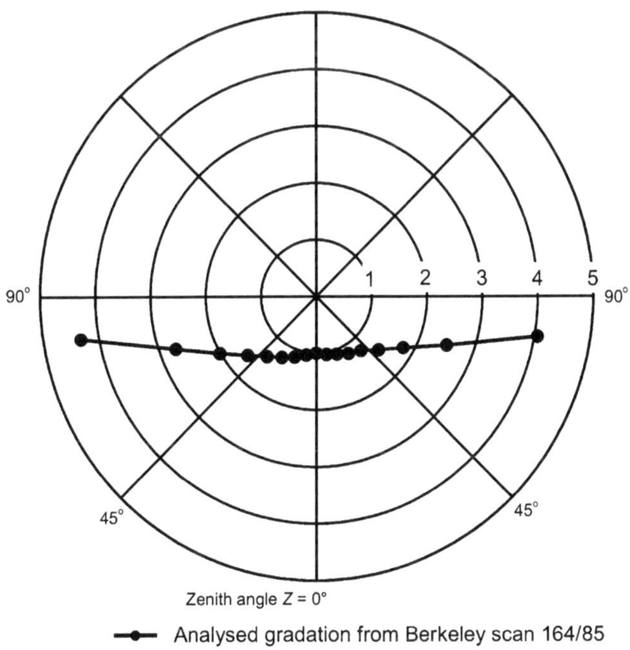

Fig. A4.6 Clear sky gradation in a polar plot

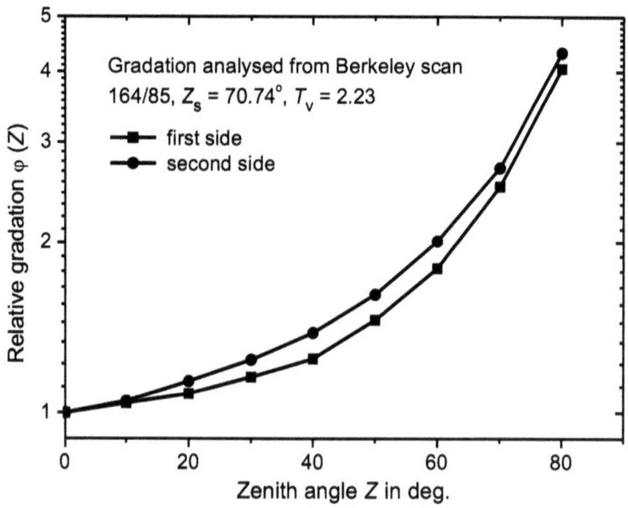

Fig. A4.7 Clear sky luminance gradation in a semilogarithmic plot

$$\frac{\varphi(\varepsilon)}{\varphi(90°)} = [1 + a_i(1 - \sin^{0.6}\varepsilon)] \text{ or } \frac{\varphi(Z)}{\varphi(0°)} = [1 + a_i(1 - \cos^{0.6}Z)], \tag{A4.5}$$

where a_i was defined as the function of the normalized globe illuminance N_{evg} (see (A4.6), (A4.7)) which roughly follows the corresponding typical range of skies from clear ($N_{\text{evg}} = 0.9$) to overcast ($N_{\text{evg}} = 0.15$).

The original normalization by Matsuzawa et al. (1997) was based on the highest correlation and lowest standard deviation between the sky luminance and global illuminance levels. This correlation was based on the model of the global illuminance under clear sky conditions and the normalization was set as the ratio

$$N_{\text{evg}} = \frac{E_{\text{vgm}}}{E_{\text{vgms}}(\gamma_s)}, \tag{A4.6}$$

where E_{vgms} as a function of solar altitude is defined in (A5.6) and E_{vgm} is the so-called relative global illuminance similar to G_v/E_v but defined as

$$E_{\text{vgm}} = \frac{mG_v}{\text{LSC}} \tag{A4.7}$$

and $a_i = -1.334N_{\text{evg}}^4 + 2.32N_{\text{evg}}^3 + 4.032N_{\text{evg}}^2 - 0.591N_{\text{evg}} - 1$.

At the same time, an analysis of various luminance scans by subtracting the gradation function was done by Kittler et al. (1997) and six typical exponential functions with parameters a and b after (A4.9) were proposed for the sky luminance patterns now adopted by the ISO/CIE (2003, 2004) standards shown in Table A4.1.

Table A4.1 Standard a and b parameters for gradation functions $\varphi(Z)$of typical homogeneous cloudiness and turbidity atmospheric conditions for use of (A4.9)

Type	Description	a	b
I	Dense layers of stratus cloudiness over dark ground surface	4.0	−0.70
II	Overcast sky with higher ground reflection	1.1	−0.8
III	Uniform sky luminance due to diffuse scattering (fog)	0.0	−1.0
IV	Partly cloudy sky	−1.0	−0.55
V	Clear and clean atmospheric conditions	−10	−0.32
VI	Clear turbid atmosphere (due to high level of water vapor, smog/aerosol content, etc.)	−1.0	−0.15

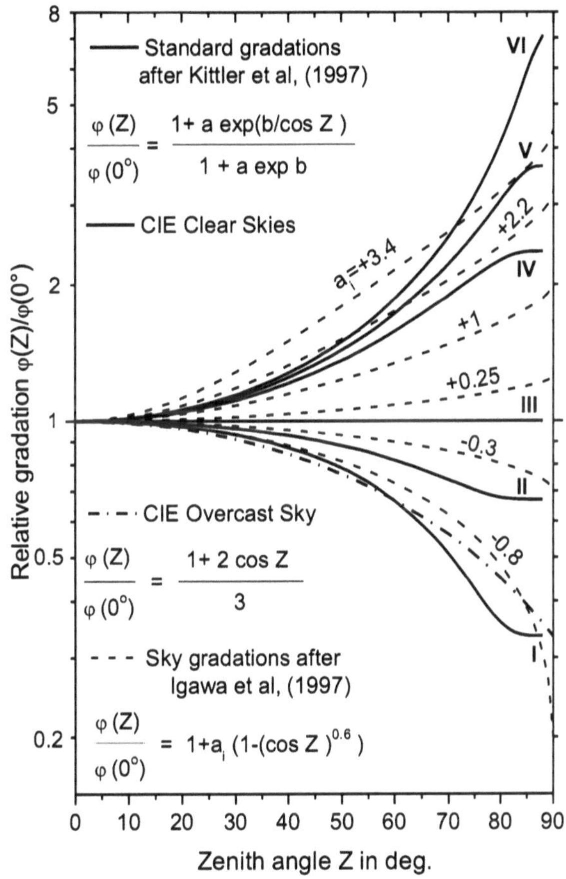

Fig. A4.8 Graphical comparison of approximated gradation functions

The comparison of three approximations is shown in Fig. A4.8, also with the simple CIE (1955) overcast sky gradation formula.

Later Igawa and Nakamura (2001) altered the gradation function concept to the exponential fit

$$\varphi(\varepsilon) = \frac{1}{1 + a_1/\exp(a_2 \sin \varepsilon)}, \qquad \text{(A4.8)}$$

where $a_1 = 5.5$ and $a_2 = 1.82N_{\mathrm{evg}}^2 - 5.82N_{\mathrm{evg}} + 2.26$..

However, the authors were probably disappointed with both previous gradation functions and therefore applied the ISO/CIE gradation function fully in Igawa et al. (2004).

All Igawa's studies indicated the link of luminance patterns to global illumination level with connection to gradation and indicatrix parameters, so if no L_{vZ} data are available to define the classification ratio $L_{\mathrm{vZ}}/D_{\mathrm{v}}$, the $G_{\mathrm{v}}/G_{\mathrm{vc}}$ ratio could serve as a sky luminance classification indicator if G_{vc} is the horizontal global illuminance under the CIE clear sky standard with $T_{\mathrm{v}} = 2$. Kittler and Darula (2000) documented that only the N_{evg} or $G_{\mathrm{v}}/G_{\mathrm{vc}}$ values without further information on the simultaneous solar altitude γ_{s} and turbidity T_{v} cannot replace fully the classification ratio $L_{\mathrm{vZ}}/D_{\mathrm{v}}$.

After studies testing gradations for every sky type by analyzing selected sky scan measurements obtained in Berkeley, in Tokyo, and in Sydney within the US–Slovak project (Kittler et al. 1998), the relative gradation function was chosen for the ISO/CIE (2003, 2004) standards after

$$\frac{\varphi(Z)}{\varphi(0^\circ)} = \frac{1 + a\exp(b/\cos Z)}{1 + a\exp b}. \qquad \text{(A4.9)}$$

The standard relative gradation and indicatrix function is respected in all standard sky distribution models. The only exception is in the Perez exponential indicatrix relation, where the missing component $-\exp(d\pi/2)$ in (4.29) distorts the normalization for the relative indicatrix $f(\chi) = L_\chi/L_{90^\circ}$ as well as the relative $f(90^\circ) = 1$ value.

References

Bemporad, A.: Zur Theorie der Extinktion des Lichtes in der Erdatmosphäre. Mitteilungen der Grossh. Sternwarte zu Heidelberg **4**, 1–78 (1904)

Bohren, C. F. and Huffman D. R.: Absorption and Scattering of Light by Small Particles. Wiley-VCH, Berlin (1998)

Born, M. and Wolf, E.: Principles of Optics: Electromagnetic Theory of Propagation, Interference and Diffraction of Light. Cambridge University Press, Cambridge (1999)

Berner, A. and Lürzer, C.: Mass size distributions of traffic aerosols at Vienna. J. Phys Chem. **84**, 2079–2083 (1980)

Cadle R. D., Particles in the Atmosphere and Space. Reinhold Publishing Corporation, New York (1966)

Chýlek, P.: Resonance structure of Mie scattering: distance between resonances. J. Opt. Soc. Am. **A 7**, 1609–1613 (1990)

Dave, J. V.: Importance of higher order scattering in a molecular atmosphere. J. Opt. Soc. Am. **54**, 307–315 (1964)

Deirmendjian, D.: Electromagnetic Scattering on Spherical Polydispersions. American Elsevier Publishing Company, Inc. New York (1969)
Draine, B. T. and Flatau, P. J.: User Guide for the Discrete Dipole Approximation Code DDSCAT7.1. eprint arXiv:1002.1505 (2007)
Horvath, H., Kreiner, I., Norek, C., Preining, O., Georgi, B.: Diesel Emissions in Vienna. Atmospheric Environment, **22**, 1255–1269 (1988)
Horvath, H., Arboledas, L. A., Olmo, F. J., Jovanović, O., Gangl, M., Kaller, C., Sánchez, C., Sauerzopf, H., Seidl, S.: Optical characteristics of the aerosol in Spain and Austria and its effect on radiative forcing. Journal of Geophysical Research, **107**, 4386 (2002)
Kasten, F., Young, A. T.: Revised optical air mass tables and approximation formula. Applied Optics **28**, 4735–4738 (1989)
Kattawar, G. W., Plass, G. N., Hitzfelder, S. J.: Multiple scattering radiation emerging from Rayleigh and continental haze layers. 1: Radiance, polarization, and neutral points. Applied Optics, **15**, 632–647 (1976)
Kerker, M., Scheiner, P., Cooke, D. D.: The range of validity of the Rayleigh and Thomson limits for Lorenz-Mie scattering. J. Opt. Soc. Am., **68**, 135–137 (1978)
Kittler, R., Perez, R., Darula, S.: A new generation of sky standards. Proceedings of the LUX Europa Conference, Amsterdam, 359–373 (1997)
Kittler, R., Darula, S., Perez, R.: A set of standard skies, characterizing daylight conditions for computer and energy conscious design. American–Slovak grant project US–SK 92 052 (1998)
Knight, R. D.: Physics for Scientists and Engineers with Modern Physics. Pearson (2004)
Kocifaj, M.: Vertical profile calculation of the attenuation coefficient of atmospheric aerosol based on the spectral radiance measurements of day-sky. Stud. Geoph. Et Geod., **36**, 376–391 (1992)
Kocifaj, M.: Sky luminance/radiance model with multiple scattering effect. Solar Energy, **83**, 1914–1922 (2009)
Kocifaj, M., Lukáč, J.: Using the multiple scattering theory for calculation of the radiation fluxes from experimental aerosol data. J. Quant. Spectrosc. Radiat. Transfer, **60**, 933–942 (1998)
Kokhanovsky, A. A., Macke, A.: The dependence of the radiative characteristics of optically thick media on the shape of particles. J. Quant. Spectrosc. Radiat. Transfer, **63**, 393–407 (1999)
Lenoble, J.: Radiative Transfer in Scattering and Absorbing Atmospheres: Standard Computational Procedures. A Deepak Publishing, Hampton (1985)
Li, H. W. D, Lau, C. C. S., Lam, J. C.: Predicting daylight illuminance on inclined surfaces using sky luminance data. Energy, **30**, 1649–1665 (2005)
Linke, F.: Transmissions – Koeffizient und Trübungsfaktor. Beiträge zur Physik der Atmosphäre, **10**, 91–103 (1922)
McCartney, E. J.: Optics of the atmosphere. John Wiley & Sons, Chichester (1977)
Mie, G.: Beiträge zur Optik trüben Medien, speziell kolloidaler Metallösungen. Ann. Phys. **25**, 377–445 (1908)
Minin, I. N.: Radiation Transfer Theory in Planetary Atmospheres (in Russian). Nauka, Moscow (1988)
Platnick, S.: Approximations for horizontal photon transport in cloud remote sensing problems. J. Quant. Spectrosc. Radiat. Transfer, **68**, 75–99 (2001)
Rayleigh, Lord: On the light from the sky, its polarization and colour. Philosophical Magazine, **41**, 107–120, 274–279 (1871)
Stroebel, L. D., Zakia, R. D. (ed.): The Focal Encyclopedia of Photography. Focal Press, (1993)
Teillet, P. M.: Rayleigh optical depth comparisons from various sources. Applied Optics, **29**, 1897–1900 (1990)
Torres, O., Bhartia, P. K., Herman, J. R., Sinyuk, A., Ginoux, P., Holben, B.: A long-term record of aerosol optical depth from TOMS observations and comparison to AERONET measurements. Journal of the Atmospheric Sciences, **59**, 398–413 (2002)
Tousey, R., Hulburt, E. O.: Brightness and polarization of the daylight sky at various altitudes above sea level. J. Opt. Soc. Am., **37**, 78–92 (1947)

Van de Hulst, H. C.: Light Scattering by Small Particles. John Willey & Sons (New York), Chapman & Hall, London (1957)
Van de Hulst, H. C.: Multiple light scattering. Academic Press (New York, London, Toronto, Sydney, San Francisco), (1980)
Wendisch, M., von Hoyningen-Huene, W.: High speed version of the method of "successive order of scattering" and its application to remote sensing. Beiträge zur Physik der Atmosphäre, **64**, 83–91 (1991)
Zakey, A. S., Abdelwahab, M. M., Makar, P. A.: Atmospheric turbidity over Egypt. Atmospheric Environment, **38**, 1579–1591 (2004)
Zuev, V. E. and Krekov, G. M.: Opticheskiye modeli atmospfery. (In Russian: Optical models of the atmosphere). Gidrometeoizdat, Leningrad (1986)

References to Appendix

CIE – Commission Internationale de l'Éclairage : Natural Daylight, Official recommendation. Compte Rendu CIE 13th Session, 2, part 3.2, 2–4 (1955)
CIE – Commission Internationale de l'Éclairage : Standardisation of luminance distribution on clear skies. CIE Publ. 22. CB CIE Paris (1973)
CIE – Commission Internationale de l'Éclairage : Spatial distribution of daylight – CIE Standard General Sky. CIE Standard S 011/E:2003, CB CIE Vienna (2003)
Hayman, S.: Luminance scans, global and diffuse illuminance data measured in Sydney-Mascot, 9th-22nd June 1992. Private Communication (1992)
Igawa, N.: Sky luminance distribution data, 12th-21st May 1992. Private Res. Report Takenaka Corp., Tokyo (1992)
Igawa, N., Nakamura, H., Matsuzawa, T., Koga, Y., Goto, K., Kojo, S.: Sky luminance distribution between two CIE standard skies. Part 2. Numerical equation for relative sky luminance distribution. Proc. Lux Pacifica Conf., E13-E18 (1997)
Igawa, N. and Nakamura, H.: All Sky Model as a standard sky for the simulation of daylit environment. Build. and Environm. **36**, 6, 763–770 (2001)
Igawa, N., Koga, Y., Matsuzawa, T., Nakamura, H.: Models of sky radiance distribution and sky luminance distribution. Solar Energy, **77**, 2, 137–157 (2004)
ISO – International Standardisation Organisation: Spatial distribution of daylight – CIE Standard General Sky. ISO Standard 15409:2004 (2004)
Karayel, M., Navvab, M., Ne'eman, E., Selkowitz, S.: Zenith luminance and sky luminance distribution for daylighting calculations. Energy and Buildings, **6**, 2–4, 283–291 (1984)
Kähler, K.: Flächenhelligkeit des Himmels und Beleuchtungsstärke in Räumen. Meteorol. Zeitschrift, **25**, 2, 52–57 (1908)
Kimball, H.H. and Hand I.F.: Sky brightness and daylight illumination measurements. Monthly weather review, **49**, 9, 481–488 (1921)
Kimball, H.H. and Hand I.F.: Sky brightness. I.E.S. Transac. **16**, 135–142 (1921)
Kittler, R.: Rozloženie jasu na bezoblačnej oblohe podľa meraní a teoretických vzťahov. (In Slovak: Luminance distribution on cloudless sky after measurements and theoretical relations.) Meteorologické zprávy, **15**, 2, 34–35 (1962)
Kittler, R.: Standardisation of outdoor conditions for the calculation of Daylight Factor with clear skies. Sunlight in buildings 273–285, Bouwcentrum, Rotterdam (1967)
Kittler, R.: A simple method of measuring and evaluating the atmospheric diffusion of sunlight when seeking the typical luminance patterns of the clear sky. Papers of the Symposium on environmental physics as applied to buildings in the tropics, **2**, D1-D14, Roorkee, U.P., India (1969)

Kittler, R.: Relative scattering indicatrix: Derivation from regular radiance/luminance distribution sky scans. Lighting Res. Technol., **25**, 3, 125–127 (1993)

Kittler, R., Perez, R., Darula, S.: Sky classification respecting energy-efficient lighting, glare and control needs. Proc.IESNA Annual Conf., Cleveland, 71–84 (1996), republished in Journ. Illum. Eng. Soc., **26**, 1, 57–68 (1997)

Kittler, R., Darula, S., Perez, R.: A new generation of sky standards. Proc. LuxEuropa Conf., Amsterdam, 359–373 (1997)

Kittler, R., Darula, S., Perez, R.: A set of standard skies characterizing daylight conditions for computer and energy conscious design. US-SK grant project. Polygrafia SAV, Bratislava (1998)

Kittler, R., Darula, S.: Determination of sky types from global illuminance. Lighting Res. Technol., **32**, 4, 187–193 (2000)

Kobav, M. B.: Uporaba digitalnega fotoaparata s širokokotnim objektivom v funkciji merilnika porazdelitve scetlosti neba. (In Slovenien: Utilising a digital photo camera with a fish-eye lens in the function of a sky luminance meter.) Proc. Razsvetljava Conf., Bled (2006)

Kobav, M. B.: Development and validation of methods used to compute time values of indoor daylight illuminances. PhD. Thesis, Faculty of Electr. Engineer., University of Ljublana (2009)

Markou, M.T.: A study on daylight conditions in European areas with different climate. (In Greek). PhD. Thesis, Department of Physics, University of Ioannina (2006)

Matsuzawa, T., Igawa, N., Nakamura, H., Koga, Y., Goto, K., Kojo, S.: Sky luminance distribution between two CIE standard skies. Part 1. Numerical equation for relative sky luminance distribution. Proc. Lux Pacifica Conf., E7-E13 (1997)

Moon, P. and Spencer, D.E.: Illumination from a non-uniform sky. Illum.Engineering, **37**, 10, 707–726 (1940)

Schramm, W.: Über die Helligkeitsverteilung am Himmel. Schriften d. Naturw. Vereins f. Schl.-Holst., **12**, 1, 81–129 (1901)

Tregenza, P.: Subdivision of the sky hemisphere for luminance measurements. Lighting Res. Technol., **19**, 1, 13–14 (1987)

Chapter 5
Sky Luminance Characteristics

5.1 Atmospheric Scattering of Sunlight Effecting Sky Luminance Distribution

Any sunbeam or directionally uniform luminous flux formed by parallel rays reaching the atmospheric border will strike tiny air molecules, aerosol particles, or water vapor droplets, which will cause its absorption, scattering, diffusion, and reflection into space. This phenomenon was studied earlier by Bouguer, but Weber (1885) was probably the first to suggest that the resulting luminance distribution into space should be characterized by an irregular luminance solid formed by directional elemental luminance in different directions relative to the original beam. It was assumed that such solids could be usually rotationally symmetrical around the original beam as an axis and could be specified by their sectional curve, called the scattering indicatrix.

Absolutely scattering diffuse media, e.g., dense fog and white overcast layers, produce a Lambertian diffuser (Lambert 1773), later called also Mie scattering (Mie 1908), when the relative indicatrix $f(\chi)$ is

$$L_\chi / L_{90^\circ} = f(\chi) = 1, \tag{5.1}$$

where the luminance in any direction to that in the direction perpendicular to the original beam is the same, thus their ratio is 1, i.e., the indicatrix can be imagined as a circle and the luminance solid forms a sphere. In the temperate climate region there are autumn days with dense fog with such a uniform unit indicatrix causing perfectly diffuse and low visibility conditions. Besides the relative indicatrix function, sometimes the absolute space indicatrix $s(\chi)$ is also specified, which in this case is

$$s(\chi) = \int_0^\pi \frac{f(\chi)}{K} \mathrm{d}\omega = 2\pi \int_0^\pi f(\chi) \sin \chi \, \mathrm{d}\chi = 1. \tag{5.2}$$

R. Kittler et al., *Daylight Science and Daylighting Technology*,
DOI 10.1007/978-1-4419-8816-4_5, © Springer Science+Business Media, LLC 2012

Thus, solving the integral when $f(\chi) = 1$ gives $K = 4\pi = 12.566$ and

$$s(\chi) = 1/12.566 = 0.07958, \tag{5.3}$$

which means that under ideal diffusion the beam luminance is spread into all directions with a 0.07958 part of the original directional luminance.

The English scientist Strutt, who later became Lord Rayleigh (1899), studying a perfect clear blue sky determined its relative scattering indicatrix function:

$$f(\chi) = 1 + \cos^2\chi. \tag{5.4}$$

The forward and backward luminance is double that in the direction perpendicular to the original beam luminance (see Fig. 4.1).

The absolute space indicatrix function (Kittler and Pulpitlova 1988) is

$$s(\chi) = \frac{1 + \cos^2\chi}{16.7552} = 0.0597\left(1 + \cos^2\chi\right). \tag{5.5}$$

This determines the equal forward and rearward scattering of molecules smaller than the spectral wavelength in the visible spectrum.

Pokrowski (1925), measuring real atmospheric scattering, tried to amend Rayleigh's function as he found a considerable elongation in the forward direction due to turbidity, but the endless elongation in the original sunbeam direction could not be corrected by the additional constants later proposed by Hopkinson (1954).

After unique measurements and visibility research of airborne objects during the war, Krat (1943) expressed the influence of turbidity on the indicatrix definition, introducing an exponential function,

$$f(\chi) = 1 + N[\exp(-3\,\mathrm{arc}\chi) - 0.009] + M\cos^2\chi, \tag{5.6}$$

where N due to turbidity or cloudiness was in the range between 0 and 17, but for clear skies was around 8–12, and formed the forward elongation of the indicatrix while M expresses the higher indicatrix tail with values between 0 and 0.5.

This function suited quite well not only Krat's measurements in Tashkent, but also those made by Pyaskovskaya-Fesenkova (1952, 1957) and by Lifshic (1965) at the Alma Ata observatory. At the same time Sobolev (1943, 1949) studied the general aspects of light diffusion in arbitrary atmospheres of Earth and other planets (Sobolev 1972).

The International Commission on Illumination (CIE) Expert Committee for daylight seeking a clear sky standard adopted two indicatrix functions to characterize it (CIE 1973) in different locations and under possible pollution expressed by T_v:

- One for a clear sky with low turbidity content (e.g., for countryside and mountainous regions) proposed by Kittler (1967)

$$f(\chi) = 0.91 + 10[\exp(-3\,\mathrm{arc}\chi)] + 0.45\cos^2\chi. \tag{5.7}$$

- The second one for a clear sky with higher turbidity (e.g., for towns and industrial regions) proposed by Gusev (1970)

$$f(\chi) = 0.856 + 16[\exp(-3\,\mathrm{arc}\chi)] + 0.3\cos^2\chi, \tag{5.8}$$

where χ is the angular distance from the original sunbeam with zenith angle Z_s to any arbitrary sky element with angular zenith coordinate Z and its azimuth angle taken from the sun meridian A_ζ, and is

$$\chi = \arccos(\cos Z_s \cos Z + \sin Z_s \sin Z \cos A_\zeta). \tag{5.9}$$

Further studies by Kittler (1985) indicated that N and M in (5.6) depend on the luminous turbidity factor T_v in the range between 1.5 and 72 (Kittler and Pulpitlova 1988) in the approximate relations

$$N = 4.3\,T_v^{\,1.9} \exp(-0.35\,T_v) \tag{5.10}$$

and

$$M = 0.71\,T_v^{-2}, \tag{5.11}$$

which fit roughly to parameters N and M in (5.6), i.e.:

- For the clear sky with low turbidity, $T_v = 2.45$.
- For the clear sky with higher turbidity, $T_v = 5.5$.
- For the overcast sky with extremely high turbidity, $T_v = 72$.

In the whole range of homogeneous skies the absolute space indicatrix function was derived (Kittler and Pulpitlova 1988, p. 27):

$$s(\chi) = f(\chi)/K, \tag{5.12}$$

where $K = 4\pi(1 + 0.041N + M/3)$.

A similar relation in a different form was published by Kondratyev (1954, p. 102), and Nagata (1971, p. 44) derived K values of 19.6 and 22.07 for CIE clear skies, which are close to those in Fig. 5.1.

Studies by Ambarcumian (1944) characterized the elongation of the indicatrix by the first coefficient of Legendre's polynomial function,

$$x_I = \frac{3}{2}\int_0^{\pi} f(\chi)\cos\chi\,\sin\chi\,\mathrm{d}\chi, \tag{5.13}$$

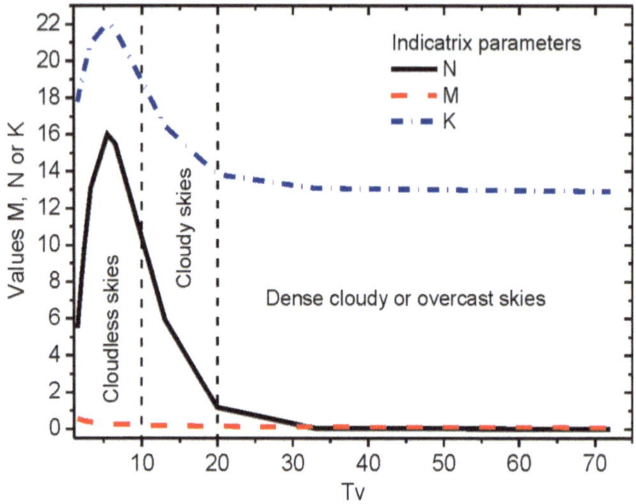

Fig. 5.1 Indicatrix parameters characterizing homogeneous atmospheres of different turbidities

which after integration yields $x_{\mathrm{I}} = 0.1153753\,N$, whereas earlier Gordov (1936) suggested a similar parameter:

$$\varepsilon_{\mathrm{I}} = x_{\mathrm{I}}/4 = 0.5 - 0.028844\,N. \tag{5.14}$$

All these parameters seem to be related to the luminous turbidity of a homogeneous atmosphere as shown in Fig. 5.1. Such an interrelation was later used to determine $\pi\,L_{\mathrm{vZ}}/D_{\mathrm{v}}$ or $L_{\mathrm{vZ}}/D_{\mathrm{v}}$ characterizing the sky luminance patterns, as will be shown later.

5.2 Luminance Distribution on the Densely Overcast Sky Vault

Whereas Bouguer studied and measured the luminance distribution on clear skies, Lambert introduced a uniform and unity sky luminance and probably believed that winter overcast skies are the critical ones. Owing to its simplicity, the latter became the basis for calculation and the concept of the daylight factor or sky factor criteria. Such a uniform luminance sky exists more seldom than realized before and is now also standardized among 15 typical skies (CIE 2003; ISO 2004).

However, luminance measurements of overcast skies in Europe by Schramm (1901) and Kähler (1908) and later in America by Kimball and Hand (1921) indicated that densely overcast skies have a gradated luminance. Kähler recommended defining it by a sine function of the elevation angle of the sky horizontal/almucantar circle ε for the ratio of luminance at this elevation to that on the horizon $L_{v\varepsilon}/L_{vh}$ in the range 2:1–3:1, i.e., in gradation with respect to horizon luminance L_{vh}

$$L_{v\varepsilon}/L_{vh} = 1 + \sin\,\varepsilon \tag{5.15}$$

to

$$L_{v\varepsilon}/L_{vh} = 1 + 2\sin\,\varepsilon. \tag{5.16}$$

A newer proposal taking zenith luminance L_{vZ} as a denominator by Moon and Spencer (1940) was accepted by the CIE (1955) for an overcast sky standard with the gradation $L_{vh}/L_{vZ} = 1:3$ using the sky element zenith angle Z instead of ε; thus,

$$L_Z/L_{vZ} = \frac{1}{3}(1 + 2\,\cos\,Z). \tag{5.17}$$

Later, Petherbridge (1955) found that (5.16) is approximately valid when the ground is covered by new snow which normalized by zenith luminance it is with gradation $L_{vh}/L_{vZ} = 1:2$, i.e.,

$$L_Z/L_{vZ} = \frac{1}{2}(1 + \cos\,Z). \tag{5.18}$$

At the same time, Fritz (1955) also realized that ground reflectance/albedo can influence the sky gradation and Steven and Unsworth (1980) introduced a more general expression:

$$L_{v\varepsilon}/L_{vZ} = \frac{1 + b_o\sin\,\varepsilon}{1 + b_o}. \tag{5.19}$$

where b_o coefficients were published by different authors in the range 1–2.

However, the recent study by Kocifaj (2010) expressing the aerosol and pollutant effects in the dense layers of the atmosphere indicated an exponential drop of luminances, rather than a cosine or sine drop, which makes a difference close to the horizon elevations mentioned also in the ISO (2004) and CIE (2003) standard documents.

It is evident that owing to multiple scattering the sun position is fully shaded by a cloud and is undetectable, i.e., the scattering indicatrix is equal to unity, and except for the horizon–zenith luminance gradation there are no azimuth deviations in sky luminance.

5.3 Sky Luminance Patterns on Arbitrary Homogeneous Skies

The first researcher who tried to define the sky luminance pattern by the indicatrix and gradation functions was probably Fesenkov and Pyaskovskaya (1934), Fesenkov (1955). Besides the scattering indicatrix influence, he assumed that owing to the rising thickness of the atmospheric layer toward the horizon the influence of the relative air mass has to be taken into account. However, his gradation function included a complex interdependence of the sunbeam penetration and air mass in the direction of the sun and sky element positions, respectively. A similar approach was followed later by Sobolev (1949, 1972) and Kittler (1986), whereas Nagata (1971, 1983) applied a more complicated version using the absolute spherical indicatrix. Some researchers tried to categorize either only clear skies (Liebelt and Bodmann 1979) or all sky patterns in a few roughly determined sky types for example (Perraudeau 1986, 1988).

Although these astronomic analytical models were derived for relatively very clear atmospheric conditions, several disadvantages opposed their practical application owing to complexity and restriction when higher turbidity and cloudiness had to be modeled.

Another approach was followed by Igawa et al. (2001, 2004) and Perez et al. (1993a, b) in trying to establish a new system of sky models that could be deduced from irradiance data and transformed via luminous efficacy to sky luminance values (see the appendices in Chap. 4 and in this chapter).

However, many cloudy skies are formed by different cloud cover and cloud type varieties and are often inhomogeneous, i.e., with the cloud layers spread nonuniformly over the entire sky vault. Because such nonuniformity occurs in many random patterns with uneven luminance distributions, the standard sky types are assumed to be quasi-homogeneous with fluent luminance patterns and even turbidity distributions. The cloud ratio D_v/G_v was suggested by Gillette and Treado (1985) to indicate the degree of conformity of a real sky luminance distribution to that of the perfect homogeneous sky model. A trial to express also luminance inhomogeneity influences was published by Perez et al. (1993b) with patchy cloudy sky luminance distributions simulated.

5.4 Standard Sky Luminance Patterns on General Skies

A simpler and more practical relative approximation was followed (Kittler 1967) to standardize first the clear sky by the CIE (1973) as well as later the ISO/CIE general sky (CIE 2003; ISO 2004) in the relative form where the sky luminance in an arbitrary sky element is normalized to the luminance at the zenith:

$$\frac{L_{\chi Z}}{L_{vZ}} = \frac{f(\chi)\varphi(Z)}{f(Z_s)\varphi(0^\circ)}, \tag{5.20}$$

where $L_{\chi Z}$ is the sky luminance of the planar element displaced from the sun position by the angle χ and from the zenith Z, L_{vZ} is the sky luminance at the zenith, $f(\chi)$ is the relative indicatrix function for the sky element, $f(Z_s)$ is the relative indicatrix function for the zenith, $\varphi(Z)$ is the gradation function for the sky element, and $\varphi(0°)$ is the gradation function for the zenith.

In the ISO/CIE standard the relative luminance sky patterns normalized by the zenith luminance as given in (5.20) are defined and the gradation functions for each sky type are determined by parameters a and b in its approximation:

$$\varphi(Z) = 1 + a \exp(b/\cos Z) \quad \text{when } 0<Z<p/2 \tag{5.21}$$

and $\varphi(\pi/2) = 1$ while

$$\varphi(0°) = 1 + a \exp b. \tag{5.22}$$

The numerical values of gradation parameters a and b are shown with curves for all 15 ISO/CIE sky types in Fig. 5.2.

The indicatrix functions for each sky type are determined by parameters c, d, and e in an equation similar to (5.6), but instead of two parameters (N and M), three parameters were applied as

$$f(\chi) = 1 + c[\exp(d\,\mathrm{arc}\chi) - \exp(d\pi/2)] + e\cos^2\chi \tag{5.23}$$

and

$$f(Z_s) = 1 + c[\exp(d\,\mathrm{arc}Z_s) - \exp(d\pi/2)] + e\cos^2 Z_s, \tag{5.24}$$

where the spherical angle between the sunbeam direction and the sky element is defined after (5.9), where A_ζ is the absolute difference of the solar azimuth A_s and the sky element azimuth A_e, i.e., $A_\zeta = |A_s - A_e|$.

All relevant angular relations determining the mutual position of the sun and sky elements are shown on Fig. 5.3, where the northern-oriented azimuth angles are presented in the lower part of the figure.

The arbitrary indicatrix value under cloudy or clear skies is determined by its angular distance from the sun as well as by the indicatrix parameters.

Indicatrix parameters c, d, and e in relation to the spherical angle χ are graphically presented in Fig. 5.4 as curves with inset parameter values. Chosen standard indicatrix types are summarized in Table 5.1.

The relative luminance distribution for any of the fifteen sky types can be calculated applying eq. (5.20) respecting the actual solar altitude. Such luminance sky pattern normalised to unity zenith luminance and solar altitude 30° are shown by Darula and Kittler (2011) in Figs. 26–28 respectively.

Fig. 5.2 Curves of the standard gradation functions

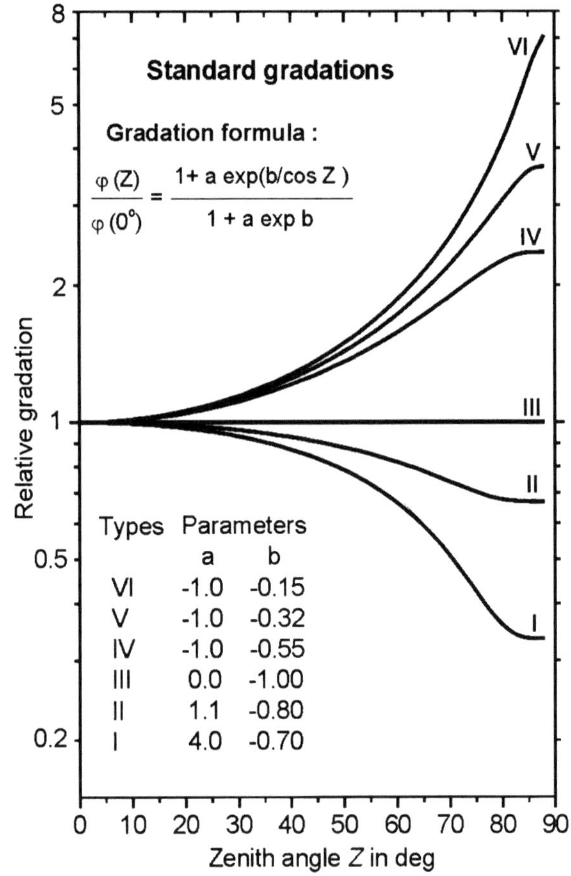

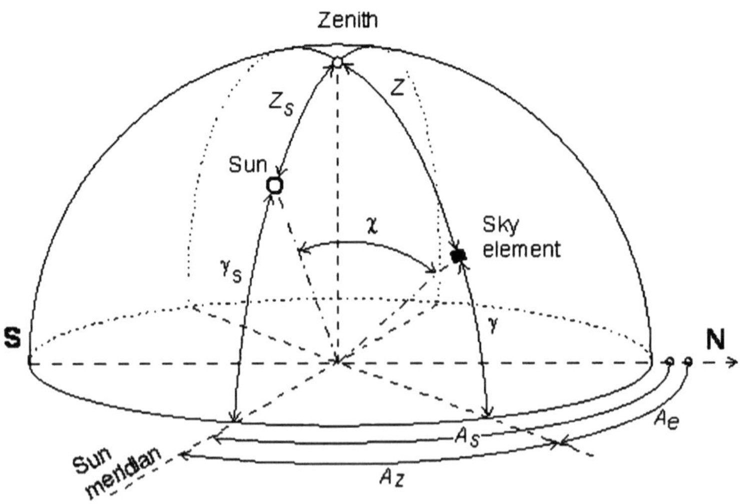

Fig. 5.3 Angular coordinates of the sun and sky element

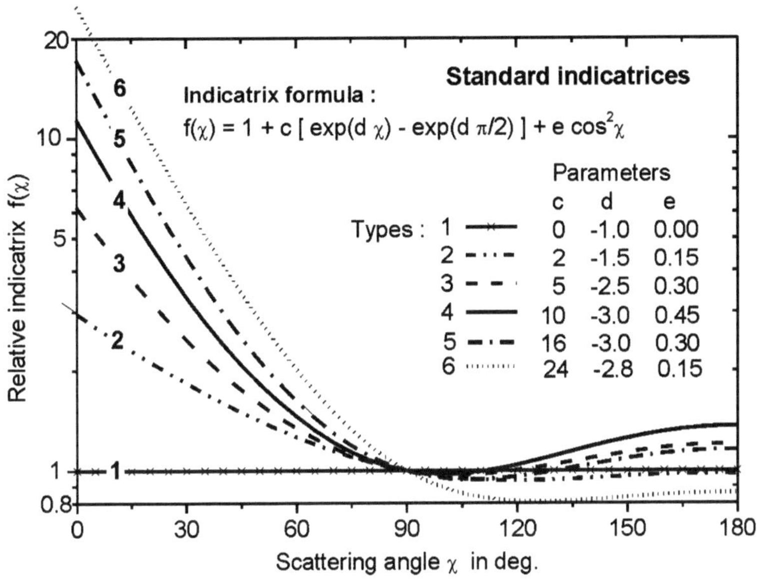

Fig. 5.4 Standard indicatrices with defined parameters $c, d,$ and e

Table 5.1 Standard *c*, *d*, and *e* parameters for indicatrix functions $f(\chi)$ of typical homogeneous cloudiness and turbidity atmospheric conditions

Indicatrix type	Description	*c*	*d*	*e*
1	Unity relative scattering indicatrix with a stable uniform course	0	−1.0	0
2	Slightly rising relative scattering indicatrix toward sun positions	2	−1.5	0.15
3	Rising trend of relative scattering indicatrix toward sun positions	5	−2.5	0.30
4	Distinct solar corona created by the relative scattering indicatrix	10	−3.0	0.45
5	Steeply rising relative scattering indicatrix toward sun positions	16	−3.0	0.30
6	Broad and high luminance solar corona caused by extensive scattering	24	−2.8	0.15

5.5 Methods for Predicting Absolute Zenith Luminance Levels

All 15 ISO/CIE sky types are standardized in relative terms, i.e., the luminance patterns are normalized to the zenith luminance L_{vZ} in accordance with (5.20). Under the assumption of homogeneous or at least quasi-homogeneous turbidity or cloudiness, the indicatrix functions are related to the average, quasi-uniform turbidity factor T_v shown in (5.6), where N and M in (5.10) and (5.11), respectively, can be determined after the T_v value. It has to be noted that the ISO/CIE sky type standardization prescribes for the indicatrices a slightly different parameterization system, i.e., roughly $c = N$ and $e = M$, but instead of a constant value of 3 in (5.6), different values of d are applied in the case of cloudy and overcast skies.

If the vectorial character of the zenith luminance L_{vZ} is related to the whole sky pattern, then its integration producing the "scalar" horizontal sky illuminance D_v and the L_{vZ}/D_v ratio can be taken as an indicator of the sky type.

The importance of the L_{vZ}/D_v parameter under clear sky conditions was probably first noticed by Fleisher, a colleague of Krochmann (1967), in a reciprocal form D_v/L_{vZ}, but later Phillips (1980) applied another form ($\pi L_{vZ}/D_v$) for characterizing the CIE (1973) clear sky standard with different solar altitudes. Valko (1986), analyzing his periodical sky radiance measurements, also found the $\pi L_{vZ}/D_v$ ratio to be relevant in connection to cloudiness quantities and types (Kittler and Pulpitlova 1988, p. 63). In the latter publication, there is a figure documenting the $\pi L_{vZ}/D_v$ ratio changes with solar altitude and turbidity, with the extreme value for the overcast sky $\pi L_{vZ}/D_v$ ratio of 1.286, which can be transformed to $L_{vZ}/D_v = 1.286/\pi = 0.40935$, which represents a constant independent of solar altitude changes. The original concept of defining both the gradation and the indicatrix functions in slightly different forms expressed the dependence on the average homogeneous luminous turbidity of the atmosphere T_v published by Kittler (1985) respecting (5.10) and (5.11) as

$$f(\chi) = 1 + 4.3\,T_v{}^{1.9}\exp(-0.35\,T_v)\,[\exp(d\,\mathrm{arc}\chi) - \exp(d\pi/2)] + \frac{0.71}{T_v{}^2}\cos^2\chi \qquad (5.25)$$

and

$$\varphi(Z) = \frac{m_\zeta}{0.094\,T_v(1-m_\zeta)}\left|1 - \frac{\exp(-a_v m_\zeta T_v)}{\exp(0.1T_v)}\right|, \qquad (5.26)$$

where m_ζ is the relative optical air mass for the zenith angle of the sky element ζ.

Integrating D_v and expressing the $\pi L_{vZ}/D_v$ ratio, Kittler and Pulpitlova (1988, p. 51) drew Fig. III.1, which indicated the importance of this ratio in the dependence of sky type classification on solar altitude and turbidity/cloudiness conditions. During a project studying the set of a new generation of sky standards (Kittler et al. 1997, 1998), the L_{vZ}/D_v ratio was chosen as an indicator of sky types, and both L_{vZ} and D_v values were used in the integration in the standardized form as

$$\frac{L_{vZ}}{D_v} = \frac{f(Z_s)\varphi(0^\circ)}{\int\limits_{Z=0}^{\pi/2}\int\limits_{A_Z=0}^{2\pi}[f(\chi)\,\varphi(Z)\,\sin Z\,\cos Z]\,dZ\,dA_Z} = KZ_v\,(\mathrm{cd/lm}). \qquad (5.27)$$

Typical curves for homogeneous skies in Fig. 5.5 were obtained after integrating (5.27). The dense overcast skies are represented by straight horizontal lines which are not dependent on solar altitudes, whereas curves rising with the sun elevation specify cloudy and clear skies (Darula et al. 2006). Because the integration of D_v values is rather complicated by the intersection of many curves in the region over 45° solar altitudes, especially for clear skies, and creates some problems in the sky type, it is better to orientate the sky type classification to lower solar altitude conditions even in the tropics, i.e., use morning and afternoon measurement data. It is evident from Fig. 5.4 that mainly due to the indicatrix influences, the constant

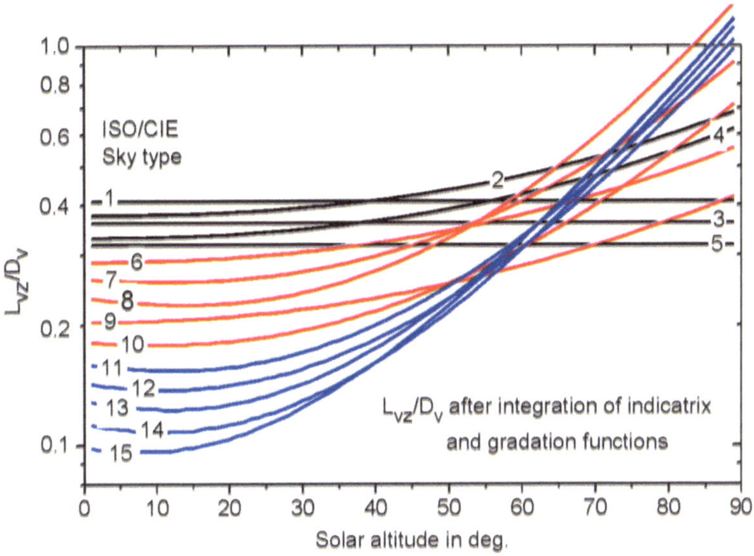

Fig. 5.5 L_{vZ}/D_v curves classifying standard sky groups. Overcast skies in *black*, quasi-cloudy skies in *red*, and clear skies in *blue*

and relativity high ratio for overcast skies decreases with low solar altitudes to ratios over 0.1, but rises with high solar altitudes.

At the same time (Kittler et al. 1998), a rather good approximation of the resultant L_{vZ}/D_v parameter was achieved, but was valid only for solar altitude under 70°:

$$\frac{L_{vZ}}{D_v} = \frac{1}{\text{LSC}}\left[\frac{B(\sin\gamma_s)^C}{(\sin\gamma_s)(\cos\gamma_s)^D} + E\right] (\text{cd/lm}) \tag{5.28}$$

or

$$L_{vZ} = \frac{D_v}{E_v}\left[\frac{B(\sin\gamma_s)^C}{(\cos\gamma_s)^D} + E\,\sin\gamma_s\right] (\text{cd/m}^2), \tag{5.29}$$

where auxiliary parameters B, C, D, and E are given in Table 5.2.

From knowledge that the interdependence between L_{vZ}/D_v and T_v is valid for clear skies only (for which it is measurable and predictable with good accuracy), an approximation formula for L_{vZ} was also found by applying further auxiliary parameters A1 and A2 in Tables 5.1 and 5.2; thus,

$$L_{vZ} = A\sin\gamma_s + \left[\frac{0.7(T_v+1)\,(\sin\gamma_s)^C}{(\cos\gamma_s)^D}\right] + 0.04\,T_v\,(\text{kcd/m}^2), \tag{5.30}$$

where A = A1 T_v + A2 (Table 5.2)

Table 5.2 Auxiliary parameters for calculating absolute zenith luminance L_{vZ} or the D_v/E_v ratio for standard sky types

	Recommended or standardized parameters							
	Typical					For L_{vZ} After T_v		
Sky type	D_v/E_v	B	C	D	E	A1	A2	Typical T_v
1	0.10	54.63	1.00	0.00	0.00	Because these sky standards are associated with no sunlight, the relation L_{vZ} after T_v is not valid		
2	0.18	12.35	3.68	0.59	50.47			
3	0.15	48.30	1.00	0.00	0.00			
4	0.22	12.23	3.57	0.57	44.27			
5	0.20	42.59	1.00	0.00	0.00			
6	0.38	11.84	3.53	0.55	38.78			
7	0.42	21.72	4.52	0.64	34.56	0.957	1.790	12.0
8	0.41	29.35	4.94	0.70	30.41	0.830	2.030	10.0
9	0.40	10.34	3.45	0.50	27.47	0.600	1.500	12.0
10	0.36	18.41	4.27	0.63	24.04	0.567	2.610	10.0
11	0.23	24.41	4.60	0.72	20.76	1.440	−0.75	4.0
12	0.15	23.00	4.43	0.74	18.52	1.036	0.710	2.5
13	0.28	27.45	4.61	0.76	16.59	1.244	−0.84	4.5
14	0.28	25.54	4.40	0.79	14.56	0.881	0.453	5.0
15	0.30	28.08	4.13	0.79	13.00	0.418	1.950	4.0

However, when the solar altitude is over 70°, approximations (5.28) and (5.29) cannot be used; therefore, the precise L_{vZ}/D_v parameter after the integration of (5.27) equal to Z_v is to be used and then

$$L_{vZ} = \frac{D_v}{E_v} KZ_v \,\mathrm{LSC} \sin \gamma_s \; (\mathrm{cd/m^2}). \tag{5.31a}$$

In a concise way, the general relationship can be formulated also by an abstract form when the luminous solar constant (LSC) is in lux:

$$L_{vZ} = (D_v/E_v)(L_{vZ}/D_v)\mathrm{LSC} \; \sin \gamma_s \; (\mathrm{cd/m^2}). \tag{5.31b}$$

The ratio L_{vZ}/D_v can be determined either using the approximate (5.28) for solar altitudes under 70° or as the result of integrating (5.27) during equatorial noon sun positions. The ratio D_v/E_v depends on momentary states of atmospheric transparency and turbidity conditions, i.e., either typical D_v/E_v ratios from Table 5.3 or turbidity-related values valid under quasi-clear conditions when usually with the rising sun height these ratios slightly decrease after Darula and Kittler (2005):

$$\frac{D_v}{E_v} = \frac{[(\mathrm{A1}\, T_v + \mathrm{A2}) \sin \gamma_s + 0.7(T_v + 1)X + 0.04\, T_v]}{BX + E \sin \gamma_s}, \tag{5.32a}$$

Table 5.3 Description of sky types and the typical D_v/E_v parameter for the 15 standard sky types

Sky type	Gradation type	Indicatrix type	Description	Typical D_v/E_v
1	**I**	**1**	**CIE standard overcast sky Steep luminance gradation toward the zenith, azimuthal uniformity**	**0.10**
2	I	2	Overcast, with steep gradation and slight brightening toward the sun	0.18
3	II	1	Overcast, moderately graded with azimuthal uniformity	0.15
4	II	2	Overcast, moderately graded and slight brightening toward the sun	0.22
5	**III**	**1**	**Sky of uniform luminance**	**0.20**
6	III	2	Partly cloudy sky, slight brightening toward the sun with uniform gradation	0.38
7	III	3	Partly cloudy sky, brighter circumsolar region with uniform gradation	0.42
8	III	4	Partly cloudy sky, clear solar corona with uniform gradation	0.41
9	IV	2	Partly cloudy sky, with obscured sun position	0.40
10	IV	3	Partly cloudy sky, with brighter circumsolar region	0.36
11	IV	4	White–blue sky with distinct solar corona	0.23
12	**V**	**4**	**CIE standard clear sky with low luminous turbidity**	**0.15**
13	**V**	**5**	**CIE standard clear sky with polluted atmosphere**	**0.28**
14	VI	5	Cloudless turbid sky with broader solar corona	0.28
15	VI	6	White–blue turbid sky with broad solar corona region	0.30

where

$$X = \frac{(\sin \gamma_s)^C}{(\cos \gamma_s)^D}. \tag{5.32b}$$

It is evident that the D_v/E_v parameter defines also the zenith luminance. Therefore, the frequency of occurrence of this ratio under different standard sky types was analyzed by Darula and Kittler (2004a, b, 2005). Because in every sky type for a range of D_v/E_v parameters an almost Gaussian distribution was found, the most frequent ratios were taken as typical respecting their mean and mode values. As can be seen in (5.29) and (5.30), there is a possibility to compare the relationship derived for clear sky conditions and the measurement data (Darula and Kittler 2005) which determine also the D_v/E_v ratio in the five standards of the clear sky group dependence on the turbidity value T_v under various solar altitudes. For orientation, both typical D_v/E_v ratios and T_v values are given in Tables 5.2 and 5.3. However, these recommended D_v/E_v values were obtained from the analysis of the long-term measurements in Bratislava recorded during year-round changes (Darula and Kittler 2004a, b, 2005), but in different daylight climate regions some deviations can exist.

For zenith luminance prediction, Perez et al. (1990) proposed an approximation model with a_i, c_i, e_i, and d_i parameters prescribed for each of the eight sky clearness ranges in a table to be used in the approximation

$$L_{vZ} = D_e(a_i + c_i \cos Z + e_i \exp(-3Z) + \Delta d_i)\ (\mathrm{kcd/m^2}). \qquad (5.33)$$

The calculation of L_{vZ} is based on the pseudo efficacy (kcd/W) to get the zenith luminance expressed in kilocandelas per square meter.

Igawa and Nakamura (2001) used a regression equation to compute zenith luminance:

$$L_{vZ} = A_I \sin^{1.3}(0.7\,\gamma_s) + B_I \tan^{1.3}(0.7\,\gamma_s) + C_I\ (\mathrm{kcd/m^2}), \qquad (5.34)$$

where A_I, B_I, and C_I are Igawa's parameters, which are dependent on the so-called normalized global illuminance N_{evg} defined in the appendix in Chap. 4 by (A4.6) or in the appendix in this chapter by (A5.4).

Recently, Kocifaj (2011) proposed reducing the number of these scaling parameters.

Note that the Lambert uniform sky is identical to sky type 5, whereas the CIE overcast sky copies sky 1 and the CIE clear sky with the so-called Kittler indicatrix (for countryside) represents sky type 12.

Besides homogeneous skies, in reality a multiple variety of actual partly cloudy skies with different sky types, cover, and placement of cloud patches exists (Perez et al. 1993b). Owing to their momentary presence and fluent movements and changes in shape and luminance patterns, such occasional states cannot be taken into account because of their rare occurrence during a year. Many trials have already been done to reduce the number of locally typical and most frequent sky types that could be recommended for year-round daylight climate changes to predict energy balance, glare control, or necessary artificial lighting enhancement (Lam et al. 1999; Tregenza 1999, 2004; Wittkopf and Soon 2007, Chirarattananon and Chaiwiwatworakul 2007, 2008).

More realistic luminance distributions of large-area sources or apertures to the sky mean a better simulation of reality, and are now manageable in spite of their complexity owing to use of computers and the software available. For instance, to describe the ISO/CIE sky patterns more vividly and instructively as well as to enable their user-friendly application, several computer tools have already been presented, e.g., Modelsky by Kocifaj and Darula (2002), SkyModeller by Roy, available from http://www.cadplan.com.au/, and Virtual Sky Domes by Wittkopf (2004). Now a sophisticated program Velux Daylight Visualizer contains possibilities of aplying all ISO/CIE sky types (http://viz.velux.com/Daylight_Visualizer/News/release032310.aspx). Also, HOLIGILM, a user-friendly freeware tool available from http://www.holigilm.info, can be used to predict different sky type luminance patterns. Here we include only an arbitrarily chosen example for San Francisco (37°47′N) applying ISO sky type 13 on May9 at 9 o'clock true solar time, when the solar altitude is 44.5° (Fig. 5.6).

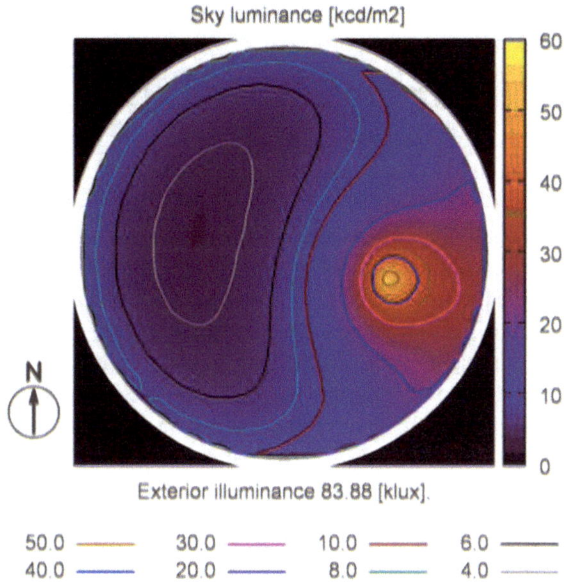

Fig. 5.6 Partly polluted clear sky pattern in San Francisco (38°N)

5.6 Partial Conclusions

Although the science and photometry of daylight sources was established and considerably developed by Bouguer and Lambert, further studies and research were needed especially in:

- Measuring the real daylight conditions with their changes in various climate regions.
- The specification of typical sky luminance patterns.
- Different seasonal, monthly, and yearly expectancy and variability of sunny, cloudy, and overcast periods linked with sunshine duration.
- Measuring or predicting scattering properties of atmospheric layers with the aim of predicting practical sky luminance distributions.

To answer these and many other questions about new and sophisticated knowledge and inventions in instrumentation and sensor technology, illuminating engineering methods based on theoretical photometry have to be revised. Current automatic recorders, sky trackers, and sky scanners enable measurements of local daylight parameters in small time steps and long-term variations as well as the determination of prevailing sky luminance patterns needed for exterior and interior illuminance and luminance prediction or glare prevention (Kittler et al. 2010). But the most revolutionary innovation is in the sky luminance measurement, determination, and prediction possibilities.

The original proposal to define and standardize typical sky distribution on the basis of the indicatrix and gradation functions by Kittler (1967) in the set of

15 standard sky types by Kittler et al. (1997, 1998) was soon followed by the Igawa et al. (1999, 2004) proposal of 20 sky types and also by a fluent series of "all weather" skies by Perez et al. (1993a). In the first case the long-accepted relative sky luminance patterns were preferred, whereas both latter definitions tried to express the absolute luminance level approximated after the diffuse and global illuminance expectances. Even nonhomogeneous, quite irregular cloudy sky patterns were proposed for approximation by Perez et al. (1993b).

Today, the currently valid ISO/CIE (2004/2003) general 15 sky types represent the prevailing range of sky luminance patterns worldwide. Of course, locally in different climate zones, in seasonal or monthly variations the probability of occurrence of any regularly prevailing sky types has to be selected after long-term measurement data. Such a selection is also influenced by different criteria according to the purpose (e.g., window design, glare problems, or year-long energy savings). However, it is also important to understand that minimal or Lambertian cases could be used as favorable simplified conditions for theoretical calculations.

Within the full range of homogeneous skies from the ideal clear sky to a very dense overcast sky, several parameters of the atmosphere inextricably depend on the atmospheric turbidity, pollution, and cloudiness:

1. The higher the luminous turbidity and air mass in the direction of sunbeams, the greater the reduction in direct sunlight illuminance at ground level.
2. Owing to the turbidity and cloudiness state of the momentary local atmosphere, both scattering and gradation redistribution of sunlight takes place, resulting in the overall sky luminance pattern changes.
3. Any turbidity rise causes the reduction of solar beam illuminance, whereas skylight gradually rises, resulting in more shadowless illumination conditions.
4. The relative illumination level of the directional sunlight (with its vectorial character) to the diffuse skylight (scalar) diminishes with the level of the extraterrestrial horizontal illuminance.
5. Whereas the relative sky luminance patterns are influenced by both the relative scattering and the gradation function normalized by the zenith luminance, the absolute luminance level of the whole sky vault is determined by the momentary extraterrestrial horizontal illuminance E_v, its penetration through the atmosphere D_v/E_v, and by its redistribution caused by turbidity and cloudiness, which characterize the sky type and the ratio of L_{vZ}/D_v.
6. Because sunlight and sky luminance patterns are spatial and directional, the light flux entering any architectural space or aperture is biased by its size and orientation.
7. When sunlight is absent, the diffuse skylight illumination is the only source of daylight in the exterior or interior spaces and is dependent on the momentary local sky luminance distribution, which can be defined by sky scanning measurements, fish-eye photography, or theoretical calculation with graphical sky images as luminance isomaps of the plan-projected sky hemisphere.

Appendix 5

Comparison of Basic Approaches and Approximations for Defining the Sky Luminance Patterns

Basic for all relevant sky models defining their luminance patterns are the gradation and indicatrix functions, which predetermine their luminance distribution under actual sun positions and the state of the atmosphere. The first luminance models of Fesenkov (1955) indicated the possible separation of both these functions assuming only the first degree of diffusion in the arbitrary volume of the atmosphere. Several refinements were proposed by Sobolev (1943, 1949, 1972) and Kittler (1967, 1985, 1986), but unfortunately the more complex formulae did not result in higher precision.

Some approximations and simplifications were accepted to determine practical sky models for daylighting purposes. Whereas under overcast conditions only the gradation function has to be simulated by different formulae as described in the appendix in Chap. 4, under cloudy and clear skies both functions are important with a multiplying effect. The indicatrix function was applied in different models in its absolute form $s(\chi)$ after (5.5) only rarely for clear sky models (Nagata 1971, 1983), whereas many others applied the relative indicatrix function normalized to the luminance perpendicular to sunbeams, i.e., $f(\chi) = L_\chi / L_{90^\circ}$ after (5.6). In this respect many relevant recent sky luminance models (Perez et al. 1993; Kittler et al. 1997, 1998; Igawa et al. 2001, 2004; ISO 2004) have applied the same exponential relations to model the gradation and relative indicatrix functions after (5.21) and (5.23), respectively. The ISO (2004) standard defines the parameters a and b for the gradation functions, c, d, and e parameters for the relative indicatrix functions for every one of the 15 standard sky types after Kittler et al. (1997, 1998), whereas in both the Perez and Igawa models these parameters are calculated after different indicators, e.g., Perez et al. (1993) specified within the eight ranges of the clearness after four model coefficients for determining the $a_p - e_p$ parameters in the relation containing also the influence of the sky brightness. So the procedure to find the suitable gradation and indicatrix functions is quite complex and tedious.

Sky luminance distribution modeling was followed by Perez et al. (1993a), applying the first measurement results from the multipurpose scanning photometer (Perez et al. 1992) recorded in Berkeley. The data set taken between June 1985 and December 1986 comprised more than 16,000 sky scans. The so-called all-weather model used the ISO/CIE gradation function and a distorted indicatrix function excluding the normalizing member $-\exp(d\pi/2)$ in (5.23). This model is based on the two-parameter description, the parameters of sky clearness $\in$ and sky brightness Δ (Perez et al. 1983), which included the horizontal diffuse irradiance D_e, the zenith angle of the sun position Z_s, the optical air mass m, the solar constant (SC) and normal parallel beam irradiance P_{en} as

$$\in = \frac{(D_e + P_{en})/D_e + 1.041\,Z_s^3}{1 + 1.041\,Z_s^3}, \tag{A5.1}$$

$$\Delta = \frac{mD_e}{SC}. \tag{A5.2}$$

After these basic two parameters, also Perez's a_P, b_P, c_P, d_P, and e_P subparameters for the gradation and indicatrix functions have to be calculated after separate relations and subcoefficients in eight different ranges of sky clearness.

For zenith luminance prediction, Perez et al. (1990) proposed an approximation model with parameters a_i, c_i, c_i', and d_i prescribed for each of the eight sky clearness $\in$ ranges in a table to be used in the approximation

$$L_{vZ} = D_e\,[a_i + c_i \cos Z + c'_i \exp(-3Z) + \Delta d_i]\,(\text{kcd/m}^2). \tag{A5.3}$$

The calculation of L_{vZ} is based on the pseudo efficacy (kcd/W) to get the zenith luminance expressed in kilocandelas per square meter.

In both the Perez et al. (1993) and Igawa et al. (2004) modeling procedures for momentarily measured irradiance, at least G_e and D_e as well as the simultaneously registered solar zenith angle Z_s have to be available.

In all Igawa's models, gradation and indicatrix functions and auxiliary relations were determined by regression analysis depending on the momentary solar altitude and the so-called normalized global illuminance N_{evg} (Igawa et al. 1997).

The normalization was set as a ratio

$$N_{evg} = E_{vgm}/E_{vgms}(\gamma_s), \tag{A5.4}$$

where E_{vgm} is the so-called relative global illuminance, which is similar to G_v/E_v but is defined as

$$E_{vgm} = m\,G_v/\text{LSC}. \tag{A5.5}$$

$E_{vgms}(\gamma_s)$ is the relative global illuminance of the CIE standard clear sky under atmospheric transmittance 0.75 estimated in the regression analysis as dependent on the solar altitude γ_s by the equation (Igawa et al. 1997)

$$E_{vgms}(\gamma_s) = 0.19 + 2.09\gamma_s - 2.581\gamma_s^2 + 1.486\gamma_s^3 - 0.323\gamma_s^4, \tag{A5.6}$$

previously determined by Matsuzawa et al. (1997) as

$$\begin{aligned} E_{vgms}(\gamma_s) &= 0.197 + 1.943\gamma_s - 2.376\gamma_s^2 \\ &+ 1.327\gamma_s^3 - 0.232\gamma_s^4 - 0.031\gamma_s^5. \end{aligned} \tag{A5.7}$$

The comparison of the Japanese gradation functions with the ISO/CIE gradation functions is given in Fig. 5.4, and indicatrix functions are compared in Fig. A5.1.

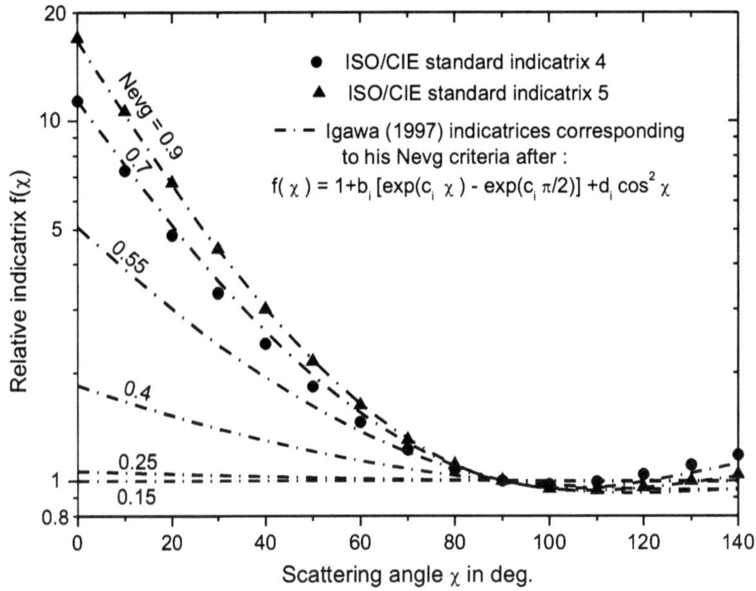

Fig. A5.1 Comparison of ISO/CIE indicatrix functions with those of Igawa et al. (1997)

Later, more measured data were obtained in Japan using a sun tracker (Fig. A5.2) and an EKO scanner (Fig. A5.3).

A new version of a model based on radiance distribution was published (Igawa et al. 2004), where differently determined parameters based on global irradiance were given (e.g., Ce, Cle, Seeg, Si), but the ISO/CIE gradation and indicatrix functions were respected. There are a few weak points in applying the latest Japanese method. The main weakness is the basic dependence on G_e or D_e, which have to be obtained either from a meteorological network or an International Daylight Measurement Programme (IDMP) station. Furthermore, the sky luminance distribution is linked more closely to D_e or D_v, influenced directly by the luminance patterns, whereas parallel sunbeams in D_v or G_v distort the problematic luminous efficacy of D_v/D_e or G_v/G_e at different solar altitudes. The complex multiple normalizing system introduced by Igawa et al. (2004) is too complicated for practical use because it produces an infinite number of sky patterns linked to momentary or daily changing G_e or G_v and D_e or D_v measured values interrelated by the luminous efficacy.

To address the question of whether luminous efficacy could be a representative parameter for sky type classification, two examples of the monthly occurrence were evaluated comparing prevailingly overcast and clear situations in already documented months, i.e., for August 2001 in Fig. A5.4 and for November 1995 in Fig. A5.5.

ISO/CIE clear sky types 11–15 differ in the various turbidities expressed by the luminous turbidity factor T_v, and great variance of the luminous efficacy was found,

Fig. A5.2 Japanese sun tracker

with the highest under $T_v = 2$, but with rising turbidity the variance is suppressed. Probably that it is an effect of the luminous efficacy of sunbeams as well as the diffusivity of the skylight, which is less sensitive to solar altitude changes in the turbid atmosphere.

Data for the global luminous efficacy $\mathrm{Eff}_g = G_v/G_e$ plotted against solar altitude γ_s under the overcast sky type (ISO/CIE sky type 1) show that there is no significant dependence on solar altitude and the values obtained vary by ± 10 lm/W around $\mathrm{Eff}_g = 115$ lm/W.

However, the dependence of these models on the inaccurate luminous efficacy is problematic and questionable.

It is true that in reality only on very seldom occasions are repeated sky luminance patterns, either homogeneous or nonhomogeneous, exactly the same, but these can result in a very similar horizontal irradiance or illuminance level measured at ground level. However, creating the sky gradation and indicatrix

Fig. A5.3 EKO sky luminance scanner. (Photo by Igawa)

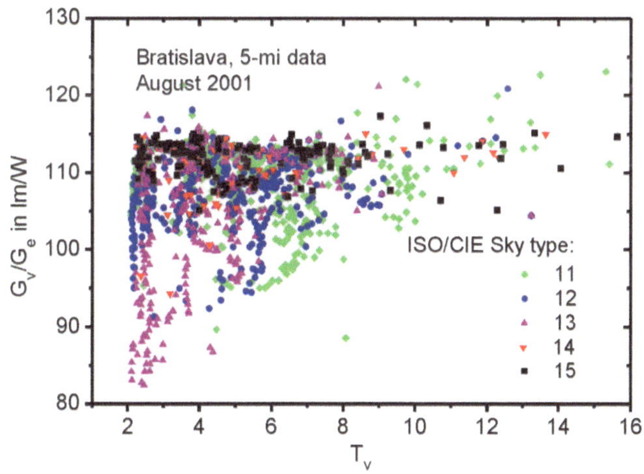

Fig. A5.4 Global luminous efficacy spread dependent on turbidity differences

functions after such scalar values is very doubtful and cannot predict typical distribution patterns.

The magnitude of skylight is the portion of extraterrestrially available sunlight scattered penetrating the atmosphere expressed by the ratio D_e/E_e in irradiance unit or D_v/E_v in illuminance units. Using the ratio D_e/E_e for predicting sky luminance

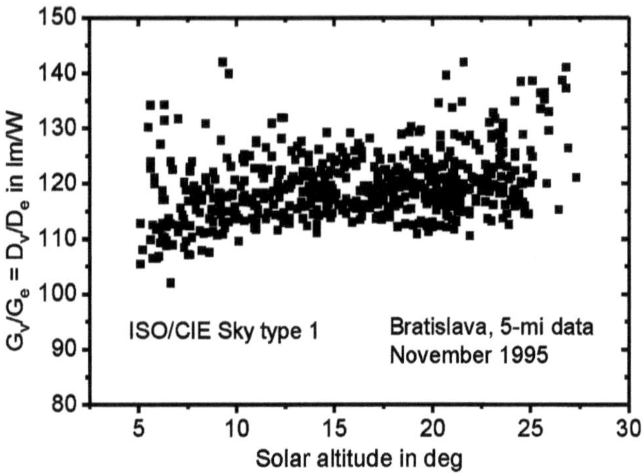

Fig. A5.5 Global luminous efficacy under overcast skies

models assumes the application of luminous efficacy recalculation, which can be inaccurate when diffuse skylight is taken as an average efficacy of 120 lm/W, whereas the extraterrestrial LSC/SC is only 97.6 lm/W (Darula et al. 2005), as

$$\frac{D_e}{E_e} = \frac{D_e}{\mathrm{SC}\,\sin\gamma_s}, \tag{A5.8}$$

$$\frac{D_v}{E_v} = \frac{D_v}{\mathrm{LSC}\,\sin\gamma_s}. \tag{A5.9}$$

In a slightly alternative form, it was used by Perez et al. (1993) as the sky's brightness after (A5.2):

$$\Delta = m\frac{D_e}{\mathrm{SC}}. \tag{A5.10}$$

The difference between (A5.8) and (A5.10) is very small except at very low solar altitudes, roughly under 15°, as the optical air mass m deviates from the $1/\sin\gamma_s$ values. The argument that parallel sunbeams penetrating the layer of Earth's atmosphere at $\gamma_s = 0°$ produce some normal irradiance reduced by the m value cannot mean that their influence on the horizontal plane is not equal to zero.

Similarly, Igawa and Nakamura (2001) used instead of D_e/E_e the so-called relative global illuminance in the first step of parameterization:

$$E_{vgm} = m\frac{G_v}{\mathrm{LSC}}. \tag{A5.11}$$

In Igawa's earlier studies, his normalization parameter was the value of N_{evg}[see (A5.4)], but later he introduced a new normalization factor – the so-called standard global illuminance, which was defined by the best-fit equation

$$S_{evg}(\gamma_s) = -36.78\gamma_s^5 + 188.79\gamma_s^4 - 375.95\gamma_s^3 + 306.2\gamma_s^2 + 15.47\gamma_s + 0.83\ (\text{klx}). \tag{A5.12}$$

In Igawa et al. (2004), a similar standard global irradiance assuming a clear sky with the Linke turbidity factor $T_L = 2.5$ was used and defined as

$$\text{Seeg} = \frac{0.84\text{SC}}{m} \exp(-0.0675m)\ (\text{W/m}^2), \tag{A5.13}$$

as a clear sky index:

$$\text{Kc} = G_e/\text{Seeg}. \tag{A5.14}$$

Thus, the basic parameter in the Igawa and Nakamura (2001) model is called normalized global illuminance:

$$N_{evg} = G_v/S_{evg}(\gamma_s), \tag{A5.15}$$

which is used to calculate the gradation and indicatrix coefficients and thus their functions.

However, all these changes in the determination of the normalization did not affect the arguments of Kittler and Darula (2000) that only the N_{evg} values without further information on the simultaneous solar altitude γ_s and turbidity T_v cannot replace fully the classification ratio L_{vZ}/D_v as well as the approximation owing to luminous efficacy of global horizontal irradiance G_e or Seeg.

Expression (5.20) for the relative luminance distribution is used in a similar form, but instead of the zenith angular distance, elevation angles from the horizon are used; thus,

$$\frac{L_{\chi Z}}{L_{vz}} = \frac{f(\chi)\,\varphi(Z)}{f(Z_s)\,\varphi(0°)} = \frac{f(\chi)\,\varphi(\gamma)}{f(\pi/2 - \gamma_s)\,\varphi(\pi/2)}. \tag{A5.16}$$

The sky radiance distribution was also determined in Igawa et al. (2004), but the gradation and indicatrix functions need an even more complex calculation procedure, as used to determine the gradation and indicatrix coefficients in Igawa and Nakamura (2001) after the sky index Si

$$\text{Si} = \text{Kc} + \text{Cle}^{0.5}, \tag{A5.17}$$

where Kc after (A5.14) is used and the cloudless index first applied in Perradeau's (1988) sky model is

$$\mathrm{Cle} = \frac{1 - D_e/G_e}{1 - \mathrm{Ce_s}}, \tag{A5.18}$$

where the mean clear sky theoretical cloud ratio value is

$$\begin{aligned}\mathrm{Ce_s} &= 0.01299 + 0.07698m - 0.003857m^2 + 0.0001054m^3 \\ &\quad - 0.000001031m^4.\end{aligned} \tag{A5.19}$$

To determine the gradation and indicatrix coefficients $a_i - e_i$, Igawa et al. (2004) proposed approximation relations dependent on the sky index Si, which is also used to classify five basic sky types in Si ranges as follows:

- Overcast sky if Si is lower than 0.3.
- Nearly overcast sky when Si is between 0.3 and 0.6.
- Intermediate sky if Si $= 0.6 - 1.5$.
- Nearly clear sky if Si $= 1.5 - 1.7$.
- Clear sky if Si is over 1.7.

Five similar categories based on the cloudless index Cle were proposed also by Perradeau (1988), who called the Cle value the nebulosity index.

It is evident that all sky models need at least some parameterization base, either the simple expected D_v/E_v ratio or measured D_e, G_e, m, or γ_s for a particular situation. Then parameterized cases can be determined for sky radiance or luminance distributions in relative or absolute units. The typology of sky luminance patterns assumes also simple and clear rules to classify sky luminance scans, with certain sky types or standards that can be identified from IDMP regularly measured data. The quasi-homogeneous cases can be selected from all sets using the L_{vZ}/D_v parameter within the narrow ±2.5% range of the particular L_{vZ}/D_v ISO standard curves. When diffuse sky illuminance or zenith luminance is not measured or is unavailable, then both the Perez and Igawa sky models, respectively, can be used if regular irradiance data measured by meteorological stations with specified time or solar altitude information can be obtained. When no local data can be found, then for practical tasks sky luminance patterns under a selected sky type can be taken from the set of 15 ISO/CIE standards. The great handicap of the Perez model is the absence of the CIE overcast sky probably owing to only the subtropical sky scans gathered in Berkeley, California. Thus, in the eighth overcast clearness category the indicatrix is not exactly $f(\chi = 1)$ and the gradation L_{vZ}/L_{vh} does not follow the 3:1 decreasing tendency. However, owing to manageable and systematically defined relationships, the Perez model was incorporated into the radiance calculation process, where it is especially suited because of its irradiance-oriented base.

References

Ambarzumian, B.: On zhe problem of the diffuse reflection of light. Journ. of Physics, **8**, 2, 65–75 (1944)
CIE – Commission Internationale de l'Éclairage: Natural Daylight, Official recommendation. Compte Rendu CIE 13th Session, 2, part 3.2, 2–4 (1955)
CIE – Commission Internationale de l'Éclairage: Standardisation of luminance distribution on clear skies. CIE Publ. 22. CB CIE Paris (1973)
CIE – Commission Internationale de l'Éclairage: Spatial distribution of daylight – CIE Standard General Sky. CIE Standard S 011/E:2003, CB CIE Vienna (2003)
Darula, S. and Kittler, R.: New trends in daylight theory based on the new ISO/CIE Sky Standard: 1. Zenith luminance on overcast skies. Build. Res. Journ., **52**, 3, 181–197 (2004a)
Darula, S. and Kittler, R.: New trends in daylight theory based on the new ISO/CIE Sky Standard: 2. Typology of cloudy skies and their zenith luminance. Build. Res. Journ., **52**, 4, 245–255 (2004b)
Darula, S. and Kittler, R.: New trends in daylight theory based on the new ISO/CIE Sky Standard: 3. Zenith luminance formula verified by measurement data under cloudless skies. Build. Res. Journ., **53**, 1, 9–31 (2005)
Darula, S., Kittler, R., Wittkopf, S.K.: Outdoor illuminance levels in the tropics and their representation in virtual sky domes. Architectural Science Review, **49**, 3, 301–313 (2006)
Darula, S. and Kittler, R.: Sunlight and skylight availability. Advances in Energy Research. **1**, 4, 133–182, Nova Publ., NY (2011)
Gusev, N. M.: Private communication to CIE TC3.2 with additional scattering indicatrix proposal. (1970)
Hopkinson, R.G.: Measurements of sky luminance distribution at Stockholm. Journ. Opt. Soc. Amer., **44**, 6, 455–459 (1954)
Fesenkov, V.G. and Pyaskovskaya J.V.: Osveshchennost dnevnogo nebesnogo svoda i indikatrisa rasseyaniya sveta atmosferoy. (In Russian. Illuminance of the daytime sky vault and the scattering indicatrix of light by the atmosphere.), Doklady AN SSSR, **4**, 3, 129–132 (1934)
Fesenkov, V.G.: O jarkosti bezoblachnogo neba pri sfericheskoy zemlye. (In Russian. About luminance of the cloudless sky for spherical Earth.), Doklady AN SSSR, **101**, 5, 845–847 (1955)
Fritz, S.: Illuminance and luminance under overcast skies. J. Opt. Soc. Amer., **45**, 10, 820–825 (1955)
Gillette, G. and Treado S.: Exploring the issue of sky consitions as applied to current daylight practice. Lighting Design and Application, **15**, 5, 23–27 (1985)
Gordov, A.N.: O vozmozhnosti primeneniya teoriyi rasseyaniya k realnoj atmosfere. (In Russian. About a possibility of applying the theory of diffusion to the real atmosphere.) Zhurn. Geofiz., **6**, 4, 295–324 (1936)
Chararattananon, S. and Chaiwiwatworakul, P.: Distribution of sky lluminance and radiance of North Bangkok under standard distributions. Renewable Energy, **32**, 1328–1345 (2007)
Chararattananon, S. and Chaiwiwatworakul, P.: Application of evolutionary computation for identification of parameters of sky luminance. Leukos, **5**, 2, 101–117 (2008)
Igawa, N., Nakamura, H., Matsuura, K.: Sky luminance distribution model for all sky conditions. Proc. CIE Session, Warsaw, Vol. l., part 2, 26–28, CIE Publ. No.133 (1999)
Igawa, N. and Nakamura H.: All sky model as a standard sky for simulation of daylight environment. Building and Environment, **36**, 763–770 (2001)
Igawa, N., Koga, Y., Matsuzawa, T., Nakamura, H.: Models of sky radiance distribution and sky luminance distribution. Solar Energy **77**, 2, 137–157 (2004)
ISO – International Standardisation Organisation: Spatial distribution of daylight – CIE Standard General Sky. ISO Standard 15409:2004 (2004)
Kähler, K.: Flächenhelligkeit des Himmels und Beleuchtungsstärke in Räumen. Meteorol. Zeitschrift, **25**, 2, 52–57 (1908)

Kimball, H.H. and Hand I.F.: Sky brightness and daylight illumination measurements. Monthly Weather Review, **49**, 9, 481–488 (1921)

Kittler, R.: Standardisation of the outdoor conditions for the calculation of the Daylight Factor with clear skies. Proc. Sunlight in Buildings, Bouwcentrum Rotterdam, 273–286 (1967)

Kittler, R.: Luminance distribution characteristics of homogeneous skies: a measurement and prediction strategy. Light. Res. and Technol., **17**, 4, 183–188 (1985)

Kittler, R.: Luminance models of homogeneous skies for design and energy performance predictions. Proc. Intern. Daylighting Conf. Long Beach, Calif., 18–22 (1986)

Kittler, R. and Pulpitlova, J.: Základy využívania prírodného svetla. (In Slovak. Basis of the utilization of daylight.) Veda Publ., Bratislava (1988)

Kittler, R., Darula, S., Perez, R.: A new generation of sky standards. Proc. LuxEuropa Conf., Amsterdam, 359–373 (1997)

Kittler, R., Darula, S., Perez, R.: A set of standard skies characterizing daylight conditions for computer and energy conscious design. US-SK grant project. Polygrafia SAV, Bratislava (1998)

Kittler, R., Darula, S., MacGowan D.: The critical window luminance causing glare in interiors. In: Selected papers of the light and lighting conference with special emphasis on LEDs and solid state lighting. Publ. CIEx34:2010 (2010)

Kondratyev, K.J.: Luchistaya energiya solnca. (In Russian. The radiation energy of the sun.) Gidrometeoizdat, Leningrad. (1954)

Kocifaj, M., and Darula, S.: Modelsky – jednoduchý nastroj pre modelovanie rozloženia jasu na oblohe. (In Slovak. Modelsky – the simple tool for sky luminance distribution modelling). Meteorol. Zprávy. **55**, 4, 110–118 (2002)

Kocifaj, M.: Overcast sky luminance is dependent on the physical state of the atmosphere below cloud level. Light. Res. and Technol., **42**, 2, 149–159 (2010)

Kocifaj, M.: CIE standard sky model with reduced number of scaling parameters. Solar Energy, **85**, 3, 553–559 (2011)

Krat, V.A.: Indikatrisa rasseyaniya sveta v zemnoy atmosfere. (in Russian. Indicatrix of light diffusion in earth atmosphere.) Astronom. J., **20**, 5–6 (1943)

Krochmann, J.: The calculation of the daylight factor for clear sky conditions. Proc. Sunlight in Buildings, Bouwcentrum Rotterdam, 286–301 (1967)

Lam, K. P., Mahdavi, A., Ullah, M. B., Ng, E., Pal, P.: Evaluation of six sky luminance prediction models using measured data from Singapore. Lighting Res. Technol., **31**, 1, 13–17 (1999)

Lambert, J.H.: Merkwürdigste Eigenschaften der Bahn des Lichts durch die Luft und über-haupt durch verschiedene sphärische und centrische Mittel. Hande und Spener Verlag, Berlin (1773)

Liebelt, C. and Bodmann H.W.: Leuchtdichte ubd Strahldichteverteilung des trüben Himmels. Archiv Met., Geoph., Bioklim. Ser 3, **27**, 325–333 (1979)

Lifshic, G.S.: Rasseyaniye sveta v atmosphere. (In Russian. Light diffusion in the atmosphere.) Nauka Publ., Alma Ata (1965)

Mie, G.: Beiträge zur Optik trüber Medien, speziell kolloidaler Metallösungen. Annalen d. Physik, **25**, 377–445 (1908)

Moon, P. and Spencer D.E.: Illumination from a non-uniform sky. Illum. Eng., **37**, 10, 707–726 (1940)

Muneer, T.: Comment 1 on 'Algorithms for calculation of daylight factors in streets'. Light. Res. and Technol., **34**, 2, 144–145 (1980)

Nagata, T.: Luminance distribution on clear skies. Transac. of Arch. Inst. Japan, 7/185, 65–71 and 8/186, 41–50 (1971)

Nagata, T.: Luminance distribution of clear sky and the resultant horizontal illuminance. Journal of Light and Visual Envir. **7**, 1, 23–27 (1983)

Perez, R., Ineichen, P., Seals, R., Michalsky, J., Stewart, R.: Modeling daylight availability and irradiance components from direct and global irradiance. Solar Energy, **44**, 5, 271–289 (1990)

Perez, R., Seals, R., Michalsky, J.: All weather model for sky luminance distribution – preliminary configuration and validation. Solar Energy, **50**, 3, 235–245 (1993a)

Perez, R., Seals, R., Michalsky, J., Ineichen, P.: Geostatistical properties and modeling of random cloud patterns for real skies. Solar Energy, **51**, 1, 7–18 (1993b)

Perraudeau, M.: Climat luminieux a Nantes. Resultats de 15 mois de mesures. Rep. Centre Scient. Techn., du Batiment, Nantes, 84 pp. (1986)
Perraudeau, M.: Luminance models. Proc. Nat. Lighting Conf., 291–292 (1988)
Petherbridge, P.: The brightness distribution of the overcast sky when the ground is snow-covered. Quarterly Journ. of the Royal Meteorol. Soc., **81**, 349, 4766–477 (1955)
Phillips, R.O.: Making the best use of daylight in buildings. Proc. Solar energy applications in design of buildings. Appl. Sci. Publ, Barking, 95–119 (1980)
Pokrowski, G.I.: Über die Lichtzerstreuung in der Atmosphäre. Zeitschr. für Physik, **34**, 9–11, 49–58 (1925)
Pyaskovskaya-Fesenkova, J.V.: Nekotorye danniye ob atmosfernoy indikatrise rasseyaniya sveta. (In Russian. Some data about the atmospheric indicatrix of light diffusion). Doklady Akad. Nauk SSSR, **86**, 5 (1952)
Pyaskovskaya-Fesenkova, J.V.: Issledovaniye rasseyaniya sveta v zemnoj atmosfere. (In Russian. Investigation of the light scattering in earth atmosphere.) Izd. Akad. Nauk SSSR. Moscow (1957)
Rayleigh, J.W.: On the transmission of light through an atmosphere containing small particles in suspension and on the origin of the blue of the sky. Phil. Mag. **47**, 5, 375–384 (1899)
Roy, G.G., Kittler R., Darula S.: An implementation of the Method of Aperture Meridians for the ISO/CIE Standard General Sky. Light. Res. and Technol., **39**, 3, 253–264 (2007)
Schramm, W.: Über die Helligkeitsverteilung am Himmel. Schriften d. Naturw. Vereins f. Schl.-Holst., **12**, 1, 81–129 (1901)
Sobolev, V.V.: Priblizhennoye resheniye zadachi o rasseyaniyi sveta v srede s proizvolnoy indikatrisoy rasseyaniya. (In Russian. The approximative solution of the task about light scattering in the media with an arbitrary scattering indicatrix.) Astronom. Zhurnal, **20**, 5–6, 14–22 (1943)
Sobolev, V.V.: O rasseyaniyi sveta v atmosferach zemli i planet. (In Russian. About light scattering in the atmosphere of earth and planets.) Uchennye zapiski LGU, ser. matem. nauk, **18**, 17–52, 116 (1949)
Sobolev, V.V.: Rasseyaniye sveta v atmosferach planet. (In Russian. Light scattering in the atmosphere of planets.) Nauka Publ., Moscow (1972)
Steven, M.D. and Unsworth M.H.: The angular distribution and interception of diffuse solar radiation below overcast skies. Quart. Journ. Roy. Met. Soc., **106**, 57–61 (1980)
Tregenza, P.:Standard skies for maritime climates. Lighting Res. Technol., **31**, 3, 97–106 (1999)
Tregenza, P.: Analysing sky luminance scans to obtain frequency distributions of CIE Standard General Skies. Lighting Res. Technol., **36**, 4, 271–281 (2004)
Valko, P.: Angular distribution of sky radiance and diffuse irradiance on inclined surfaces. I. Cloudless skies. II. Cloudy and overcast skies. In: Course on physical climatology for solar and wind energy. Intern. Centre for Theoret. Physics, Trieste (1986)
Weber, L.: Intensitätsmessungen des diffusen Tageslicht. Annal. der Physik und Chemie, **26**, 11, 374–389 (1885)
Wittkopf, S. K.: A method to construct Virtual Sky Domes for the use in standard CAD-based light simulation software. Architectural Science Review, **47**, 3, 275–286 (2004)
Wittkopf, S. and Soon L.K.: Analysing sky luminance scans and predicting frequent sky patterns in Singapore. Lighting Res. Technol., **39**, 1, 31–51 (2007)

References to Appendix

Darula, S., Kittler, R., Gueymard, C.: Reference luminous solar constant and solar luminance for illuminance calculations. Solar Energy, **79**, 4, 559–565 (2005)
Fesenkov, V.G., Pyaskovskaya, J.V.: Osveshchennost dnevnogo nebesnogo svoda i indikatrisa rasseyaniya sveta atmosferoy. (In Russian. Illuminance of the daytime sky vault and the scattering indicatrix of light by the atmosphere.), Doklady AN SSSR, **4**, 3, 129–132 (1934)

Fesenkov, V.G.: O jarkosti bezoblachnogo neba pri sfericheskoy zemlye. (In Russian. About luminance of the cloudless sky for spherical Earth.), Doklady AN SSSR, **101**, 5, 845–847 (1955)

Igawa, N., Nakamura, H., Matsuzawa, T., Koga, Y., Goto, K., Kojo, S.: Sky luminance distribution between two CIE standard skies. Part 2. Numerical equation for relative sky luminance distribution. Proc. Lux Pacifica Conf., E13-E18 (1997)

Igawa, N. and Nakamura, H.: All Sky Model as a standard sky for the simulation of daylit environment. Build. and Environm. **36**, 6, 763–770 (2001)

Igawa, N., Koga, Y., Matsuzawa, T., Nakamura H.: Models of sky radiance distribution and sky luminance distribution. Solar Energy **77**, 2, 137–157 (2004)

ISO – International Standardisation Organisation: Spatial distribution of daylight – CIE Standard General Sky. ISO Standard 15409:2004 (2004)

Kittler, R.: Luminance models of homogeneous skies for design and energy performance predictions. Proc. Intern. Daylighting Conf. Long Beach, Calif., 18–22 (1986)

Kittler, R. and Pulpitlova, J.: Základy využívania prírodného svetla. (In Slovak Basis of the utilization of daylight.) Veda Publ., Bratislava. (1988)

Kittler, R., Darula S., Perez R.: A new generation of sky standards. Proc. LuxEuropa Conf., Amsterdam, 359–373 (1997)

Kittler, R., Darula S., Perez R.: A set of standard skies characterizing daylight conditions for computer and energy conscious design. US-SK grant project. Polygrafia SAV, Bratislava (1998)

Kittler, R., Darula, S.: Determination of sky types from global illuminance. Lighting Res. Technol., **32**, 4, 187–193 (2000)

Matsuzawa, T., Igawa, N., Nakamura, H., Matsuzawa, T., Koga, Y., Goto, K., Kojo, S.: Sky luminance distribution between two CIE standard skies. Part. 1. Proc. Lux Pacifica Conf., E-7-E-13 (1997)

Nagata, T.: Luminance distribution on clear skies. Transac. of Arch. Inst. Japan, 7/185, 65–71 and 8/186, 41–50 (1971)

Nagata, T.: Luminance distribution of clear sky and the resultant horizontal illuminance. Journal of Light and Visual Envir., **7**, 1, 23–27 (1983)

Perez, R., Ineichen, P., Seals, R., Michalsky, J., Stewart, R.: Modeling daylight availability and irradiance components from direct and global irradiance. Solar Energy, **44**, 5, 271–289 (1990)

Perez, R., Seals, R., J. Michalsky, J.: Modeling Skylight Angular Distribution from Routine Irradiance Measurements. Proc. IES Annual Conference, San Diego (1992)

Perez, R., Seals, R., Michalsky, J.: All weather model for sky luminance distribution – preliminary configuration and validation. Solar Energy, **50**, 3, 235–245 (1993)

Perradeau, M.: Luminance model. Proc. Nat. Lighting Conf., 291–292 (1988)

Sobolev, V.V.: Priblizhennoye resheniye zadachi o rasseyaniyi sveta v srede s proizvolnoy indikatrisoy rasseyaniya. (In Russian. The approximative solution of the task about light scattering in the media with an arbitrary scattering indicatrix.) Astronom. Zhurnal, **20**, 5–6, 14–22 (1943)

Sobolev, V.V.: O rasseyaniyi sveta v atmosferach zemli i planet. (In Russian. About light scattering in the atmosphere of earth and planets.) Uchennye zapiski LGU, ser.matem. nauk, **18**, 17–52, 116 (1949)

Sobolev, V.V.: Rasseyaniye sveta v atmosferach planet. (In Russian. Light scattering in the atmosphere of planets.) Nauka Publ., Moscow (1972)

Chapter 6
Simulation of Seasonal Variations in the Local Daylight Climate

6.1 Basic Characteristics of the Local Daylight Climates

The natural human biorhythm was mainly influenced by periodic changes of daytime activities and nighttime rest. In the original evolution of the human environment in the equatorial region, the regular daily and yearly changes of sunrise and sunset influenced considerably also vital behavior of the body for permanent 12 h of daytime and the same nighttime period. Migration to northern and southern territories during prehistoric time brought experience of seasonal daylight changes. These were linked with the daytime interval varying in the yearly cycles. It is evident that owing to the geographical altitude of the locality and the yearly solar declination changes the sunrise H_{sr} and sunset H_{ss} hours follow (6.1) if solar altitude $\gamma_s = 0$, i.e.,

$$H_{sr} = \frac{1}{15°} \arccos(\tan \varphi \tan \delta)\,(\mathrm{h}), \tag{6.1}$$

where φ is the geographical altitude and δ is the solar declination, and

$$H_{ss} = 24 - H_{sr}\,(\mathrm{h}). \tag{6.2}$$

So, the significant period for sunlight exposure of Earth's surface is determined by these two times. Then the maximum possible daily duration of the presence of the sun, the so-called astronomical sunshine duration S_a, is calculated after:

$$S_a = H_{ss} - H_{sr} = \frac{1}{7.5°} \arccos(-\tan \varphi \tan \delta)\,(\mathrm{h}). \tag{6.3}$$

The length of the day is shortest in locations in the northern hemisphere during the winter solstice, whereas the length is with maximum at the summer solstice. In Fig. 6.1 measured global illuminance and diffuse illuminance are compared during typical sunny winter, spring, and summer days. These illuminance

R. Kittler et al., *Daylight Science and Daylighting Technology*,
DOI 10.1007/978-1-4419-8816-4_6,

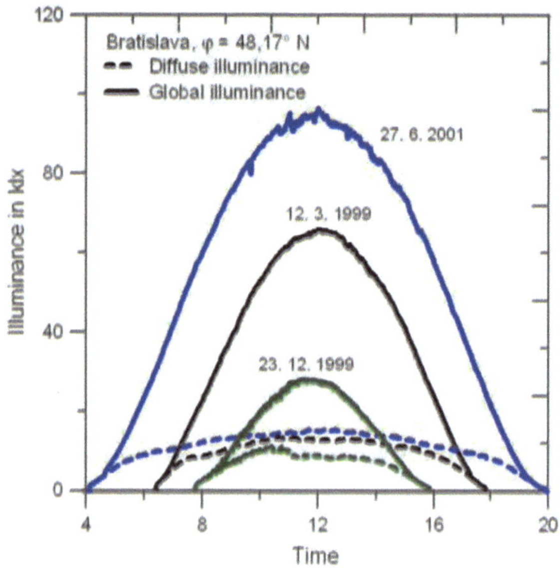

Fig. 6.1 Seasonal changes of day lengths and illuminance levels. *Solid lines* global illuminance, *dashed lines* diffuse illuminance

courses document seasonal changes of day lengths and illuminance levels outdoors for a locality at 48.17°.

The day length changes significantly influence the occurrence of skylight and sunlight, both depending also on the variability of cloudiness and atmospheric turbidity during the year. Coupling continual changes of solar altitude with different absorption and scattering properties of the atmosphere causes unique nonrepeating conditions. Therefore, real exterior daylight conditions differ locally and daylight climate has to be studied, assessed, and evaluated after long-term measurements of its momentary and typical situations.

6.2 Methods for Defining Local Daylight Conditions Based on Daylight Measurements

The first measurements of sky luminance performed by Schramm (1901) and Kähler (1908) resulted in models of the sky luminance distribution under overcast sky conditions with luminance gradation 1:3, later proposed by (Moon and Spencer 1942) and standardized by the International Commission on Illumination (CIE) and ISO.

New possibilities to measure and determine local daylight conditions regularly by more precise and digital measurement equipment enable one to record several quantities continually and simultaneously in shorter or long periods. So, in 1991 the CIE Technical Daylight Committee initiated the International Daylight Measurement Programme (IDMP), which started with three types of stations measuring daylight parameters at several universities and research institutes (http://idmp.entpe.fr/). At basic class stations only global illuminance and irradiance are measured, while at the general class stations besides global and diffuse horizontal illuminance and irradiance also illuminance on vertical planes facing the four cardinal points and zenith luminance are registered in 1-min steps during daytime. At the research class stations, furthermore data are collected on direct normal sun illuminance and irradiance and sky luminance or radiance distributions by scanners, sometimes supplemented by sunshine duration, relative humidity, or photosynthetically active radiation (PAR) data. Depending on the measuring instruments, instantaneous data in 1-, 5-, or 15-min steps or integrated values taken from the same intervals were recorded. The operation of measurements and processing of the data obtained were coordinated after CIE guide 108-1994 (CIE 1994). A total of 48 stations operated around the world in 1994. The CIE IDMP station activities depend on the local support and staff possibilities. Therefore, some stations have already ended their activities or the equipment has been moved to other localities. However, new stations have also been established, e.g., in Hong Kong at the Chinese University of Hong Kong, in Arcavacata di Rende at Università della Calabria in Italy, and the station in Osaka, Japan was rebuilt as a research and daylight calibration center.

Measured data stored at CIE IDMP stations reflect local specifications and can be used in the definition of local daylight climate and to refine general analytic models for example investigating:

- Parametric characteristics
- Statistical occurrence frequency
- Simulation possibilities
- Resemblance to artificial neural networks applicability
- Combined methods evaluating sets of measured data

The first measurements were focused on the description of the dependence of exterior global and diffuse illuminance on the solar altitude under homogeneous and simply defined atmospheric conditions. Two such types were investigated – especially overcast and clear skies – and later results of measurements gathered during cloudy days were published. Occasionally, measurements allowed statistical methods or fitting techniques to be applied and also allowed empirical formulae and the range of illuminance levels in various climatic zones to be proposed. In that time daylight research was also focused on the description of the importance of atmospheric conditions such as turbidity, cloud types, and cloud cover with respect to the documentation of ground illuminance levels.

The exterior illuminance was usually expressed by the sine function of the solar altitude and is significantly influenced by ground reflection as shown, for example, by Krochmann (1963), Hopkinson et al. (1966), (Kittler 1968), Feitsma (1971), and Dogniaux (1978) (Fig. 6.2). So, very high differences of illuminance values can be found for situations without snow and with snow-covered ground. Values of the global illuminance under clear sky conditions are generally higher in the case of the sun shining in Fig. 6.3, for example, after Ogisso (1965), Kittler (1968, 1972),

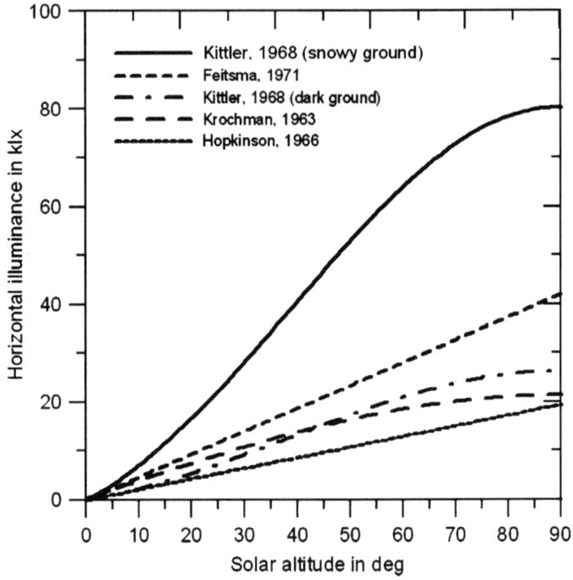

Fig. 6.2 Dependence of exterior illuminance on the solar altitude during an overcast sky

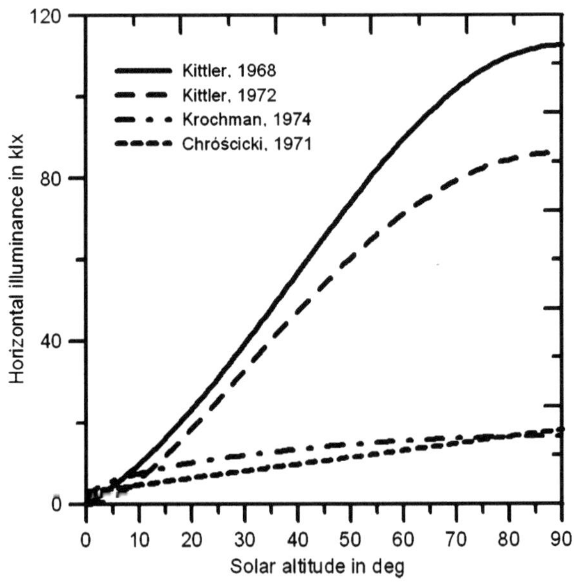

Fig. 6.3 Dependence of changes of exterior illuminance on the solar altitude during a clear day

Chróścicki (1971), Krochmann and Seidl (1974), and Ruck (1982). The lower curves in Fig. 6.3 represent diffuse illuminance levels while higher curves document global level under favorable conditions.

More complex daylight conditions occur during cloudy situations when clouds cover some parts of the sky and the sun is shining irregularly. The data obtained by Chróścicki (1971), Gillette et al. (1984), Ruck (1985), Tregenza (1986), Ruck and Selkowitz (1986), and Kittler and Pulpitlová (1988) cover a wide range within

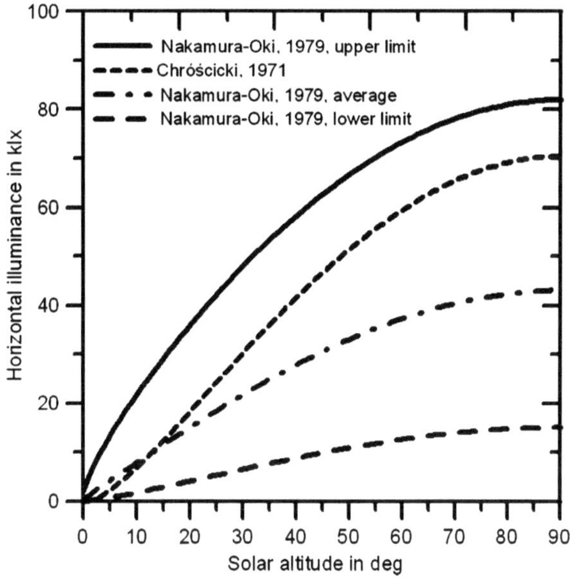

Fig. 6.4 Possible ranges of exterior illuminance levels during cloudy days

identified ranges (Nakamura and Oki 1979). The lower and upper limits of possible measured illuminance values are plotted in Fig. 6.4.

Measured data obtained in that period of irregular measurements have shown a high spread of values because the atmospheric conditions were not specifically determined.

Geographical location and orographic configuration of the specific place seemed to be quite important in daylight local conditions. Several studies of daily, monthly, and annual availability based on measured data were published. Similar to solar engineering practice, authors expressed hourly or monthly values by several parameters, such as the diffuse clearness index $k_{dd} = D_v/E_v$, the sunbeam index $k_b = P_v/E_v$, the clearness index $k_t = G_v/E_v$, the diffuse ratio or cloudiness index $k_d = D_v/G_v$, and the cloud ratio CR as published by Perradeau and Chauvel (1986), Kittler (1989), Kittler and Darula (1999), Kambezidis et al. (2002), and Muneer (2004).

The relation of D_v/E_v to P_v/E_v and G_v/E_v presented in Chap. 3 was illustratively shown in PGD diagrams. These relations usually represent either the momentary or the short-term daylight occurrences, but can characterize roughly the general daylight situations from data recorded during a long-term period (e.g., month, year, several years). The first analyses of daylight parameters were based on hourly or daily averages of measured daylight parameters, e.g., by Soler (1990), Corbella (1997), Kinghorn and Muneer (1998), and Muneer (2004), whereas later studies and analyses were directed to evaluate instantaneous data or data averaged from 1-min readings to 5-min averages, e.g., by Kittler (1989), Perez et al. (1990), Kittler et al. (1992), Muneer and Angus (1993), Dumortier et al. (1994), and Kittler and Darula (1997).

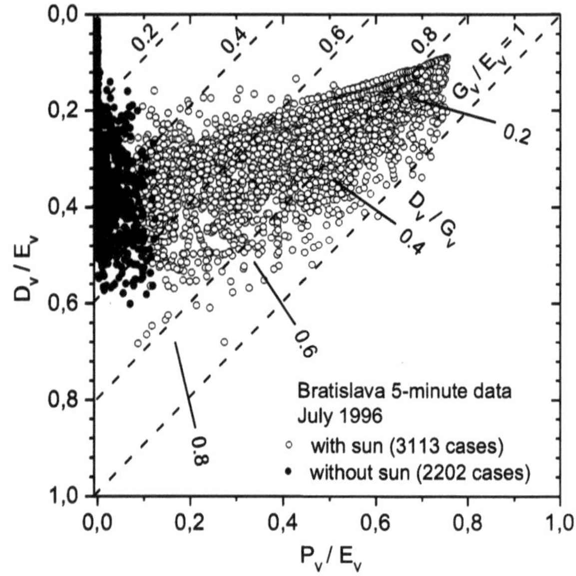

Fig. 6.5 Identification of daylight situations after clearness and cloudiness

Instantaneous illuminance and luminance data measured continuously and regularly register more precisely real daylight characteristics, including short time changes of levels and occurrences of P_v/E_v and D_v/E_v ratios. This possibility to use such parameterization of light transmittance through the atmosphere was presented in Fig. 3.9, and another example is given in Fig. 6.5. The beam index $k_b = P_v/E_v$ evidently separates a sunny situation from an overcast situation and gives information about the rate of attenuation of sunlight during July 1996 in Bratislava. The overcast situations were in the interval $P_v/E_v<0.1$ and the range of $0<D_v/E_v<0.5$. Clear sky conditions can be identified in the small area bordered by $0.7<G_v/E_v<0.95$ and $0.6<P_v/E_v<0.75$.

Another graphical system is based on the parameter $k_t = G_v/E_v$, expressing the quantity of light penetrating through the atmosphere compared with the extraterrestrial quantity (Kittler 1989). A similar monthly analysis of global and diffuse illuminance can be done using the diagram shown in Fig. 6.6. Here, overcast situations are placed in the upper-left part above the curve $k_b<0.1$, whereas sunny situations are concentrated diagonally in the strip directed toward the bottom corner to $k_b = P_v/E_v = 1$.

Kittler et al. (1997, 1998) introduced a model for calculation of the luminance distribution and sky type classification as already explained in Chaps. 3 and 5 (Fig. 3.12). The parameter L_{vZ}/D_v allows one to identify the participation of overcast, cloudy, and clear daylight conditions within a particular month. This parameter is quite constant under overcast skies, as is evident in Fig. 6.7, but is influenced by solar altitude under a clear sky daylight climate (Fig. 6.8). However, the sky type prevalence during the day, month within a year is distinctly shown. In the example in Fig. 6.7, the prevailing occurrence of overcast skies is in the

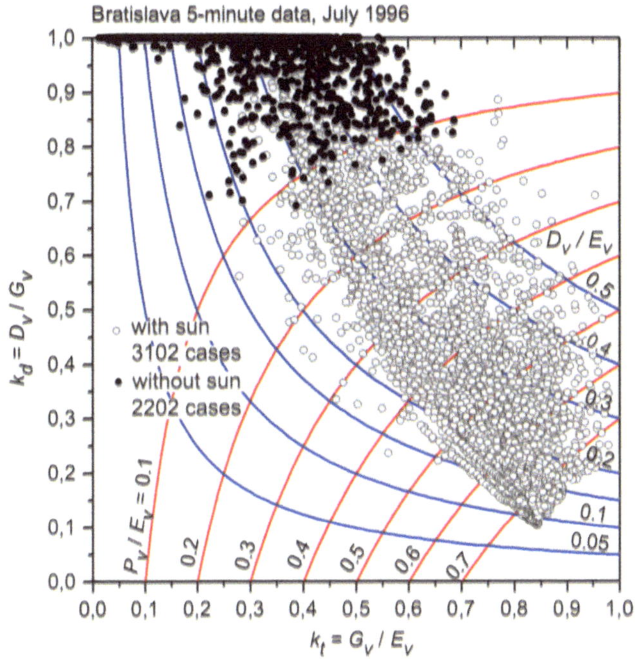

Fig. 6.6 Identification of daylight situations, Bratislava, July 1996

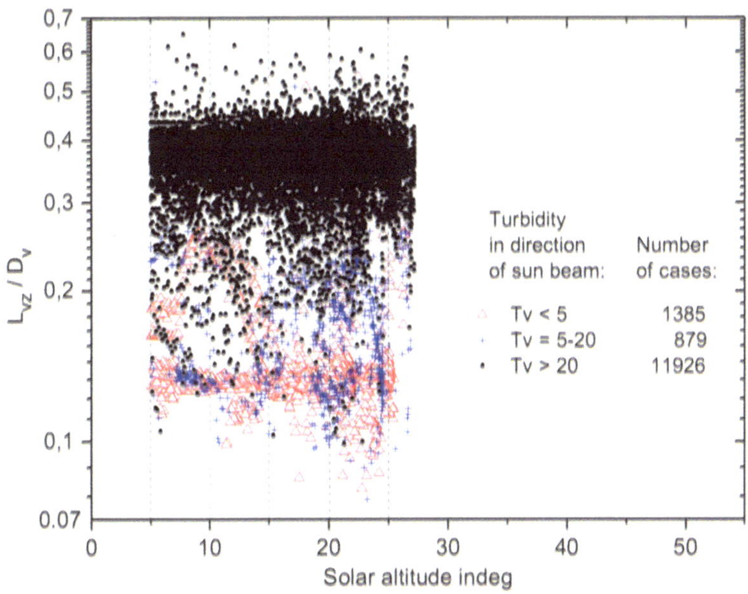

Fig. 6.7 Prevailing occurrence of overcast skies during winter in Bratislava, November 1995

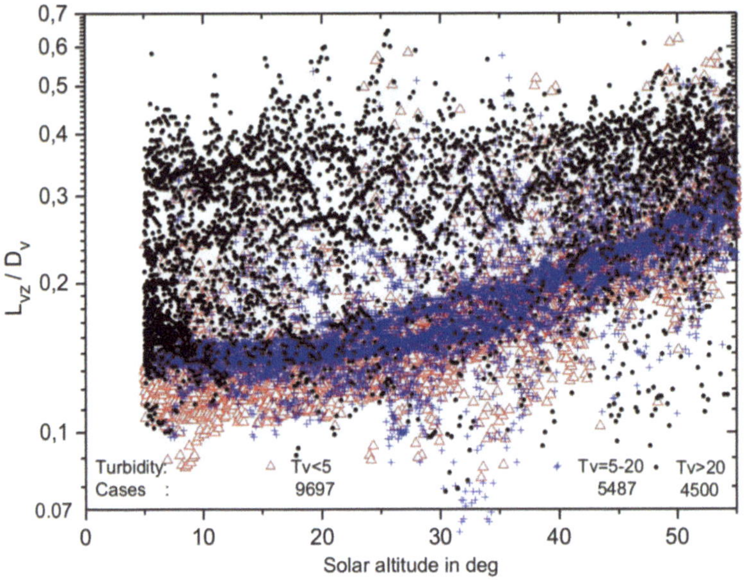

Fig. 6.8 Typical concentration of clear sky situations in the curved strip in the lower part of the diagram, Bratislava, July 1995

horizontal strip with range $0.27<L_{vZ}/D_v<0.43$ in November, with disproportionally fewer cases of clear situations in the range $0.12<L_{vZ}/D_v<0.18$ during a typical winter in Bratislava. Also, lower values of luminous turbidity factors T_v are associated with very clear atmospheres. For daylight during summer months, the measured L_{vZ}/D_v parameters in Fig. 6.8 are concentrated in the narrow curved strip rising from $L_{vZ}/D_v = 0.12$ at $\gamma_s = 5°$ to $L_{vZ}/D_v = 0.28$ at $\gamma_s = 55°$.

Another relation of L_{vZ}/D_v to the clearness index $k_t = G_v/E_v$ can give overall information about attenuation of light received at ground level and the distribution among sky types (Figs. 6.9–6.12). In the temperate daylight climate (Bratislava), the winter period (November to February) as in January 1994 in Fig. 6.9 is characterized by prevailing overcast skies, i.e., dominance of sky situations in the ranges $0.0<G_v/E_v<0.3$ and $0.25<L_{vZ}/D_v<0.43$, whereas sky situations in the range $0.5<G_v/E_v<1$ are scarcer and correspond to sun presence and low solar altitudes. In contrast, the summer period in Fig. 6.10 is rich with sunny and clear sky days with a high efficiency rate G_v/E_v over 0.4 linked with rising L_{vZ}/D_v values indicating many sunny periods during June to September. In the Mediterranean climate, the prevailing availability of sunlight and the scarce occurrence of sunless situations were identified in Athens even in the whole winter period (November to February 1994), as shown in Fig. 6.11, and especially during a typical Mediterranean summer period from June to September 1994 (Fig. 6.12).

Classical statistical techniques or advanced methods such as cluster analysis and factor analysis can be used for the determination of diurnal variations of illuminance levels and other daylight climate parameters (Markou et al. 2009).

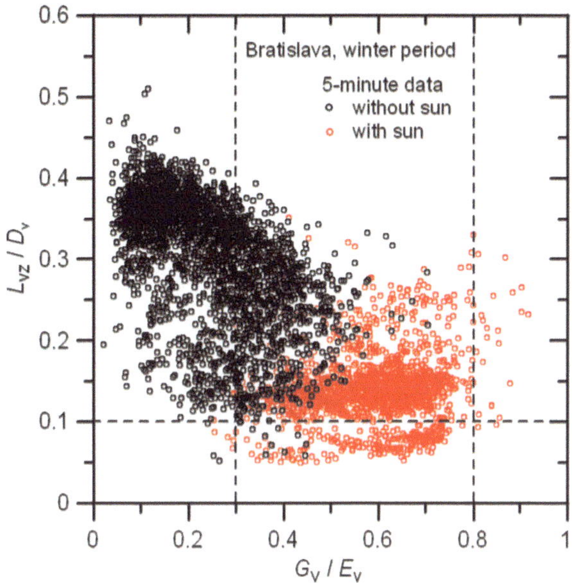

Fig. 6.9 Occurrence of sky types during January 1994 in Bratislava

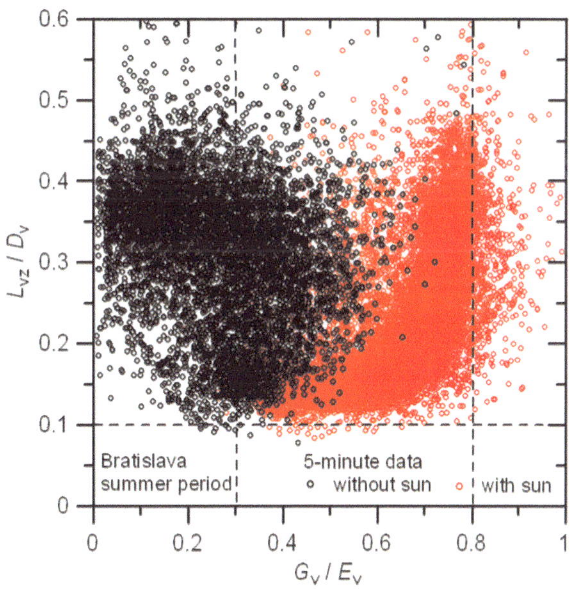

Fig. 6.10 Occurrence of sunny sky types during June 1994 in Bratislava

Good daylighting design in buildings requires information about the availability and intensity of natural light. As interiors are illuminated by sunlight and skylight, the availability statistics, such as average, maximum, minimum, standard deviation, cumulative distribution, and probability of occurrences of diffuse and global horizontal or vertical illuminances have been published (Nakamura and Oki 1979; McCluney and Bornemann 1986; Robins 1986; Darula and Kittler 1995;

Fig. 6.11 Even in the winter period in Athens the weather is very sunny

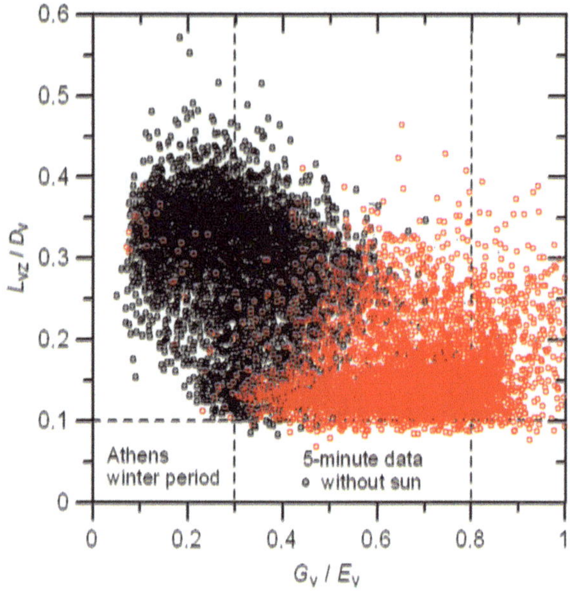

Fig. 6.12 There are few overcast and cloudy periods in Athens during summer and sunny periods prevail

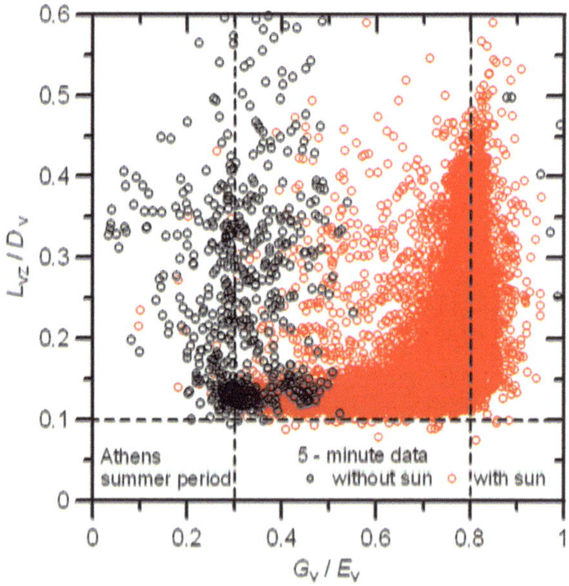

Ullah 1996; Darula 1997; Muneer and Kinghorn 1998; Ruiz et al. 2002; Perez-Burgos et al. 2007).

The statistical averages of global horizontal illuminance measured in Bratislava during 1994 and 1995 are given in Table 6.1. Average values in each month and calculated standard deviations from instantaneous measurements are documented

Table 6.1 Averages and standard deviations of the global illuminance in kilolux for Bratislava based on data measured in 1994 and 1995

	Average			Standard deviation		
Month	1994	1995	1994, 1995	1994	1995	1994, 1995
January	13.572	11.885	12.688	10.475	9.405	9.964
February	20.163	19.299	19.730	14.195	14.645	14.429
March	23.578	23.100	23.339	19.750	20.426	20.092
April	32.443	32.288	32.365	25.309	26.324	25.821
May	37.653	41.082	39.366	28.643	31.060	29.923
June	39.771	40.389	40.079	29.691	31.158	30.433
July	44.942	50.135	47.541	29.304	31.188	30.373
August	39.600	39.547	39.573	27.507	28.696	28.107
September	32.178	29.125	30.651	22.095	24.206	23.225
October	21.575	24.864	23.219	15.796	15.996	15.981
November	11.960	9.049	10.505	10.315	8.627	9.619
December	10.191	7.318	8.755	7.848	5.834	7.062

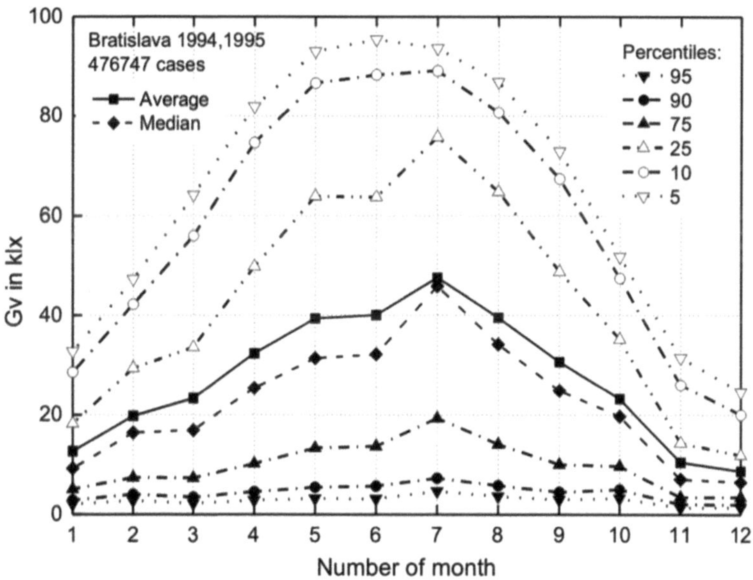

Fig. 6.13 Statistical monthly courses of global horizontal illuminance in Bratislava for 1994 and 1995. *Solid line* monthly average values, *dashed line* median values, *dot-dashed line* percentiles

in Table 6.1, whereas maximum and median values correspond with monthly changes of solar altitudes. These statistics are plotted in Fig. 6.13, where significant differences between seasons during a year in Bratislava are shown. The values of all statistics are similar for spring and autumn periods.

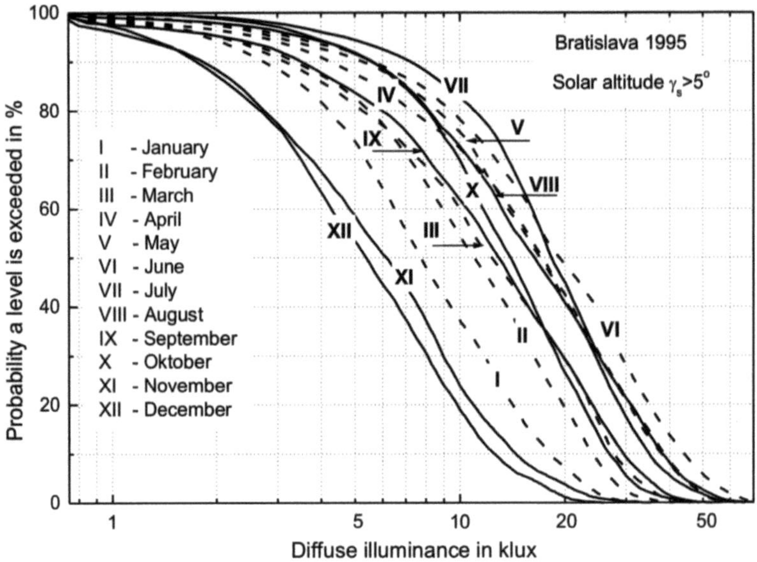

Fig. 6.14 Availability of diffuse illuminance levels during 1995 in Bratislava

The diffuse and global illuminance availability curves from data measured in Bratislava during 1995 are plotted in Figs. 6.14 and 6.15. Diffuse illuminance levels up to 20 klx can occur during winter months, when higher levels are seldom observed. However, owing to higher solar altitudes and white clouds often occurring during all other seasons, the diffuse illuminance can rise to 50 klx. When global illuminance includes also direct sunlight, its values can be several times higher than those of the diffuse illuminance. Figure 6.15 shows that in the winter period global illuminance can rise to 50 klx, whereas during summer it can rise even to 100 klx. However, owing to frequent fog and low darker cloudiness during November and December drop to 32% of global illuminances compared with values in Bratislava in January and February is observed. This asymmetry to the fictitious equinox axis is observed also in the availability of diffuse illuminance.

Dumortier et al. (1994) analyzed exterior illuminances measured during March 1993 to February 1994 in Lyon, France. He recognized only two seasons, i.e., winter, representing the period from October to March, and the summer season around the summer solstice (April to September). Winter periods in Lyon are characterized by illuminance about 5–15 klx higher than in Bratislava, whereas during the summer period illuminance can be 20–75 klx.

Navvab et al. (1989) presented results of evaluating daylight parameters derived form regular measurements in Ann Arbor, USA, comparing 4-, 10-, 20-, and 60-min averages of direct illuminance obtained during clear days and partly cloudy days. They found that 10-min integrated data fully represent the sky conditions. Similar analysis was published by Walkenhorst et al. (2002) in a study comparing the results of simulated annual indoor illuminances based on 1-min and hourly

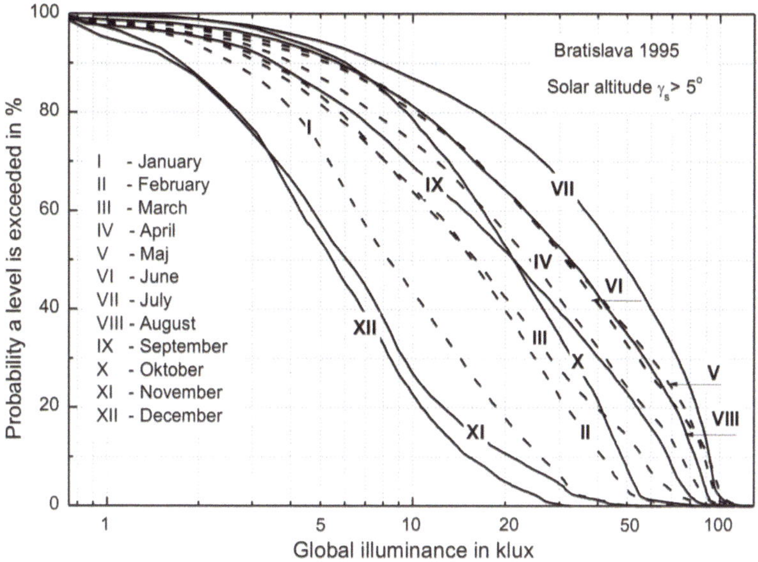

Fig. 6.15 Availability of global illuminance during 1995 in Bratislava

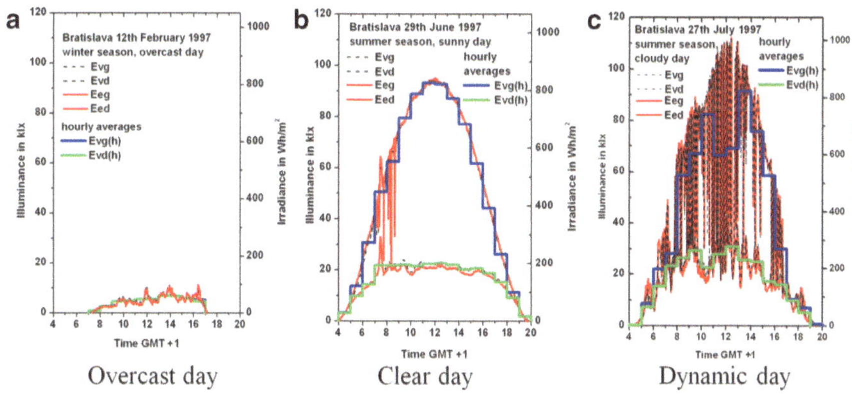

Fig. 6.16 Comparison of daily courses based on the instantaneous and hourly data

averaged data. Two different automated daylight-dependent artificial lighting strategies to predict annual artificial lighting demand were investigated. Determination of such demand based on hourly averages is systematically underestimated by up to 27% compared with 1-min-average measured data.

Skylight and sunlight as daylight sources were analyzed by Darula and Kittler (2010) with respect to their occurrence on daily illuminance courses and the composition of data for common evaluations. Comparison of 1-min instantaneous data with hourly averages indicates differences especially between the occurrence

Table 6.2 Selected computer programs for daylight simulation

Name	URL
ADELINE	http://www.ibp.fhg.de/wt/adeline/
Daylight	http://www.archiphysics.com/programs/daylight/daylight.htm
DAYSIM	http://www.daysim.com/
EnergyPlus	http://apps1.eere.energy.gov/buildings/energyplus/cfm/reg_form.cfm
Leso DIAL	http://www.ibp.fraunhofer.de/software/lesodial.html
Radiance	http://radsite.lbl.gov/radiance/HOME.html
SkyVision	http://www.nrc-cnrc.gc.ca/eng/projects/irc/optical-characteristics.html

of maximal levels within the different daily courses (Fig. 6.16). During overcast and clear days (Fig. 6.16a, b) owing to fluent smaller changes the consecutive illuminances and hourly averages follow more or less the overall daily trend. Totally different situations occur during days with rapid dynamic illuminance changes (Fig. 6.16c), when the averaging method eliminates real maximum peaks within hourly periods. However, resulting stepped hourly illuminance courses of global and diffuse illuminance hide any information about the daylight dynamic character. In consequence, large differences between 1-min and hourly maximums have to be expected.

In extreme cases, the evaluations based on the hourly averages can reduce the maximal instantaneous global illuminance to about 26%, whereas a reduction of up to 40% in diffuse illuminance was found.

Scientists have recently adopted computer techniques also for daylight research and analysis. The first simulations based on the calculation of illuminance distributions in rooms were extended to visualization, investigation of the indoor luminous environment, and energy aspects of design and interior performance (Ward 1994; Reinhart and Herkel 1999; Mardaljevic 2000; Janák et al. 2003; Li et al. 2004). There are many programs and packages available cost-free and for payment. Some selected programs and World Wide Web addresses are listed in Table 6.2. However, some of these program results might be misleading because average hourly illuminances are applied also for dynamic days.

Daylighting design and solution for sun energy utilization require methods and algorithms expressing characteristic daylight changes during the day and the predetermination of typical courses of global, diffuse, and direct illuminance. A specific solution to derive synthetic short-term or long-term sequences of daylight parameters by Darula and Kittler (2008) enables one to predict daily synthetic courses of diffuse and global illuminances based on D_v/E_v and T_v monthly values respecting local specific conditions.

Artificial neural networks also represent a powerful method for categorization of various climatic parameters and for local specification. Çolak and Onaygil (1999) presented a study of the calculation of illuminance (artificial and daylight) preference on the desk in offices. Kazanasmaz et al. (2009) modeled indoor illuminance at a reference point using 13 input weather, room dimension, and time parameters. Li et al. (2010) classified daylight conditions and calculated the frequency of occurrence of the 15 CIE standard skies by applying a probabilistic neural network with five input variables, i.e., zenith luminance, global, direct, and diffuse illuminance, and solar altitude.

Satellite data are an alternative information database for irradiance and illuminance evaluations. The SATEL-LIGHT Internet server (http://www.satel-light.com) offers daylight illuminances and solar radiation for localities in western and central Europe. These data were derived from images produced by the *METEOSAT* satellite in 1996–2000. The available database covers fully 3 years (1998–2000) of 30-min data with a spatial resolution of 2.5 km (Dumortier and Van Roy 2003; Dumortier 2005). The results of daily course computation of global horizontal illuminance from horizontal irradiance using luminous efficacy were presented in Dumortier (2005). Calculation of horizontal illuminances with respect to sunshine duration also based on luminous efficacy was described by Olseth and Skartveit (2001). Janjai et al. (2003, 2008a, b) introduced the derivation of illuminance data from satellite images. He and Ng (2010) analyzed Hong Kong satellite data and illuminance measurements at their IDMP station and described a method for predictions respecting higher solar altitudes in subtropical regions.

6.3 Using Meteorological Data to Estimate Year-Round Daylight Availability

Year-round changes of daylight climate and exterior illuminance represent one of the most important factors of the human and built environment. In reality, cloud cover and cloud types, atmospheric content, and sun position in the hemisphere influence variables related to meteorological situations. Perradeau and Chauvel (1986), Kittler and Pulpitlová (1988) Koga et al. (1999), and Matuszawa et al. (1999) studied significant influences of the cloud type and cloud cover on natural illuminance levels. Cloud cover, sunshine duration, and atmospheric humidity are the main meteorological variables determining the weather situation related to daylight availability. Meteorological services offer data from measurements and observations such as horizontal irradiance (hourly, daily, or monthly values), sunshine duration, cloud type and cover, and atmospheric humidity.

Because common available meteorological data indirectly reflect photometric sky parameters, they can be used to express approximately daylight conditions by simple empirical formulae or diagrams.

Muneer (1997, 2004) discussed and compared several solar radiation and daylight models using the meteorological parameters mentioned. The hour–month rectangular diagrams of annual availability of global illuminance availability in Teddington/Kew, Eskdalemuir, and Lewick together with tables of averaged values (1933–1939) for Teddington are documented in Hopkinson et al. (1966). Kambezidis et al. (1998) presented the availability of global and diffuse horizontal illuminances for Athens in hour–month diagrams. Li et al. (2002) studied the interrelation between measured horizontal irradiance and illuminance in

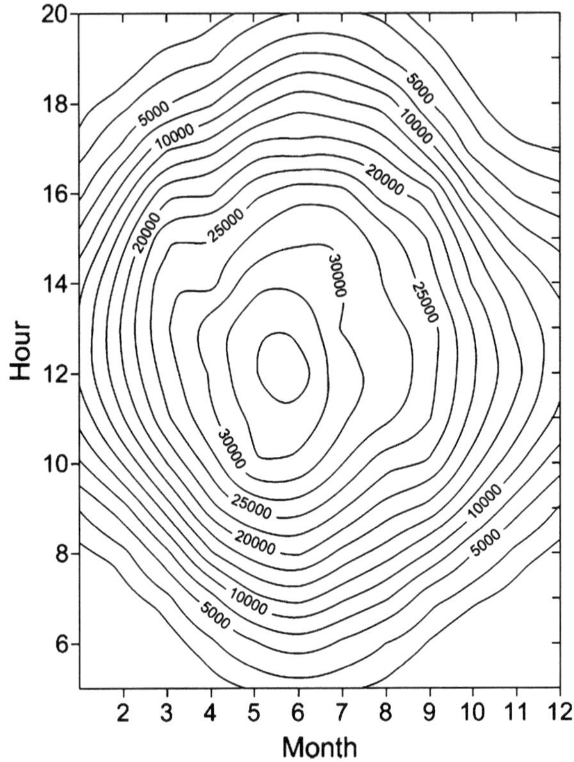

Fig. 6.17 Hourly values of diffuse horizontal illuminance in lux, Nantes 1993

Hong Kong. Similar diagrams of annual availability of horizontal diffuse and global illuminance for Bratislava (Kittler and Pulpitlova 1988) and for Lyon (Dumortier 1994) were developed. CSTB (1993) collected hourly illuminance data measured in 1993 and 1994 at Nantes in tables. These data are plotted here in hour–month diagrams in Figs. 6.17 and 6.18 as examples of the yearly illuminance distribution.

Comparison of Figs. 6.17 and 6.18 with Figs. 6.14 and 6.15 demonstrates significant differences between instantaneous and averaged illuminance levels. Whereas instantaneous global illuminance levels higher than 65,000 lx can occur frequently during summer in Bratislava (Fig. 6.14), hourly averages eliminate higher values and global illuminances are restricted to 65,000 lx in Nantes, the geographical latitude of which is lower than that of Bratislava.

The standard meteorological database which contains hourly values for a full year used by Mardaljevic (2008) in climate-based daylight analysis resulted in a method for the criterion of annual utilization of daylight in interiors under various outdoor conditions. This method is based on an acceptable/designed working-plane illuminance level and prediction of indoor illuminance using the daylight coefficient method and the simulation program Radiance. The new evaluation criterion called useful daylight illuminance (UDI) was proposed to replace the daylight factor and was introduced in this method (Mardaljevic 2000; Nabil and Mardaljevic 2005; Kleindienst et al. 2008).

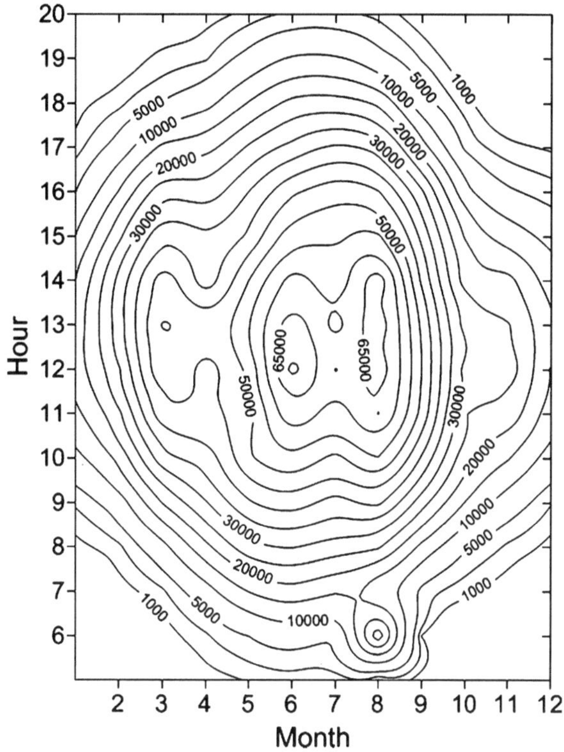

Fig. 6.18 Hourly values of global horizontal illuminance in lux, Nantes 1993

6.4 Sunshine Duration as a Local Daylight Climate Generator

After John Francis Campbell invented the sunshine recorder in 1853 and George Gabriel Stokes modified it in 1879, regular measurements of sunshine duration started and sunshine duration is now one of the basic parameters regularly measured by meteorological services. The first serious information about sunshine duration in days is from the UK. The Greenwich Observatory has measured sunshine duration since approximately 1870 (Whipple 1878) and it has been measured across the USA since 1890. Long-term statistical evaluations of sunshine duration were processed mainly with respect to utilizing sun radiation. Establishing CIE IDMP stations gave a new opportunity to study the influences on sunlight and skylight (Kittler and Pulpitlova 1988). Darula and Kittler (2004) found that a dependence between relative sunshine duration and daily global and diffuse illuminance profiles exists. Statistical analyses of sunshine duration occurrences based on data recorded in Bratislava (1994–2001) and Athens (1992–1996) are documented in Tables 6.3 and 6.4 and allow one to derive formulae for the predetermination of sequential types of clear, overcast, cloudy, and dynamic situations of outdoor daylighting. In this sense, one can apply either daily relative sunshine duration or morning and

Table 6.3 Monthly averages of relative sunshine duration after measured 5-min data in Bratislava, 1994–2001

Month	Year 1994	1995	1996	1997	1998	1999	2000	2001	Mean
January	0.298	0.185	0.157	0.098	0.313	0.170	0.218	0.207	0.206
February	0.351	0.343	0.407	0.444	0.556	0.238	0.368	0.505	0.402
March	0.321	0.327	0.342	0.490	0.477	0.454	0.359	0.303	0.384
April	0.446	0.390	0.493	0.477	0.442	0.515	0.655	0.466	0.486
May	0.461	0.548	0.442	0.579	0.569	0.547	0.716	0.668	0.566
June	0.531	0.428	0.602	0.521	0.549	0.519	0.736	0.495	0.548
July	0.687	0.722	0.555	0.431	0.505	0.573	0.415	0.511	0.550
August	0.603	0.540	0.502	0.706	0.626	0.561	0.739	0.750	0.628
September	0.512	0.444	0.212	0.715	0.356	0.568	0.486	0.259	0.444
October	0.398	0.479	0.436	0.538	0.313	0.414	0.460	0.404	0.430
November	0.224	0.117	0.295	0.275	0.291	0.156	0.319	0.331	0.251
December	0.247	0.090	0.237	0.113	0.140	0.261	0.176	0.289	0.194

Table 6.4 Monthly averages of relative sunshine duration after measured 5-min data in Athens, 1992–1996

Month	Year 1992	1993	1994	1995	1996	Mean
January	0.499	0.554	0.516	0.477	0.210	0.451
February	0.526	0.417	0.476	0.609	0.370	0.480
March	0.417	0.584	0.654	0.661	0.265	0.516
April	0.599	0.681	0.668	0.637	0.628	0.643
May	0.585	0.599	0.748	0.738	0.662	0.666
June	0.672	0.781	0.856	0.808	0.866	0.797
July	0.770	0.880	0.873	0.807	0.891	0.844
August	0.890	0.874	0.884	0.765	0.858	0.854
September	0.816	0.844	0.870	0.661	0.725	0.783
October	0.528	0.742	0.591	0.682	0.484	0.605
November	0.476	0.287	0.518	0.415	0.593	0.458
December	0.382	0.543	0.481	0.263	0.332	0.400

afternoon half-day values, which can characterize also daylight situation changes when relative sunshine durations are different in morning and afternoon hours (Darula et al. 2004; Darula and Kittler 2008a, 2011). Thus, the relative sunshine duration in any period is the ratio of the measured duration of sunshine to the astronomically possible duration within the same period assuming cloudless weather (after (6.3)).

An example of a diagram with formulae for predicting illumination situations based on monthly sunshine duration values for mornings is shown in Fig. 6.19. A similar diagram was developed for afternoon time that was roughly valid for any locality. The definition curves in Fig. 6.19 document the fact that maximal values of

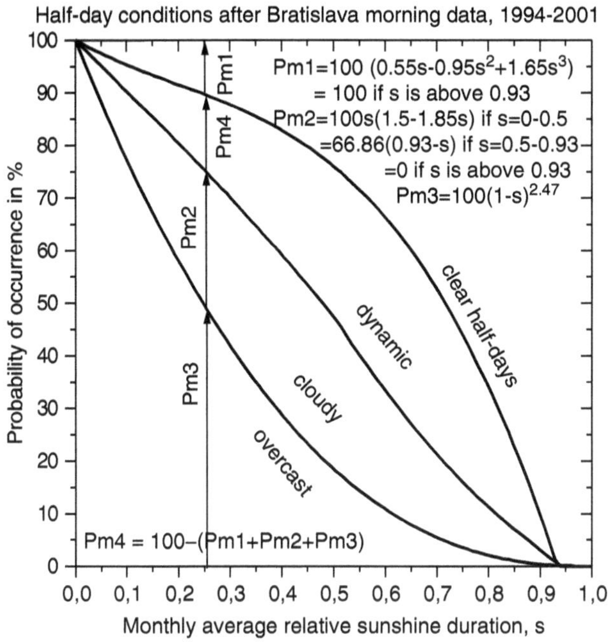

Fig. 6.19 Redistribution of daylight type courses for morning half days

sunshine duration are associated with clear situations, whereas minimal values have to be expected during overcast days. So, these occurrences and their monthly changes can be modeled and predicted. Averaged monthly values of sunshine duration are summarized in tables or often modeled by trigonometric functions; thus, typical daylight situations can be determined. Good similarity and agreement between measured and calculated values is shown in Fig. 6.20, the results for years with extreme differences are plotted (Darula and Kittler 2005).

Because sunshine duration is the most frequent parameter recorded by meteorological stations worldwide and is available in many places where no other parameter is available, it was also applied for identification of sky situations in specific localities. For instance, Nakamura and Oki (1987) investigated monthly values of sunshine duration during 1979–1982 representing daylight climate in Japan. They selected three sky types as relevant: clear, overcast, and intermediate. Kittler (1997) analyzed the irradiance, illuminance, and zenith luminance measured in Bratislava in 1995 and found a significant dependence between sunshine duration and the luminous turbidity factor. Rahim et al. (2004) analyzed data from 1995–2000 measured at the meteorological observatory in Ujung Pandang (formerly Makassar) and classified three sky conditions for each month of the year.

A different method based on derivation of hourly global and diffuse irradiances from satellite images was presented in Olseth and Skartveit (2001) and Good (2010). First, the normal beam irradiance is calculated and then the sunshine duration is obtained from the number of periods with normal irradiance exceeding 120 W/m^2.

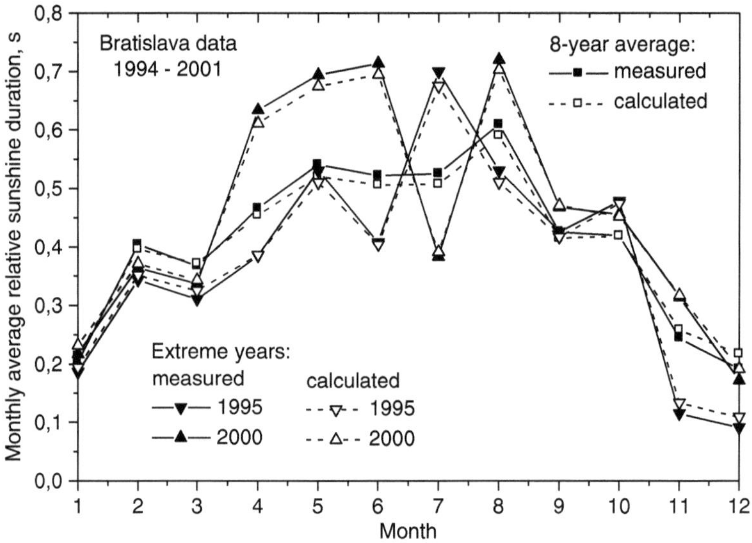

Fig. 6.20 Monthly sunshine duration occurrences in Bratislava

6.5 Daylight Reference Years for Simulating Long-Term Yearly Variations in Daylighting Availability

One important problem in building design and solar engineering is energy consumption and conservation, which depends also on the accurate prediction of long-term availability of daylight. The precise simulations use meteorological variables as inputs and computer methods to describe daylight availability with yearly changes. These offer a variety of results that are more or less applicable in practice. To better utilize daylight in buildings, it is necessary to find and classify typical conditions that are locally applicable. In solar engineering several methods were developed characterizing typical weather conditions during months in a reference year. There are several alternatives in the determination of daylight climate conditions:

- Composition of prevailing conditions of daylight parameters, their frequency, and duration during a year
- Grouping of usually occurring relevant characteristics which represent typical seasonal or yearly variations

Darula et al. (2001) analyzed daylight climate characteristics such as horizontal illuminance and sky patterns in Bratislava and Athens from simultaneous measurements in both localities. Seasonal variations in central Europe and the Mediterranean region differ in the specific occurrence of clear and overcast skies during the winter and the summer period, respectively.

Table 6.5 Example of daylight reference year results for Bratislava and Athens

Month	Bratislava	Athens
January	1999	1993
February	1995	1995
March	1997	1993
April	1999	1993
May	1999	1996
June	1996	1995
July	1998	1996
August	1998	1994
September	2000	1993
October	1994	1993
November	1999	1995
December	1998	1994

Changes of local climate depend on annual cycles, geographical latitude, and topographical specifications. Data from meteorological and IDMP measurements allow the generation of typical available values of illuminances, their daily courses, sunshine duration and other variables. Classic meteorological parameters, such as air temperature, relative humidity, wind velocity, and solar radiation, are utilized in the generation of meteorological reference years applicable mainly in energy analysis and studies. Formulation of a daylight reference year conditions would be based on photometric variables and their measurements. Petrakis et al. (1996) created a typical illuminance year for Athens by applying the method used in generating of typical meteorological year, but based on daylight measurements. Markou et al. (2007) developed and explained the detailed generation of four daylight reference years (DRYs) based on the photometric variables and common methods, i.e., the Danish Festa-Ratto method and the modified Sandia National Laboratory (SNL) method with two variations of weighting factors. They applied them for two localities with different climates, Athens in Greece and Bratislava in Slovakia. The 5-min global and diffuse horizontal illuminance, global and diffuse horizontal irradiance, zenith luminance, luminous turbidity factor, Linke's turbidity factor, and relative sunshine duration data gathered during 1994–2001 in Bratislava and during 1992–1996 in Athens served as an input to statistical calculations. Monthly composite DRY data represent typical daylight conditions which have shown in the selection the smallest root-mean-square deviations or other statistical criteria. Of course, in the dependence on the methods and their statistical procedures all DRYs are not identical. Although there are similarities, there are also significant differences. For Bratislava, the Festa-Ratto and modified SNL-Var1 methods give comparable results in 6 months, whereas for Athens the Festa-Ratto and modified SNL-Var2 methods give comparable results in 7 months of the 12 months. The results of the typical months selected are documented in Table 6.5 for Bratislava and Athens. Finally, each month was assigned the original instantaneous and 5-min averaged illuminance measured values, so the DRY works with the real photometric data. Such studies show that the final selection of the best DRY depends also on the method chosen, the parameters processed, and the availability and capacity of the database.

6.6 Energy Savings due to Better Daylight Utilization

The transport and the energy required to service the building performance consume most of energy related to human activities. Artificial lighting is estimated to account for up to 40% of all energy used in buildings depending on the type and utilization of indoor spaces. There are continual intentions to reduce consumption of electricity associated with artificial lighting. Using new, more efficient lamps, control systems, and better design strategies to promote effective daylight utilization can considerably reduce energy consumption and reduce maintenance costs in buildings. It is important to note the substantial differences between daylighting and artificial lighting. Daylight levels in the exterior are continually changing during the day, predominantly influenced by solar altitude, the luminance distribution on the sky vault under different cloud cover, and the turbidity of the atmosphere. Indoor daylight illuminance is strongly associated with the sky luminance distribution in the window solid angle seen from interior points; therefore, the dependence of room orientation to the cardinal points is crucial. However, artificial lighting is designed to produce constant interior levels respecting night or daylight operation regimes. Moreover, often spectral properties of artificial sources markedly differ from the changing daylight spectrum. Some solutions utilizing daylight in the interior are based on the space zoning and control of levels of additional artificial lighting. Choi and Mistrick (1999) described an integrated daylight and automatic lighting control system which follows the demand for energy savings and environmental protection. Krarti et al. (2005) presented a method for the energy saving estimation of artificial lighting when daylighting is operated by a dimming control system. Li and Lam (2001) proposed and experimentally tested a method of lighting control of indoor illuminance after measured illuminance changes on an external vertical surface. Kurian et al. (2008) described three computational models suitable for the optimum integration of visual comfort, thermal comfort, and energy consumption when daylight and artificial light are integrated. They identified the lighting control strategy, considering the reduction of glare, uniformity increase, and thermal comfort using an adaptive predictive control for artificial light dimming. Darula and Kittler (2008c) investigated daylight monthly utilization in offices with various working times under exterior illuminances of 5,000, 7,500, and 10,000 lx based on Bratislava data gathered in the period 1995–2004. An example of the times needed in 1995 for operation of luminaires during day and night in an office occupied from 7 a.m. to 4 p.m. is presented in Table 6.6. In this table are given exterior illuminance in lux needed to achieve suficient interior daylighting in three cathegories (Darula and Kittler 2008c). Summertime is included in and weekends are excluded from this calculation.

Ihm et al. (2009) described a method for evaluating the potential of daylighting to reduce energy use associated with electrical lighting. The impact of both dimming and sophisticated daylighting controls and their settings was also investigated. The electronic control systems in buildings are often connected to devices reducing or redirecting sunlight and skylight, such as movable blinds, louvers, and prismatic or dimmable glazing. For example, Bhavani and Khan (2009) presented a model for controllable blind systems respecting designed illuminance levels, blind position, occupant comfort, and benefits of daylighting in the interior.

Table 6.6 Operating time in hours of luminaires in offices in Bratislava with working time 7:00–16:00

	Daytime t_D			
	Exterior illuminance needed (lux)			Night
Month	5,000	7,500	10,000	time t_N
January	60.08	93.45	113.25	15.83
February	32.00	52.32	68.28	2.80
March	24.92	38.72	51.62	0.00
April	14.57	26.38	35.17	0.00
May	3.53	6.87	12.83	0.00
June	1.17	4.65	9.85	0.00
July	1.83	3.13	4.98	0.00
August	6.57	15.13	22.67	0.00
September	16.77	25.70	34.43	0.00
October	25.65	35.72	47.22	4.91
November	85.10	109.33	132.17	3.09
December	86.38	110.58	128.08	14.52
1995	358.57	521.98	660.55	41.15

6.7 Partial Conclusions

Weather and daylight conditions periodically change in annual cycles in each locality with more or fewer deviations; therefore, daylight climate is also logically associated with yearly, seasonal, or shorter periods. In building interiors, daylight is utilized at the moment of its occurrence, i.e., with the momentary exterior level and for the momentary human activity in the interior. To predict or model correctly such annual daylight conditions outdoors, photometric data such as luminance (for discomfort glare prevention through use of appropriate models), illuminance, irradiance, zenith luminance, and sunshine duration are needed. Several CIE IDMP stations have been in operation since 1991, where basic daylight parameters are regularly measured worldwide. At these stations instantaneous luminous variables and supplementing radiative variables after CIE guide 108-1994 (CIE 1994) are recorded and archived, but few radiometric variables with cloudiness and sunshine duration are registered at meteorological stations after World Meteorological Organization rules. Generally, the hourly data or data representing longer periods can be provided by meteorological services, but when daylight conditions for interior design with specific visual requirements are needed, then instantaneous data should be considered. Hourly averaged data suppress the momentary dynamics of daylight and eliminate peak extremes.

Daylight annual profiles can be based on the relation between sunshine duration and exterior illuminance parameters, thus allowing possibilities for approximate modeling of the characteristic or typical day or half-day daylight situations in any locality if no reliable data are available.

When simple methods for modeling daylight availability are expected in practical cases, daylight climate conditions can be expressed by data of the DRY, offering real measured illuminances in selected months representing typical conditions. However, such DRYs were recently generated only for Bratislava and Athens.

Daylight penetrates into interiors but not necessarily without costs. However, energy expenditure can be considerably reduced without inducing visual comfort by educated design and effective utilization of daylight, supplemented when necessary by electric lighting and/or adequate video display task illumination (see Chap. 12).

Appendix 6

Possibilities to Simulate Year-Round Changes of the Local Daylight Climate

Soon after the set of sky standards had been published by Kittler et al. (1997), Tregenza (1999) tried to analyze available sky scan data in several maritime locations and noticed some similarities in the frequency of sky types. In Singapore, Garston, and Sheffield, sky types 1, 3, 4, 8, 11, and 13 seemed to be most relevant. Later, Tregenza (2004) recommended his method of sky scan analysis to obtain the sky type frequency distribution. Thereafter, several IDMP stations used their long-term data to find the most frequent locally occurring sky types, as reported by Ng et al. (2007) or Li and Tang (2008) for Hong Kong, Chirarattananon and Chaiwiwatworakul (2007) for Bangkok, Wittkopf and Soon (2007) for Singapore, and Torres et al. (2010) for northern Spain. However, luminance scan measurements are available only in IDMP research stations, so other approaches to identify prevailing sky types or the daylight climate were sought, such as the classification parameter L_{vZ}/D_v(Kittler et al. 2001; Bartzokas et al. 2005), the vertical sky component, i.e., the ratio of vertical to horizontal diffuse illuminance ($VSC = D_{vv}/D_v$) (Alshaibani 2008, 2011; Li et al. 2011) or analyzing the daylight climate by applying several meteorological parameters (Markou et al. 2009).

Daylight climate in the temperate region is characterized by almost all sky types as documented by the Bratislava seasonal distribution (winter overcast and summer clear, shown in Fig. A6.1). This seasonal effect increases with the distance of the locality from the ocean (Kittler et al. 2001). In the subtropical or Mediterranean climate, e.g., in Athens, clear sky types (Fig. A6.2) prevail. A more illustrative analysis based on the L_{vZ}/D_v parameterization following the $\pm 2.5\%$ strip along each sky type L_{vZ}/D_v curve from long-term Bratislava IDMP data which show sunless cases is shown in Fig. A6.3 and sunny cases in Fig. A6.4, same presentation for Athens documents considerable different climate conditions in Figs. A6.5 and A6.6. Owing to the $\pm 2.5\%$ strip selection in this analysis, the overall number of cases taken into consideration in Bratislava was reduced to 113,473 within the range of solar altitudes 5–70°. In the original study (Darula et al. 2001), a comparison was made by taking ± 1 and $\pm 2.5\%$ strips to detect differences. However, good agreement with the previously documented data resulted.

Fig. A6.1 Seasonal sky type distribution after long-term 1994–1998 data gathered in Bratislava

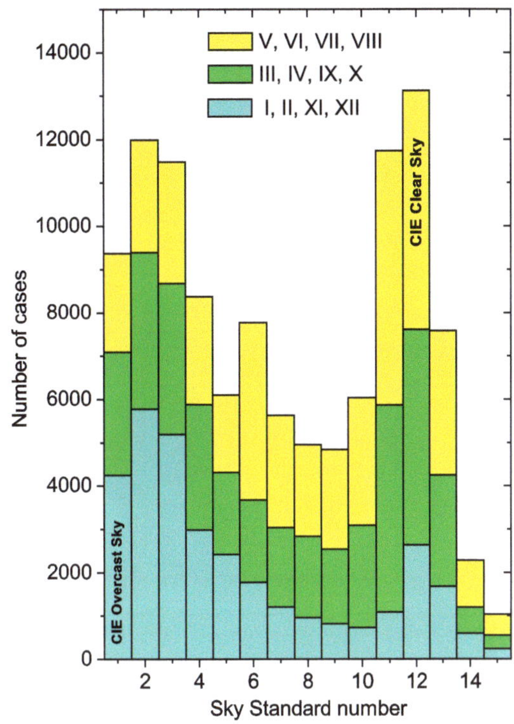

Fig. A6.2 Seasonal sky type distribution after long-term 1992–1996 data gathered in Athens

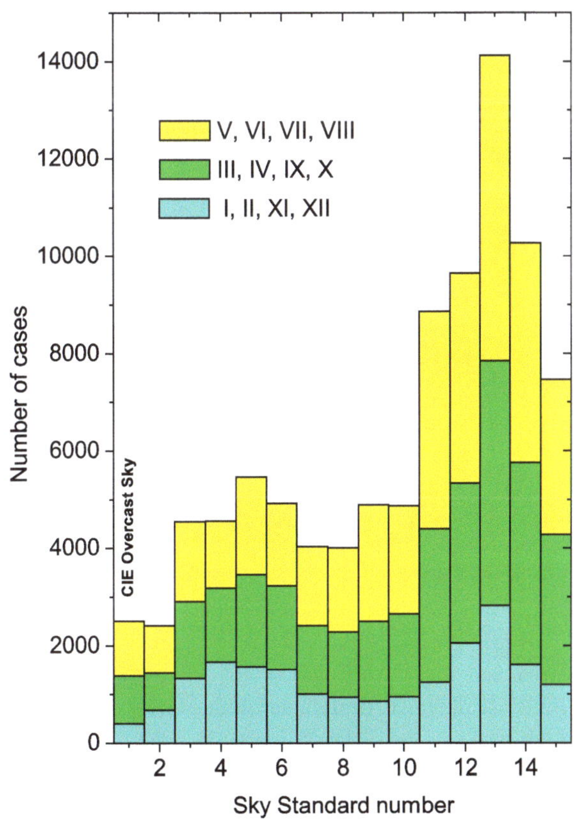

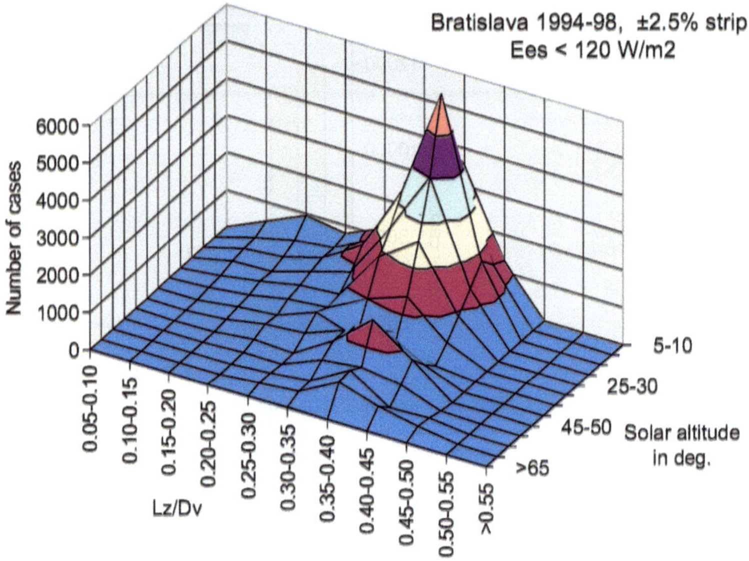

Fig. A6.3 Situations without sun represent prevailing overcast skies in Bratislava

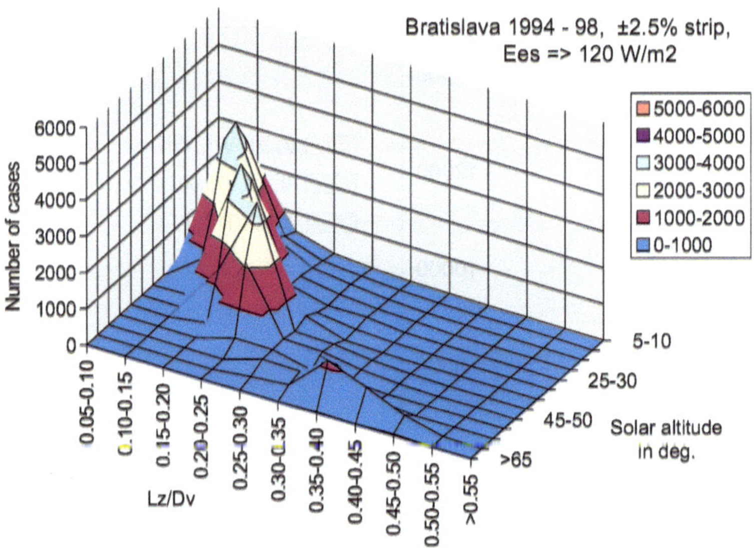

Fig. A6.4 Situations with sun documenting clear and partly cloudy skies in Bratislava

Bratislava and Athens sky type frequencies are thus compared in Figs. A6.3 and A6.4 and Figs. A6.5 and A6.6, where high-frequency peaks are shown in the overcast range in Bratislava, with almost the opposite prevalence and extremely high peaks in the clear sky range in Athens.

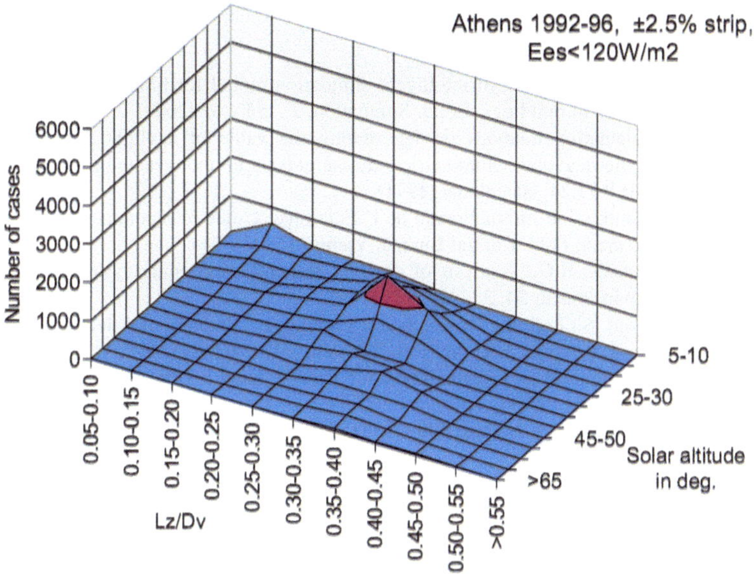

Fig. A6.5 Situations without sun represent cloudy and overcast skies in Athens

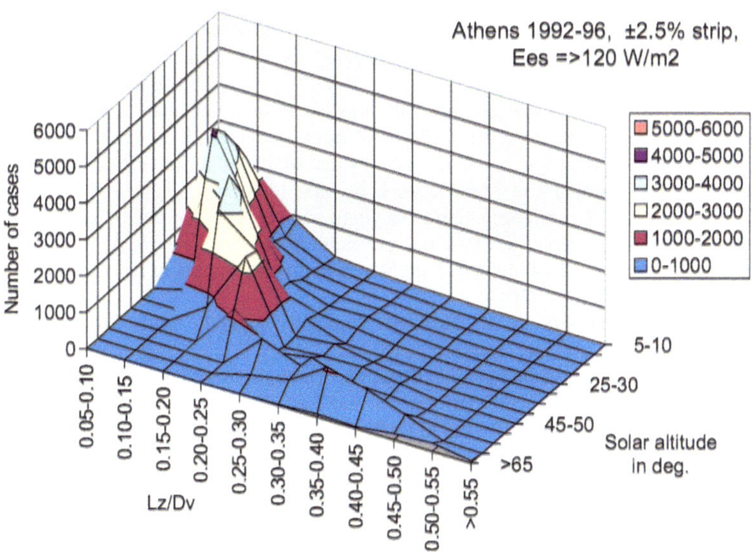

Fig. A6.6 Situations with sun documenting prevailing clear skies in Athens

References

Bhavani, R.G. and Khan. M.A.: An intelligent simulation model for blind position control in daylighting schemes in buildings. Build. Simulation, **2**, 253–262 (2009)

Chróścicki, W. Calculation methods of determining the value of daylight intensity on the ground of photometrical and actinometrical measurements in unobstructed planes. Proc. C.I. E. Session, Paper P.71.24, Barcelona, (1971)

CIE 108–1994.: Commission Internationale de l' Eclairage: Guide to recommended practice of daylight measurement. Publ. Central Bureau, Vienna, (1994)

Choi, A.S. and Mistrick, R.G.: Analysis of daylight responsive dimming system performance, Building and Environment, **34**, 3, 231–243 (1999)

Çolak, N. and Onaygil, S.: Lighting Research and Technology. Prediction of the artificial illuminance using neural networks. Lighting Res. Technol., **31**, 2, 63–69 (1999)

Corbella, O.D.: Sky luminosity categorisation for Rio de Janeiro City, Brazil. Lighting Res. Technol., **29**, 2, 69–79 (1997)

CSTB Centre de Recherche de Nantes. 1993 CSTB Data. Hourly values. Private riport Joule II, CSTB Nantes (1993)

Darula, S. and Kittler, R.: Daylighting availability after one-minute measurements for energy conscious design. Proc. Building Physics Symposium, Budapest. p. 243–248 (1995)

Darula, S.: Štatistické charakteristiky exteriérovej osvetlenosti v Bratislave podľa meraní počas rokov 1994 a 1995. (In Slovak. Statistical characteristics of exterior illuminance measured during 1994 - 1995). Světelná technika, **30**, 3–4, 42–46 (1997)

Darula, S., Kittler, R., Kambezidis, H. and Bartzokas, A.: Reference daylight conditions for energy-saving design in buildings. Final Technical Report, NOA Athens, ICA SAS Bratislava, 126 p (2001)

Darula, S. and Kittler, R.: Sunshine duration and daily courses of illuminances in Bratislava. International Journal of Climatology, **24**, 14, 1777–1783 (2004)

Darula, S., Kittler, R., Kambezidis, H.D. and Bartzokas, A.: Frequency probabilities of daylight illuminance courses due to sunshine duration. Proc. 1st Symposium Technological developments of daylight applications, Athens, 26–29 (2004)

Darula, S. and Kittler, R.: Monthly sunshine duration as a trustworthy basis to predict annual daylight profiles. Proc. 10th European Lighting Conf. Lux Europa 2005, Berlin, 141–144 (2005)

Darula, S. and Kittler, R.: Occurrence of standard skies during typical daytime half-days. Renewable Energy, **33**, 3, 491–500 (2008a)

Darula, S. and Kittler, R.: Contribution to the modelling of illuminance and irradiance daily courses. Przeglad Elektrotechniczny (Electrical Review), **84**, 8, 11–14 (2008b)

Darula, S. and Kittler, R.: Lighting energy savings due to daylight: time effectiveness based on Bratislava data. Building Res. Journal, **56**, 4, 241–253 (2008c)

Darula, S, and Kittler, R.: Determination of daylight sources. Proc. of the 1st central European symposium on building physics. Research on Building Physics, Cracow. p. 465–469 (2010).

Darula, S. and Kittler, R.: Sunlight and skylight availability. Advances in Energy Research. **1**, 4, 133–182, Nova Publ., NY (2011)

Dogniaux, R.: Disponibilité de l'éclairement lumineux naturel. Édité par I'lnstitut royal méteorologique de Belgique, Brussels (1978)

Dumortier D., Fontoynont M. and Avouac-Bastie P.: Daylight availability in Lyon, Proc. of the European Conference on Energy Performance and door Climate in Buildings. Ecole Nationale des Travaux de l'Etat Lyon, 1315–1320 (1994)

Dumortier, D. and Van Roy, F.: Daylight information throught Europe from the SATEL—LIGHT and SODA internet servers. Proc. Conf. 25th Session of the CIE, San Diego, 1, D3–4 – D3–7 (2003)

Dumortier, D.: Estimating global irradiances ad illuminances from the second generation of METEOSAT satellites – Results from the HELIOSAT-3 European project. Proc Conf. LUX EUROPA 2005, Berlin, 224–226

Feitsma, B.: Daylight illumination measurements at Katwijk (I), NTO, Delft, report E51 (1971)
Gillette, G., Pierpoint, W., Treado, S.: A general illuminance model for daylight availability. Journal of IES, **13**, 4, 330–340 (1984)
Good, E.: Estimating daily sunshine duration over the UK from geostationary satellite data. Weather, **65**, 12, 324–328 (2010)
He, J. and Ng, E.: Using satellite-based methods to predict daylight illuminance for subtropical Hong Kong. Lighting Res. Technol., **42**, 2, 135–147 (2010)
Hopkinson, R.G., Petherbridge, P., Longmore, J.: Daylighting. Heinemann, London (1966)
Ihm, P., Nemri, A. and Krarti, M. Estimation of Lighting energy savings from daylighting. Build. Environment, **44**, 3, 509–514 (2009)
Janák, M., Hraška, J., Straňák, Z.: Dynamic Assessment of Daylight. Proc. of Workshops Newly Associated States EnerBuild RTD. January 27 - 28, 2003, Prague, Technology Center of the Academy of Science, Czech Republic, p. 28–37 (2003)
Janjai, S., Jantarach, T., Laksanaboonsong, J.: A model for calculating global illuminance from satellite data. Renewable Energy, **28**, 15, 2355–2365 (2003)
Janjai, S., Tohsing, K., Nunez, M. and Laksanaboonsong, J.: A technique for mapping global illuminance from satellite data. Solar Energy, **82**, 6, 543–555 (2008a)
Janjai, S., Masiri, I., Nunez, M. and Laksanaboonsong, J.: Modeling sky luminance using satellite data to classify sky conditions. Building and Environment, **43**, 12, 2059–2073 (2008b)
Kambezidis, H.D., Muneer, T., Tzortzis, M. and Arvanitaki, S.: Global and diffuse horizontal solar illuminance: Month-hour distribution for Athens, Greece in 1992. Lighting Res. Technol., **30**, 2, 69–74 (1998)
Kambezidis, H.D., Oikonomou, Th., Zevgolis; D.: Daylight climatology in the Athens urban environment: guidance for building designers. Lighting Res. Technol., **34**, 4, 297 (2002)
Kähler, K.: Flächenhelligkeit des Himmels und Beleuchtungsstärke in Räumen. Meteorolgische Zeitschr., **25**, 2, 52–57 (1908)
Kazanasmaz, T., Günaydina, M. and Binola, S.: Artificial neural networks to predict daylight illuminance in office buildings. Building and Environment, **44**, 8, 1751–1757 (2009)
Kinghorn, D. and Muneer, T.: Daylight illuminance frequency distribution: Review of computational techniques and new data for UK locations. Lighting Research and Technology, **30**, 4, 139–150 (1998)
Kittler, R., Kittlerová, L.: Design and evaluation of daylighting. (in Slovak. Návrh a hodnotenie denného osvetlenia). ALFA Bratislava (1968)
Kittler, R.: Standardisation of the solar radiation with regard to the prediction of insolation and shading of buildings. Teaching the teachers and building climatology conference, Stockholm, **2**, 45 (1972)
Kittler, R. and Pulpitlová, J.: Basis of the utilization of daylight. (in Slovak. Základy využívania prírodného svetla). VEDA Bratislava (1988)
Kittler, R.: Expected daylight climate standards based on the evaluation of the IDMP data. Proc. Conf. Daylight and solar radiation measurement, Berlin, 244–253 (1989)
Kittler, R., Hayman, S., Ruck, N. and Julian, W.: Daylight measurement data: methods of evaluation and representation. Lighting Res. Technol., **24**, 4, 173–187 (1992)
Kittler R. and Darula S.: Prevailing sky conditions: Identifying simple parameters for definition Lighting Research and Technology, **29**, 2, 63–68 (1997)
Kittler, R., Perez, R. and Darula, S.: A new generation of sky standards. Proc. The 8th European Lighting Conf. Lux Europa 1997, Amsterdam, 359 – 373 (1997)
Kittler, R.: The relation of sky types to relative sunshine duration. Build. Res. Journal, 45, 1, 41–59 (1997)
Kittler, R., Perez, R. and Darula, S.: A new generation of sky standards. Final Report. Polygrafia, Bratislava (1998)
Kittler, R., Darula, S.: Specification of the local daylight climate and the importance of overcast sky conditions. Light and Engineering, **7**, 4, 25–27 (1999)

Kleindienst, S., Bodart, M. and Andersen, M.: Graphical Representation of Climate-Based Daylight Performance to Support Architectural Design, LEUKOS, **5**, 1, 39–61 (2008)

Koga, Y., Anai, K. and Nakamura, H.: Cloud Amount and Daylight Availability, Balkan light '99, 124–129, (1999)

Krarti, M., Erickson, P.M. and Hillman, T.C.: A simplified method to estimate energy savings of artificial lighting use from daylighting. Build. Environment, **40**, 6, 747–754 (2005)

Krochmann, J.: Über die Horizontalbeleuchtungsstärke der Tagesbeleuchtung, Lichtechnik, **15**, 559 (1963)

Krochmann, J. and Seidl, M.: Quantitative data on daylight for illuminating engineering. Lighting Research and Technology, **6**, 3, 165–171 (1974)

Kurian, C.P., Aithal, R.S., Bhat, J. and George, V.I.: Robust control of optimisation of energy consumption in daylight-artificial light integrated schemes. Lighting Res. Technol., **40**, 7–24 (2008)

Li, D.H.W and Lam J.C.: Evaluation of lighting performance in office buildings with daylighting controls. Energy and Buildings, **33**, 793–803 (2001)

Li, D.H.W., Lam, J.C., Lau, C.C.S.: A Study of Solar Radiation Daylight Illuminance and Sky Luminance Data Measurement for Hong Kong. Architectural Science Review, **45**, 21–30 (2002)

Li, D.H.W., Lau, C.C.S., Lam, J,C.: Predicting daylight illuminance by computer simulation techniques. Lighting Res. Technol., June, **36**, 2, 113–128 (2004)

Li, D.H.W., Tang, H.L., Lee, E.W.M., Muneer, T.: Classification of CIE standard skies using probabilistic neural networks. Int. J. Climatol., **30**, 2, 305–310 (2010)

Mardaljevic, J.: Daylight Simulation: Validation, Sky Models and Daylight Coefficients. PhD thesis. (2000)

Mardaljevic, J.: Simulation of annual daylighting profiles for internal illuminance. Lighting Res. Technol., **32**, 3, 111–118 (2000)

Mardaljevic, J.: Climate-Based Daylight Analysis for Residential Buildings. Impact of various window configurations, external obstructions, orientations and location on useful daylight illuminance. (2008)

Markou, M.T., Kambezidis, H.D., Bartzokas, A., Darula, S., Kittler, R.: Generation of daylight reference years for two European cities with different climate: Athens, Greece and Bratislava, Slovakia. Atmospheric Research, **86**, 3–4, 315–329 (2007)

Markou, M.T., Bartzokas, A., Kambezidis, D.H.: Daylight climatology in Athens, Greece: types of diurnal variation of illuminance levels. International Journal of Climatology. **29**, 14, 2137–2145 (2009)

Matuszawa, T., Koga, Y., Nakamura, H., Anai, K.: Daylight illuminance by cloud type and amount. CIE 24th Session, Warsaw, **1**, 2, 56–58 (1999)

McCluney, R. and Bornemann, H.J.: The Time Rate of Changing Daylight. Proc. International Daylighting Conference, Long Beach, CA, 36–44 (1986)

Moon, P. and Spencer. D.E.: Illumination from a non-uniform sky. Illum. Eng., **37**, 10, 707–726 (1942)

Muneer, T. and Angus, R.C.: Daylight illuminance models for the United Kingdom. Lighting Res. Technol., **25** 3, 113–123, (1993)

Muneer, T. and Kinghorn, D.: Predicting daylight availability for the UK using cumulative frequency curves. Proc. Conf. Daylighting '98. Ottawa, 165–170 (1998)

Muneer, T.: Solar Radiation and Daylight Models: For the Energy Efficient Design of Buildings. Butterworth-Heinemann (1997, 2004)

Nakamura, H., Oki, M.: Study on the statistic estimation of the horizontal illuminance from unobstructed sky. J. Light. and Vis. Envir., **3**, 1, 23–31 (1979)

Nakamura, H. and Oki, M.: Investigation on the relation between the probability of occurrence of the three skies and relative sunshine duration. Proc. of the CIE 21st Session, Venice, **1**, 228–229 (1987)

Nabil, A. and Mardaljevic, J.: Useful daylight illuminance: a new paradigm for assessing daylight in buildings, Lighting Res. Technol., **37**, 1, 41–59 (2005)

Navvab, M., Love, J., Ne´eman, E.: Daylight and solar availability data for Ann Arbor, Michigan U.S.A. Proc. Conf. Daylight and solar radiation measurement. Berlin, 61–69 (1989)

Ogisso, S.: Study on daylight sources. Journal of the Fac. Eng., Univ. of Tokyo (B), **28**, 2, 103–118 (1965)

Olseth, J.A. and Skartveit, A.: Solar irradiance, sunshine duration and daylight illuminance derived from METEOSAT data for some European sites. Theoretical and Applied Climatology, **69**, 3–4, 239–252 (2001)

Perez, R., Ineichen, P., Seals, R., Michalsky, J., R. Stewart, R.: Modeling Daylight Availability and Irradiance Components from Direct and Global Irradiance. Solar Energy, **44**, 271–289 (1990)

Perez-Burgos, A., Bilbao, J., Argimoro de Miguel, A.: An evaluation of illuminance measurements at Valladolid (Spain). Journal of Atmospheric and Solar-Terrestrial Physics, **69**, 8, 939–946 (2007)

Perradeau, M. and Chauvel, P.: One year´s measurements of luminous climate in Nantes. Proc. International Daylight Conference, Long Beach, CA, p. 83–88 (1986)

Petrakis, M., Lykoudis, S., Kassomenos, P., Assimakopoulos, D.N.: Creation of a typical meteorological year for Athens based on daylight measurements. Proc. of the 7th Conference of Union Hellenic of Physicists and Union Cyprus of Physicists. Heraklion, Crete, Greece (in Greek), (1996)

Rahim, R., Mulyadi, B., Mulyadi, R.: Classification of daylight and radiation data into three sky conditions by cloud ratio and sunshine duration. Energy and Buildings, **36,** 660–666 (2004)

Reinhart C.F. and Herkel, S.: An evaluation of RADIANCE based simulations of annual indoor illuminance distributions due to daylight. Proc. of Building Simulation 1999, II. 563–570, Kyoto (1999)

Robins, C.R.R.: Daylighting. Design and analysis. Van Nostrand Reinhold Co., New York (1986)

Ruck, N.C.: Daylight availability in Australia. Proc. Daylighting and Energy Conservation, Fac. of Arch., Univ. of NSW, 4.1–4.16 (1982)

Ruck, N.C.: Representation of skylight availability in Australia in different climatic zones using regression procedures. Lighting Res. Technol., **17**, 2, 72–78 (1985)

Ruck, N.C. and Selkowitz, S.: A review of measured skylight availability data. Proc. of International Daylighting Conference, Long Beach, CA, 67–71 (1986)

Ruiz, E., Soler, A., Robledo, L.: Statistical assessment of a model for global illuminance on inclined surfaces from horizontal global illuminance. Energy Conversion and Management. **43**, 5, 693–708 (2002)

Schramm, W.: Uber die Verteilung des Lichtes in der Atmosphäre. Schr. Naturwiss. Ver. Schlesw.-Holst., **12**, 1, 81–129 (1901)

Soler, A.: Global and diffuse illuminances: Estimation of monthly average hourly values. Lighting Res. Technol., **22**, 4, 193–196 (1990)

Tregenza, P.R.: Measured and calculated frequency distributions of daylight illuminance. Lighting Res. and Technol., **18**, 2, 71–74 (1986)

Ullah, M.B.: International Daylight Measurement Programme — Singapore data III: Building energy savings through daylighting. Lighting Res. Technol., **28**, 2, 83–87 (1996)

Walkenhorst, O., Luther, J., Reinhart, C., Timmer, J.: Dynamic annual daylight simulations based on one-hour and one-minute means of irradiance data. Solar Energy, **72**, 5, 385–395 (2002)

Ward, G.J.: The RADIANCE Lighting Simulation and Rendering System. Proc. of '94 SIGGRAPH conference, Computer Graphics, 459–72 (1994)

Whipple, G.M.: On the relative duration of sunshine at the royal observatory, Greenwich, and at the Kew observatory, during the year 1877. Quarterly Journal of the Royal Meteorological Society, **4**, 28, 201–207 (1878)

References to Appendix

Alshaibani, K.: Determination of CIE Standard General Sky from horizontal and vertical illuminance/irradiance. Proc. World Renewable Energy Congresss, Glasgow (2008)

Alshaibani, K.: Finding frequency distribution of CIE Standard General Skies from sky illuminance or irradiance. Lighting Res. Technol., **43**, in print (2011)

Bartzokas, A., Darula, S., Kambezidis, H. D., Kittler, R.: Comparison between winter and summer sky luminance distribution in Central Europe and in the Eastern Mediterranean. Journ. Atmosph. and Solar Terrestr. Physics, 67, 7, 709–718 (2005)

Chirarattananon, S. and Chaiwiwatworakul, P.: Distributions of sky luminance and radiance of North Bangkok under standard distributions. Renewable Energy, 32, 8, 13328–1345 (2007)

Darula, S., Kittler, R., Kambezidis, H., Bartzokas, A.: Reference daylight conditions for energy-saving building design. Final Res. Report of the Greek-Slovak project, ICA SAS, Bratislava (2001)

Kittler, R., Darula, S., Perez, R.: A new generation of sky standards. Proc. LuxEuropa Conf., Amsterdam, 359–373 (1997)

Kittler, R., Darula, S., Kambezidis, H., Bartzokas, A.: Daylight climate specification based on Athens and Bratislava data. Proc. Lux Europa, 442–449, Reykjavik (2001)

Li, D.H.W. and Tang H.L.: Standard skies classification in Hong Kong. Journ. Atmosph. and Solar Terrestr. Physics, 70, 8–9, 1222–1230 (2008)

Li, D.H.W., Cheung, K.L., Tang, H.L., Cheng, C.C.K.: Identifying CIE Standard skies using Vertical Sky Component. Journ. Atmosph. And Solar Terrestr. Physics, **73**, 13, 1861–1867 (2011)

Markou, M.T., Bartzokas, A. and Kambezidis, H.D.: Daylight climatology in Athens, Greece: types of diurnal variation of illuminance levels. International Journal of Climatology, 29, 14, 2137–2145, (2009)

Ng, E., Cheng, V., Gadi, A., Mu, J., Lee, M.: Defining standard skies for Hong Kong. Build. and Envir., 42, 2, 866–876 (2007)

Torres, J.L., Blas, M., García, A., Gracia, A. and deFrancisco, A.: Sky luminance distribution in the North of Iberian Peninsula during winter. Journal of Atmospheric and Solar-Terrestrial Physics, **72**, 1147–1154 (2010)

Tregenza, P.R.: Standard skies for maritime climates. Lighting Res. Technol., 31, 3, 97–106 (1999)

Tregenza, P.R.: Analysing sky scans to obtain frequency distributions of CIE Standard General Skies. Lighting Res. Technol., **36**, 4, 271–281 (2004)

Wittkopf, S.K. and Soon, L.K.: Analysing sky luminance scans and predicting frequent sky patterns in Singapore. Lighting Res. and Technol., 39, 1, 31–51 (2007)

Chapter 7
Fundamental Principles for Daylight Calculation Methods

7.1 Skylight Availability on Horizontal Unobstructed Planes Outdoors

Historically, the sky vault in hemispheric form was the geometrical image used to define angular sun coordinates. The same sky hemisphere concept was employed to describe the luminance distribution over the vast endless upper half-space and to visually represent particular solid angles on that sky. Therefore, any object, the sun, building, window aperture, etc., could be projected from an illuminated point or surface element onto the sky hemisphere and thus the spherical area would represent the solid angle and, if the radius of the hemisphere is equal to unity, the solid angle would be in steradians. And so the sky hemisphere could instructively show its overall luminance pattern as well as the luminance of each of its surface elements that pass the scattered skylight in all directions to its center. Thus, in contrast to parallel sunbeam illumination directed from the sun position, the skylight is distributed to ground level from the whole upper half-space. The advantage of this hemispheric sky model is that all surface elements have their normal directed to the sphere center, i.e., have the most efficient position indicating also a defined most effective solid angle.

However, the apex of the solid angle cone can be imagined either in the light source element or in any planar element of the illuminated surface. Both the plane angle and the solid angle are independent of the length of lines or rays. Therefore, real or even infinite surfaces or space objects can be characterized by their angular extent.

Many researchers in the seventeenth and eighteenth centuries investigated candle light, which was assumed to be a point source, and tried to study the light ray fluxes in small solid angles restricted by circular apertures or board holes. Sometimes, even purely geometric line analysis or luminous pyramids were imagined to describe the light flow from a point source to a larger illuminated plane. The first trials to understand area sources were reported by Zahn (1686, 1702) and influenced by his telescopic observations of the sky. Although he did not express the

R. Kittler et al., *Daylight Science and Daylighting Technology*,

DOI 10.1007/978-1-4419-8816-4_7,

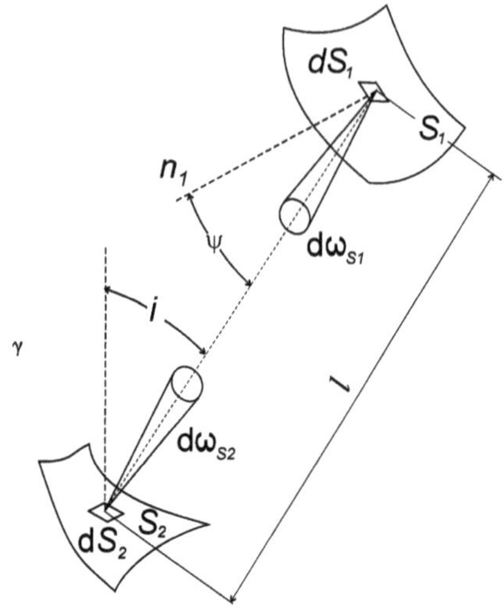

Fig. 7.1 The element of the planar source and the illuminated element

illuminance from the sky, he certainly helped Lambert identify the main influences on the intensity of illumination falling on any body as:

- The size of the light source
- Its distance from the illuminated object
- Its luminance or brightness
- The perpendicular position of the illuminated surface in relation to the direction of the incident rays

However, Lambert (1760) was the first to explicitly use the solid angle (*angulus solidus*) in the photometry of skylight.

Lambert, realizing that the luminance pattern of the sky could be of some importance in future, divided the planar light source into surface elements dS_1 with luminance L_{dS1} projecting toward the arbitrary illuminated planar element dS_2 and a partial luminous flux $d\Phi_1$ after Fig. 7.1:

$$dE_2 = \frac{d\Phi_1}{dS_2} = L_{dS1}\, d\omega_{S1}\, \cos i = L_{dS1}\, d\omega_p \,(\mathrm{cd/m^2}). \quad (7.1)$$

This formula in the exact mathematical form, sometimes referred to as the Lambert and Beer formula (1854), is basic for any planar sky sources such as windows, inclined apertures, or top lights if their dimensions and luminance patterns are known. However, owing to the lack of information about sky luminance during Lambert's lifetime, he suggested an overcast sky with uniform and unity luminance over the total sky vault as critical; thus, $L_{dS1} = 1$. This assumption simplified considerably the calculations. Furthermore, he realized the advantage of the solid angle, which is the projected area of the source onto the unit hemisphere, i.e.,

$$d\omega_{S1} = \frac{dS_1 \cos \psi}{l^2} (\mathrm{sr}), \quad (7.2)$$

which is of particular significance (see the appendix in Chap. 3).

At ground level the illuminance is

$$dE_2 = L_{dS1}\, d\omega_{S1}\, \cos i\, (\mathrm{lx}), \tag{7.3}$$

where cos i is the cosine of the incidence angle for sloped illuminated planes and, for an arbitrary sky element, the luminance is defined by the angular spherical distance from the sun position χ and the zenith angle of the sky element Z; thus, the skylight illuminance on any arbitrary sloped plane is

$$E_2 = \int L_{\chi Z}\, \cos i\, d\omega_{S1}\, (\mathrm{lx}). \tag{7.4}$$

It should be noted that Lambert also overcame the problem of measuring the sky luminance as well as the distance between the sky and a surface element placed on the ground by using applied spherical geometry. The virtual hemisphere with its radius equal to 1 is not only the sky vault image for a horizontal plane at ground level, but in general it can represent the half-space seen from any arbitrary sloped or vertical illuminated plane. The advantage of this theoretical concept was enormous, as will be explained in the following and later text.

When the horizontal skylight illuminance from the unobstructed sky outdoors is taken as normalizing, then for the whole hemisphere its projected solid angle onto the horizontal plane is $\omega_{ph} = \int \cos i\, d\omega_{S1} = \pi$, which was termed the "absolute illuminance" by Lambert (1760), i.e., the normalizing maximum exterior horizontal illuminance E_{vh} and the sky factor (often used later as the sky component of the daylight factor in absolute values or as a percentage) was already defined for the unity sky luminance as

$$\mathrm{SF} = \frac{E_{vi}}{E_{vh}} = \frac{\omega_p}{\pi} \tag{7.5}$$

or

$$E_{vi} = \mathrm{SF}\, E_{vh}\, (\mathrm{lx}) \tag{7.6}$$

where SF is the sky factor, E_{vi} is the interior illuminance of a planar element taken on the horizontal working plane from the sky alone, i.e., without any reflected skylight, and E_{vh} is the horizontal exterior illuminance from an unobstructed sky of uniform unity luminance.

It is evident that outdoors when the whole sky vault has no external obstruction, the half-sphere projected solid angle $\omega_{ph} = \pi$; and then the horizontal illuminance from a sky of unity luminance is π, too. Thus, $\mathrm{SF}(1:1) = 1$ or 100% and logically for any vertical plane (façade) $\mathrm{SF}(1:1) = 1/2$ or 50% without any dependence on its orientation.

When the gradated luminance $L_{vh}/L_{vz} = 1:3$ on the (CIE) overcast sky is standardized by CIE (1955) or $L_{vh}/L_{vz} = 1:2$, the different horizontal sky rings and the zenith cap on the hemisphere have different validities in relation to their

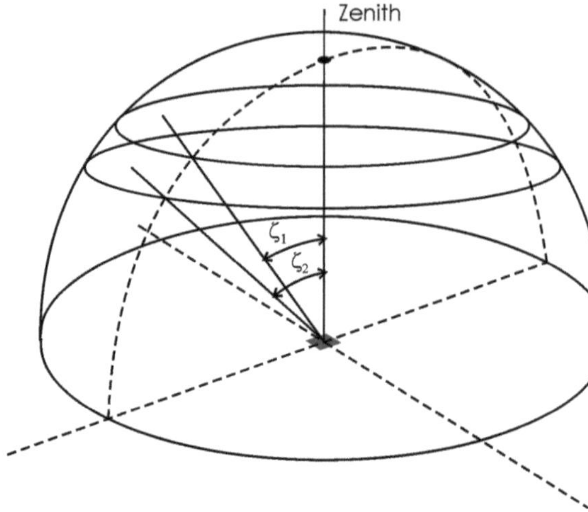

Fig. 7.2 Exterior horizontal illuminance calculated after circular rings

luminance, but the total effect of the unobstructed sky is the same because the projection of the solid angle (hemisphere) onto the horizontal plane is still a circular area with unity radius; thus,

$$\mathrm{SF}(1:1) = \mathrm{SF}(1:3) = \mathrm{SF}(1:2) = 1\,. \tag{7.7}$$

In the case of the unobstructed outdoor hemisphere illuminating a horizontal plane, the spherical geometry can be applied in three ways:

1. Taking horizontal circular bands with the zenith cap into account, the exterior skylight illuminance $D_\mathrm{v} = E_\mathrm{vh}$ after Fig. 7.2 following the classical integration is

$$D_\mathrm{v} = \frac{1}{2}\sum_{i=1}^{k_\zeta}\int_{A=0}^{A=2\pi} L_{\chi Z}\left(\sin^2\zeta_{i+1} - \sin^2\zeta_i\right)\mathrm{d}A\ (\mathrm{lx}) \tag{7.8}$$

or following the numerical summation instead, i.e., the total skylight illuminance on a horizontal planar element is the sum of the products of the mean luminance of each sky band element and the subtended area forming the element solid angle, is

$$\begin{aligned} D_\mathrm{v} &= \frac{\pi}{k_A}\sum_{i=1}^{i=k_\zeta}\sum_{n=1}^{n=k_A} L_{\chi Z}(\sin^2\zeta_{i+1} - \sin^2\zeta_i) \\ &= \frac{\pi}{k_A}\sum_{j=1}^{j=k_\varepsilon}\sum_{n=1}^{n=k_A} L_{\chi Z}(\cos^2\varepsilon_j - \cos^2\varepsilon_{j+1}), \end{aligned} \tag{7.9}$$

where n is the index of an azimuth element within the band with the final number of elements k_A, i.e., the higher the number of circular bands and their zones or elements, the more precise is the summation. From a comparison of (7.8) and (7.9), it is evident that the step in the azimuth is $2\pi/k_A$. Each zenith angle $\zeta_{i+1}>\zeta_i$ and each elevation angle $\varepsilon_{j+1}>\varepsilon_j$ are taken from the horizon.

A subtle subdivision in the computer program KISOB (Kittler and Pulpitlova 1988, p. 96) with a 5° step in elevation and azimuth zones contained 18 times 72 elements, totaling 1,296 tiny zones 5° by 5°. Tregenza and Sharples (1993) recommended a grid of 145 sky zones with the hemisphere subdivision in eight elevation bands roughly 12° (in fact $\pi/15$ rad) wide.

The contribution of different ring strips can be calculated with respect to their luminance uniformity, i.e., SF(1 : 1) or gradations SF(1 : 2) or SF(1 : 3) (Kittler et al. 1962):

- For the zenith cap with the downward-bordering circle in elevation ε_o over the horizon
  - With unity luminance,

$$\mathrm{SF}(1:1) = \cos^2\varepsilon_1 \tag{7.10}$$

  - With 1:3 luminance gradation,

$$\mathrm{SF}(1:3) = \frac{3}{7}\cos^2\varepsilon_1 + \frac{4}{7}(1 - \sin^3\varepsilon_1) \tag{7.11}$$

  - With 1:2 luminance gradation,

$$\mathrm{SF}(1:2) = \frac{3}{5}\cos^2\varepsilon_1 + \frac{2}{5}(1 - \sin^3\varepsilon_1) \tag{7.12}$$

- For the circular ring with the downward-bordering circle in elevation ε_1 and the upward ε_2 over the horizon (Fig. 7.2)
  - With unity luminance,

$$\mathrm{SF}(1:1) = \sin^2\varepsilon_2 - \sin^2\varepsilon_1 \tag{7.13}$$

  - With 1:3 luminance, gradation

$$\mathrm{SF}(1:3) = \frac{3}{7}(\sin^2\varepsilon_2 - \sin^2\varepsilon_1) + \frac{4}{7}(\sin^3\varepsilon_2 - \sin^3\varepsilon_1) \tag{7.14}$$

  - With 1:2 luminance, gradation

$$\mathrm{SF}(1:2) = \frac{3}{5}(\sin^2\varepsilon_2 - \sin^2\varepsilon_1) + \frac{2}{5}(\sin^3\varepsilon_2 - \sin^3\varepsilon_1) \tag{7.15}$$

When the sky has an unobstructed horizon, then $\varepsilon_1 = 0°$ in (7.13)–(7.15) and the influence of the bottom sky ring can be calculated and compared with the effect of the whole hemisphere, for which $\mathrm{SF}(1:1) = \mathrm{SF}(1:3) = 1$. This possibility was used by Walsh (1961) soon after the CIE overcast sky 1:3 gradation had been standardized. He documented the loss of exterior illuminance when the sky horizon is obscured under an angle of altitude 10° when the fractional loss was almost exactly 3% for the uniform sky, but the reduction for $\mathrm{SF}(1:3)$ was only 1.6 % owing to low luminances at the lowest horizon strip.

Note that this system of dividing the hemisphere into horizontal rings is applicable only when calculating sky illuminance or sky factors produced by unglazed apertures outdoors or obstructions formed around the illuminated point (center of a round square or atrium encircled by a round obstruction or house fronts). In comparison with the current ISO/CIE (2003, 2004) standard general sky for overcast sky types 1 and 3, the gradations 1:3 and 1:2 are approximated by exponential functions instead of cosine functions as before.

Owing to the prevailingly horizontal and vertical grid of architectural objects in the urban space, for the calculation of solid angles the image of spherical lunes with their vanishing points either at the zenith or on the horizon is more suitable. The whole sky division following the almucantar horizontal rings is suitable to calculate only the outdoor illuminance from the whole hemisphere. The lune division is applicable also in the case of outdoor obstructions of the skyline by walls, house fronts, or any barriers in the urban space. If the lune system is applied to calculate the solid angle of the whole hemisphere as the sum of the spherical lune areas, then the angular extent either in azimuth φ_0 for a vertical half-lune net is used or the zenith angle Z_0 or the elevation angle ε_0 is relevant for the horizontal lune net.

2. Thus, in the simplest case of an unobstructed virtual sky hemisphere with unity radius having uniform unit luminance, its horizontal illuminance at ground level can be calculated either by the integration of vertical half-lunes φ_0 in 360° all around the horizon or following the numerical summation.

   Thus, by assuming vertical half-lunes with the basic widest angle at the horizon φ_0 with the zenith pole as a vanishing point in Fig. 7.3,

$$D_v = \int_{\zeta=0}^{\zeta=\pi/2} \int_{\varphi_0=0}^{\varphi_0=2\pi} L_{\chi Z} \sin\zeta \cos\zeta \, d\zeta \, d\varphi_0 \ (\mathrm{lx}) \tag{7.16}$$

   or following the numerical summation instead of integration, the total skylight illuminance on a horizontal planar element is the sum of the products of the mean luminance of each sky lune element and its elemental solid angle calculated for each n element:

$$D_v = \frac{\pi}{2} \sum_{\varphi_0=0°}^{\varphi_0=90°} \sum_{n=1}^{n=k} L_{\chi Z}\left(\frac{\varphi_0}{n}\right) \cos\zeta \ (\mathrm{lx}). \tag{7.17}$$

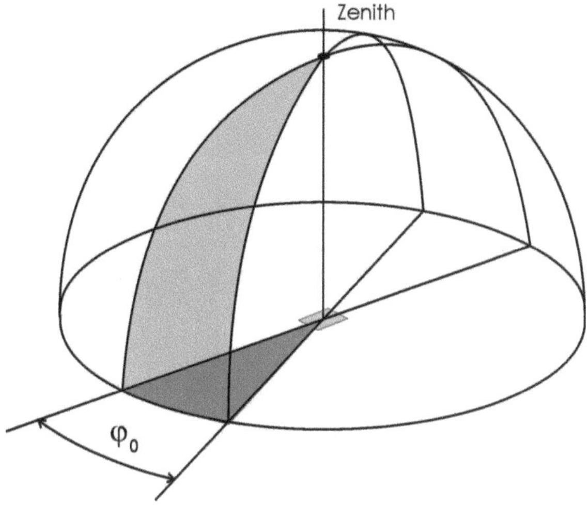

Fig. 7.3 Exterior horizontal illuminance calculated after vertical half-lunes

In the study by Kittler et al. (1962), the classical integration was applied to calculate both SF(1 : 3) and SF(1 : 2) under an unobstructed sky hemisphere, i.e., φ_0 equal to 2π, as

$$\mathrm{SF}(1:1) = \frac{\varphi_0}{2\pi} = 1, \tag{7.18}$$

$$\mathrm{SF}(1:3) = \frac{3}{7}\frac{\varphi_0}{2\pi} + \frac{2}{7}\frac{\varphi_0}{\pi} = \frac{3}{7} + \frac{4}{7} = 1, \tag{7.19}$$

$$\mathrm{SF}(1:2) = \frac{3}{5}\frac{\varphi_0}{2\pi} + \frac{1}{5}\frac{\varphi_0}{\pi} = \frac{3}{5} + \frac{2}{5} = 1. \tag{7.20}$$

3. Taking into account the horizontal lunes declining from the zenith on both sides toward the horizon after Fig. 7.4,

$$D_\mathrm{v} = \frac{\pi}{2} + \int\limits_{\zeta_0=0}^{\zeta_0=\pi/2} \int\limits_{\varphi=0}^{\varphi=\pi} L_{\chi Z} \sin\zeta_0 \cos\zeta_0 \, \mathrm{d}\zeta_0 \, \mathrm{d}\varphi \; (\mathrm{lx}). \tag{7.21}$$

An alternative method is to follow the numerical summation instead of integration; then, the total skylight illuminance on a horizontal planar element is the sum of the products of the mean luminance of each sky lune element and its elemental solid angle calculated for each n element:

$$D_\mathrm{v} = \frac{\pi}{2} + \pi \sum_{\zeta_0=0^\circ}^{\zeta_0=90^\circ} \sum_{n=1}^{n=k} L_{\chi Z}\left(\frac{\zeta_0}{n}\right) \cos\zeta \, (\mathrm{lx}). \tag{7.22}$$

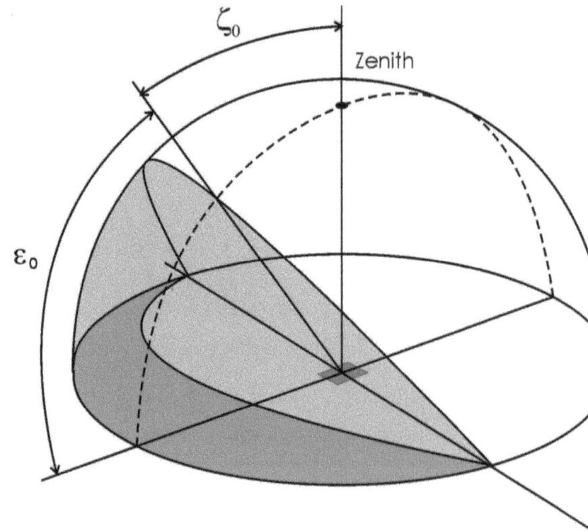

Fig. 7.4 Exterior horizontal illuminance calculated after horizontal lunes

It is evident that the solid angle of extensive horizontally stretched openings such as endless streets, courtyards, etc. can be advantageously envisaged and calculated using the horizontal spherical lune grid.

The classical integration was applied to calculate both SF(1 : 3) and SF(1 : 2) under an unobstructed sky hemisphere for the CIE (1955) overcast sky, i.e., for SF(1 : 3) as well as for the gradation SF(1 : 2), with the following results after case 22.0 in Kittler et al. (1962), if $\varepsilon_0 = 0°$:

$$\mathrm{SF}(1:1) = \frac{1}{2}(1 + \cos\,\varepsilon_0) = 1, \tag{7.23}$$

$$\mathrm{SF}(1:3) = \frac{3}{7}\left[\frac{1}{2}(1 + \cos\varepsilon_0)\right] + \frac{4}{7\pi}\left[\frac{\pi}{2} + \arcsin(\cos\varepsilon_0) + \frac{1}{2}\sin 2\varepsilon_0\right] = \frac{3}{7} + \frac{4}{7} = 1, \tag{7.24}$$

$$\mathrm{SF}(1:2) = \frac{3}{5}\left[\frac{1}{2}(1 + \cos\varepsilon_0)\right] + \frac{2}{5\pi}\left[\frac{\pi}{2} + \arcsin(\cos\varepsilon_0) + \frac{1}{2}\sin 2\varepsilon_0\right] = \frac{3}{5} + \frac{2}{5} = 1. \tag{7.25}$$

Lambert had already realized the advantages of the latter lunes when applied to calculations in the case of vertical windows with their rather simple projected solid angles, but he did not study the luminance distribution on the different skies caused by atmospheric scattering.

7.2 Daylighting of Urban Spaces and Obstructed Horizontal Outdoor Surfaces

Any open space (e.g., free horizontal terrain or a roof) can be obstructed by an endless house front or wall which reduces the sky solid angle from any horizontal planar element by shading elevation angles ε or their equivalent zenith angles ζ. The first studies of daylighting in urban spaces, namely, streets and courtyards, were published by Burchard (1919a, b), who noted that both these shading angles reach their maximum in the normal direction to the house front, i.e., in section as shown in Fig. 7.5. As these normal angles decrease sideways to the vanishing points, the sky factor or the projected angle onto the horizontal plane can be taken as the horizontally projected half sky hemisphere with $90° = \pi/2$ on one side plus the sky part reduced by a lune equal to $\pi/2(\sin\zeta_0)$ or $\pi/2(\cos\varepsilon_0)$. Thus, an open terrain shaded by an endless wall or façade on one side presents the situation determined by Kittler et al. (1962) as case 22.0 and if ε_0 is the actual shading angle, then (7.23)–(7.25) are valid:

$$\mathrm{SF}(1:1) = \frac{1}{2}(1 + \cos\varepsilon_0), \tag{7.26}$$

$$\mathrm{SF}(1:3) = \frac{3}{7}\left[\frac{1}{2}(1 + \cos\varepsilon_0)\right] + \frac{4}{7\pi}\left[\frac{\pi}{2} + \arcsin(\cos\varepsilon_0) + \frac{1}{2}\sin 2\varepsilon_0\right], \tag{7.27}$$

$$\mathrm{SF}(1:2) = \frac{3}{5}\left[\frac{1}{2}(1 + \cos\varepsilon_0)\right] + \frac{2}{5\pi}\left[\frac{\pi}{2} + \arcsin(\cos\varepsilon_0) + \frac{1}{2}\sin 2\varepsilon_0\right]. \tag{7.28}$$

In a reversed system, when an obstruction of unlimited length is studied (Cox 1871), then the "quantity of light subtracted" is cited in Swarbrick (1930, p. 25) as

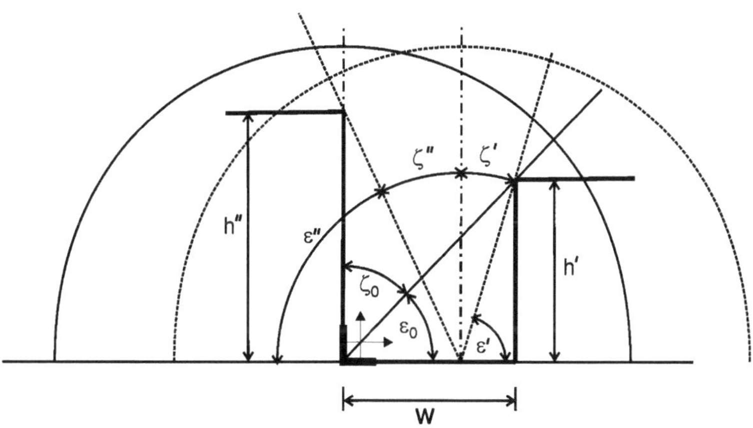

Fig. 7.5 Shading angles in the section of the street

$$\mathrm{SF}(1:1) = \frac{1}{2}(\cos\,\varepsilon_0)\,, \tag{7.29}$$

with the "quantity of light admitted"

$$\mathrm{SF}(1:1) = \frac{1}{2}(1 - \cos\,\varepsilon_0)\,. \tag{7.30}$$

In urban spaces, streets with either infinite or finite length and rectangular squares, courtyards, and atria with unglazed top apertures have to be expected. In contrast to horizontal planes obstructed only on one side, two different shading angles can be expected with the same vanishing points on the horizon. It is evident that such zenithal lunes (indicated by dashed in Fig. 7.5) have the widest angle in the normal section of an endless street with shading elevation angle ε'' on one side of the house front and ε' on the other side; thus,

$$\mathrm{SF}(1:1) = \frac{1}{2}(\cos\,\varepsilon'' + \cos\varepsilon')\,, \tag{7.31}$$

$$\mathrm{SF}(1:3) = \frac{3}{7}[\mathrm{SF}(1:1)] + \frac{4}{7\pi}\left[\arcsin(\cos\varepsilon'') + \arcsin(\cos\varepsilon') + \frac{1}{2}(\sin 2\varepsilon_0 + \sin 2\varepsilon')\right], \tag{7.32}$$

$$\mathrm{SF}(1:3) = \frac{3}{5}[\mathrm{SF}(1:1)] + \frac{2}{5\pi}\left[\arcsin(\cos\varepsilon'') + \arcsin(\cos\varepsilon') + \frac{1}{2}(\sin 2\varepsilon_0 + \sin 2\varepsilon')\right]. \tag{7.33}$$

When both house fronts are equally high and the horizontal illuminated element is placed in the center between them, $\varepsilon'' = \varepsilon'$ is a valid simplification.

Recently, these problems in streets were studied also by Matusiak et al. (1999) and Matusiak and Aschenhoug (2002), but they used instead of the classical integration within the sky lune a computer numerical integration based on the sky rings. Both approaches were compared in the discussion within Comment 2 by Kittler (2002). In connection with the atrium buildings, vertical sky components (SCv) were originally derived by Seshadri (1960), and with an additional interreflection distribution by Littlefair (2002) and by Sharples and Lash (2004). Further Radiance computer simulations tested in atria model measurements were published recently by Du and Sharples (2010).

Because sunlight is absent, densely overcast skies have the same luminance in all orientations, and the azimuth of the obstructing house front is irrelevant. However, under clear or cloudy skies, the momentary position of the sun with respect to the shading house front becomes important because of both the sunlight and the irregular sky luminance pattern, which could be shaded. The first checks can be very instructive especially for architects or solar engineers. The locally valid sun-path and droop-line diagrams can quickly show either a favorable design solution or shading consequences in existing or design situations. Such tasks as trials to find the best place for solar collectors, photovoltaic panels, or outdoor swimming pools can be easily tested by use of either the stereographic sun-path plan diagrams (Markus and Morris 1980) or the vertical unfolded cylindrical diagrams of Waldram (1950), both with appropriate droop lines.

Unfortunately all the equations above are valid only for uniform and densely overcast skies, so in the case of clear and cloudy skies computer summation of sky elements with different luminance values and with additional direct sunlight have to be taken into account in numerical integration within the appropriate solid angles.

The numerical integration is usually performed on the basis of a general formula:

$$E_2 = \int_{\zeta'}^{\zeta''} \int_{A=0}^{2\pi} L(\zeta, A)\, sin\zeta\, \cos\zeta\, d\zeta\, dA, \tag{7.34}$$

where $L(\zeta, A)$ is the luminance of a sky element with the zenith angles ζ and azimuth A. To achieve the required accuracy, the lattice spacing dA must be small enough, in particular, smaller than 5°. The more conservative criterion $d\zeta < 2°$ is usually applied for grid size related to the zenith angle. This is necessary owing to rapid changes of sky luminance along a meridian slice. The numerical accuracy can be easily verified as follows:

- In the first step, the computation is performed assuming some values of dA and dζ.
- In the next step, the same calculation is realized assuming d$A/2$ and d$\zeta/2$.

If both results differ by more than a predefined error margin (e.g., 5%), d$A/4$ and d$\zeta/4$ will be considered in the next run, etc. Because of invariance, the calculation results converge to the same value independently of whether the outer integration runs over dζ or dA.

7.3 Utilization of Daylight in Solar Facilities and Photovoltaic Panels on Vertical or Inclined Building Surfaces

Solar collector and photovoltaic panels are only rarely placed horizontally on roofs, except in tropical regions, where very high sun positions are expected. In subtropical and temperate climate zones, a sloped and vertically inclined orientation is quite common and therefore the cosine of the incidence angle has to be applied. The direct sunbeam illuminance on the sloped or vertical illuminated element was defined in Chap. 3.

In solar and photovoltaic facilities, the greatest gains are achieved under clear sky conditions. Therefore, ISO/CIE sky types should be taken into account as well as the obstructions in the vicinity. Owing to uneven luminance sky patterns on the clear skies with diurnal as well as annual changes, the directional formula for illuminance calculation has to follow the numerical integrations with respect to the actual slope and orientation of the illuminated panel after the basic (7.4).

This also means there are quite complicated cases that can be solved only by computer programs, as indicated by Matusiak and Aschenhoug (2002) for overcast skies and vertical illuminated planes. Owing to various clear and cloudy sky luminance patterns, the computer summation program has to follow a net dividing the whole sky vault, e.g., respecting the 145 sky zones or the subdivision of spherical triangles proposed by Tregenza and Sharpless (1993). In some algorithms (Kastendeuch and Najjar 2009) the radiative transfer in urban space or in the so-called urban canyons is expressed by the form factor between the sky mesh and the terrestrial mesh (Pianykh et al. 1998) respecting the sky radiance because of which the calculation has to be run in diurnal time steps and within specified solid angles, i.e., form factors.

Whereas the daylight illuminance on horizontal surfaces outdoors such as streets or squares is today less important and low atrium houses are scarce, sloped and vertically placed collectors and photovoltaic panels are gradually becoming more prevalent. However, because the effective illuminance on these collectors is more useful under clear or cloudy exterior situations than under overcast skies, it is logical that for energy efficiency the prime orientation of collector surfaces should be found for any collector slope or orientation by following the local year-round sun paths. The approximate rules are to follow the geographical latitude of the location, e.g., the advantageous slope for gains in the winter months (December to April) is $\beta = \varphi$, for the summer season (May to September) $\beta = \varphi/2$, and for the year-round gains $\beta = \varphi/2 + 10^\circ$, i.e., slightly more vertical slope (Kittler and Mikler 1986).

7.4 Partial Conclusions

The illuminance from the sky on the outdoor unobstructed horizontal plane depends on the luminance distribution pattern and the solid angle of the whole sky vault projected onto the illuminated plane. When the sky luminance pattern is either uniform or has a vertically progressing gradation, the sky horizontal elemental rings can be summed up in their projected solid angles to calculate the total horizontal illuminance level, which is comparable to exterior measurements of diffuse sky illuminance D_V. In architectural designs of open urban spaces as well as in the design of buildings and interiors, the rectangular system of vertical walls with horizontal obstruction edges as well as rectangular window frames is most common. Therefore, the decreasing obstruction droop lines or angles influence the resulting sky illuminance on outdoor and indoor horizontal surfaces. It is important to apply computer-based numerical integration within respective solid angles with the effective luminance distributions seen through the aperture opening which vary considerably with the aperture orientation under the clear and partly cloudy skies.

Appendix 7

Comparison of Basic and Approximate Formulae for Defining the Window Solid Angle

The most extensive light source present everywhere during daytime is the vast sky extending from the horizon to the zenith. The theoretical image of the sky vault is a hemisphere which represents the whole upper half-space that can be measured in solid angle units, i.e., steradians (see in Appendix 2).

Originally, the whole space was simulated by the surface of a sphere with unity radius, $A_0 = \omega_s = 4\pi$; thus, the sky hemisphere has area $A = 6.283185$ sr.

The basic rule for defining illuminance from the area source on any sloped surface element dE_{vi} is determined by the elemental solid angle of the source plane projected onto the horizontal illuminated plane ω_p multiplied by the source luminance L_{vs}, i.e.,

$$dE_{vi} = L_{vs}\, d\omega_p (\mathrm{lx}). \tag{A7.1}$$

As shown in detail in Chaps. 3–5 the typical sky luminance distributions are now standardized in the ISO/CIE general sky, including the simplest Lambertian uniform sky as sky type 5 in relative terms. If $L_{vs} = 1$, the task is to determine the projected solid angle

$$d\omega_p = d\omega_s \, cos\, i \, (\mathrm{sr}). \tag{A7.2}$$

where $d\omega$ is the solid angle element in which the area light source illuminates an element of the plane with any slope and azimuth orientation given by the incidence angle i after (3.24).

The horizontal plane was taken as the most frequent working plane of visual tasks (e.g., for handwork on a bench or reading on a table). Then its normal point to the zenith, $i = Z = 0°$ or $i = \varepsilon = \pi/2 = 90°$, if either the zenith angle Z or the elevation angle ε is taken into account.

So, the Lambert sky hemisphere or the ISO/CIE sky type 5 with a certain unity uniform luminance L_{vs} will illuminate the outdoor horizontal plane in the whole sky solid angle and its projection onto the plane coincides with the area of the hemisphere base; thus, $\omega_{ph} = \pi$ and

$$E_{vi} = D_v = 3.1415926\, L_{vs} (\mathrm{lx}). \tag{A7.3}$$

Such cases sometimes happen in reality, e.g., during dense foggy situations under an overcast sky when $L_{vs}/D_v = 1/\pi = 0.31831$.

Owing to the common architectural design and building practice involving vertical and horizontal grids of daylight obstructions and apertures, the classical division of solid angular nets cuts the hemisphere in vertical and horizontal lunes as shown e.g., in Figs. A3.3 and A3.5. Such lunes form spherical double-angle areas determined by the widest angle either on the horizon or on the section meridian of the virtual hemisphere with equal solid angles:

$$d\omega = d\varphi_0 \int_{\varepsilon=0}^{\pi/2} \sin \varepsilon \, d\varepsilon = d\varphi_0 \times 1\,(\mathrm{sr}) \tag{A7.4}$$

or

$$d\omega = d\varepsilon_0 \int_{\varphi=0}^{\pi} \sin \varphi \, d\varphi = d\varepsilon_0 \times 2(\mathrm{sr}) \tag{A7.5}$$

In both equations above and in all further relations 1 sr = rad^2 because the double-sided lune is $A_\mathrm{p} = 2\varphi_0$or $2\,d\varepsilon_0$ (note that if φ_0 is in degrees then $A_\mathrm{p} = \pi\varphi_0{}^\circ/90^\circ$).

It is evident that φ_0 lunes coincide with solid angles of endlessly high rectangular openings or vertical obstructions, whereas $2\,\varepsilon_0$ lunes are valid for endlessly long obstruction barriers on both sides (walls or house fronts) and illuminating aperture strips such as rows of windows, sawtooth or zenith top lights, and street or atrium openings.

Lambert realized that the old Maurolyco's pyramid concept is in fact the image of the solid angle of any rectangular light source in an upright position like a window. That was a logical approach to define its illuminance influence on the horizontally placed surface element. But this pyramid has to be imagined for a vertical window in the simplest configuration with its peak point M and one base corner of the pyramid P placed on a normal line from the bottom corner of the window (Fig. A7.1). Thus, such a pyramid does not have the form of a true pyramid, but it is in fact a pyramid with its peak at the side corner of a rectangle. Although Lambert used a coordinate system different from current one, using the coordinate system in Fig. A7.2 with the dimensions x, y, z when the diagonal is $d_\mathrm{r} = \sqrt{x^2 + y^2 + z^2}$, one can define the dominant (normal) maximal angles of the lune by any trigonometric function of either the azimuth angle φ_0 or the elevation angle ε_0, i.e.,

$$\tan \varphi_0 = y/x \quad \text{or} \quad \tan \varepsilon_0 = z/x, \tag{A7.6}$$

$$\cos \varphi_0 = \frac{x}{\sqrt{x^2 + y^2}} \quad \text{or} \quad \cos \varepsilon_0 = \frac{x}{\sqrt{x^2 + z^2}}, \text{etc.} \tag{A7.7}$$

Knowing the perspective principle that two parallel lines intersect at infinity at the vanishing point, one can imagine a double system of lune solid angles on the virtual sky hemisphere with unity radius. Specifically:

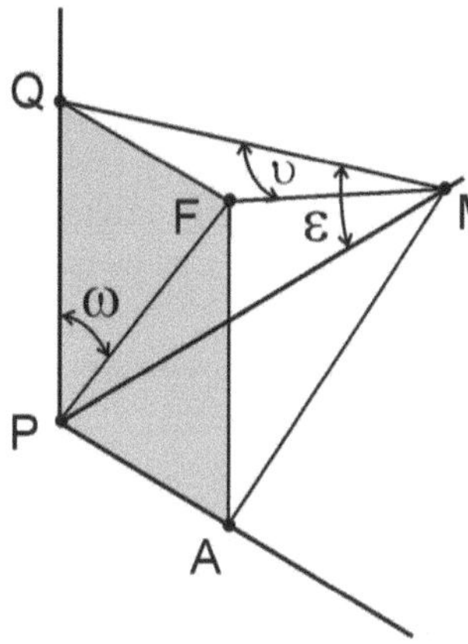

Fig. A7.1 Simplified Lambert scheme of a rectangular light source

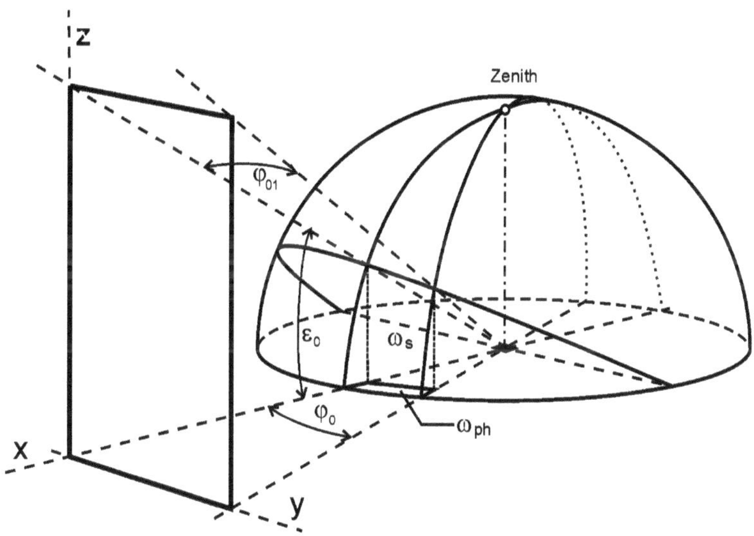

Fig. A7.2 Solid window angle and its projection onto the horizontal plane and coordinates

- One solid angle formed by an endlessly high vertical window with maximum azimuth angle φ_0 which has the half-lune area equal to the solid angle $\omega_s = \varphi_0/2$.
- Another horizontal lune for an endlessly wide (on both sides) horizontal aperture projected on the sky hemisphere has area, i.e., solid angle equal to, $\omega_s = \varepsilon_0$.

Finally, Lambert's mathematical formula as well as Wiener's (1884) geometrical relation is valid for a window solid angle projected onto the horizontal plane with the bottom window frame in the illuminated plane:

$$\omega_{ph} = \frac{1}{2}\,(\phi_0 - \phi_{01}\ \cos\,\varepsilon_0). \tag{A7.8}$$

The vanishing tendency of both lunes from φ_0 or ε_0 to zero cannot be represented by the reduction in relation to only the cosine angle as Lambert (1760, paragraph 181) suggested the form expressing $\upsilon = \varphi_{01}$ as its tangent value:

$$\tan\upsilon = \tan\varphi_{01} = \tan\varphi_0\,cos\varepsilon_0 = \frac{y}{\sqrt{x^2+z^2}}. \tag{A7.9}$$

Thus, an approximation ignoring a slight reduction of y due to the ε_0 droop line and thus the following classical formula is in the form:

$$\omega_{ph} = \frac{1}{2}\left(\arctan\frac{y}{x} - \frac{x}{\sqrt{x^2+z^2}}\arctan\frac{y}{\sqrt{x^2+z^2}}\right). \tag{A7.10}$$

Moon and Spencer (1946) derived the same formula using the vector method, but Moon (1936, 1961) presented a similar form

$$\omega_p = \frac{1}{2}\left(\arctan\frac{y}{x} - \frac{x}{\sqrt{x^2+y^2+z^2}}\arcsin\frac{y}{\sqrt{x^2+y^2+z^2}}\right), \tag{A7.11}$$

which was probably influenced by Yamauti (1924), whom he regarded as the first to derive it.

Of course, in those days all researchers were investigating the measuring or graphical possibilities to simplify the "direct daylight factor," i.e., sky factor calculations, and user-friendly graphical tools for its estimation and did not consider any sky luminance distributions obeying Lambert's assumption $L_{vs} = 1$. The so-called Lambert formula was derived using the interrelation in (A7.9) and (A7.10) after Fig. A7.1:

$$E_{vi}{=}L_{vs}\frac{(\varphi_0 - \varphi_{01}\cos\varepsilon_0)}{2} = L_{vs}\frac{[\varphi_0 - \cos\varepsilon_0\arctan(\cos\varepsilon_0\ \tan\varphi_0)]}{2}\,(\mathrm{lx}). \tag{A7.12}$$

Several researchers followed this formula and considered the interdependence between ε and φ angles defined in (A7.9) in several trials to find some further approximations for a window without a sill in the case of a horizontal or a vertical illuminance after the classical integration or the vector method for the uniform rectangular source (Herman 1900; Higbie 1925) as well as circular sources (Jones 1909, 1910). Walsh (1961), Moon and Spencer (1946) praised "the masterly presentation of the basic methods by Stevenson" (1933b), who summarized the mathematical and graphical estimations of the solid angle measurements and grilles, mentioning also some errors and approximations. In an earlier study, Stevenson (1931) treated in addition to both vertical section planes with divisions of the sky hemisphere in orthogonal projection, also the horizontal plane and stated the basic relation

$$E_{\mathrm{vi}} = L_{\mathrm{vs}} \int_0^{\varphi_0} \int_0^{\varepsilon_0} \sin\varepsilon \, \cos\varepsilon \, \mathrm{d}\varepsilon \, \mathrm{d}\varphi \, (\mathrm{lx}). \tag{A7.13}$$

This integration yielded him the relation

$$E_{\mathrm{vi}} = \frac{L_{\mathrm{vs}}\,\varphi_0}{2} (\sin^2 \varepsilon_0) = \frac{L_{\mathrm{vs}}\,\varphi_0}{4} [1 - \cos(2\varepsilon_0)] \, (\mathrm{lx}) \tag{A7.14}$$

but left out the interdependence between the lune intersection, i.e., the relation between ε_0 and φ_0. So, this was the simplest approximation formula with the desired separation of independent influences of ε_0 and φ_0 angles which suited the construction of diagrams by Waldram and Waldram (1923) and Waldram (1980) or the design of protractors by Stevenson (1932, 1933a), Dufton (1940) and Danilyuk (1941). All mentioned some approximations but defined none. However, and without just cause, as will be explained, Moon and Spencer (1946) declared the Waldram method "worthless" for a long rectangular source as it "completely ignores the correct equations which are employed by the rest of the world." So the question is what is the correct method.

The exact integration has to follow the original equation:

$$E_{\mathrm{vi}} = L_{\mathrm{vs}} \int_{\varphi=0}^{\varphi_0} \left[\int_0^{\varepsilon_0(\varphi)} \sin\varepsilon \, \cos\varepsilon \, \mathrm{d}\varepsilon \right] \mathrm{d}\varphi$$

$$= \frac{L_{\mathrm{vs}}}{4} \int_{\varphi=0}^{\varphi_0} \{ 1 - \cos[2\varepsilon_0(\varphi)] \} \mathrm{d}\varphi \, (\mathrm{lx}) \tag{A7.15}$$

when

$$\tan \varepsilon_0(\varphi) = \tan \varepsilon_0 \, cos \, \varphi. \tag{A7.16}$$

Thus, the equation to be integrated is

$$E_{\mathrm{vi}} = \frac{L_{\mathrm{vs}}}{4} \int_{\varphi=0}^{\varphi_0} \{1 - \cos[2 \arctan(\cos\varphi \, \tan \varepsilon_0)] \} \mathrm{d}\varphi \, (\mathrm{lx}), \tag{A7.17}$$

which is analytically unsolvable, but can be relatively precisely solved by numerical integration, i.e., summation of tiny elements $\mathrm{d}\varphi = (\varphi_2 - \varphi_1)/1{,}000$. The computing time is usually some milliseconds for a typical PC (operating at a frequency above 1 GHz). The numerical error is then below 0.1% for any type of integration method (including the simplest one – the trapezoidal integration method).

A further simplification of (A7.17) can be done by assuming that for very small ε_0 and φ_0 angles their tangents are equal to their radian values; thus,

$$\mathrm{Evi} = \frac{L_{vs}\,\varepsilon_0^{\,2}}{4}\left[\varphi_0 + \frac{\sin(2\varphi_0)}{2}\right] = \frac{L_{vs}\,\varepsilon_0^{\,2}}{4}\left[\varphi_0 + \sin\varphi_0\cos\varphi_0\right](\mathrm{lx}). \qquad \text{(A7.18)}$$

The results obtained after use of all three main simplification formulae, i.e., Lambert's (A7.12), Stevenson's (A7.14), and exact integration assuming small angles (A7.18), and the exact results after (A7.17) were also published (Kittler et al. 2010), where the derivation was done for a window placed sideways from the coordinate zero point.

The approximation errors can be followed here with a simplified case with regard to the different window width angle φ_0 in the relative illuminance horizontal level for selected window elevation angles $\varepsilon_0 = 20°, 50°$, and 75° in Fig. A7.3. It is evident that the exact calculation after numerical integration after (A7.17) gives results almost identical to those obtained with the traditional Lambert formula (A7.12) in all cases with an error of less than 0.1%.

Although the differences seem to be quite small, some errors might be significant, e.g.:

- If the window is relatively narrow (φ_0 under 45°) even if it is quite high (ε_0 over 50°), e.g., owing to the illuminated point being close to the window, all approximate formulae give good results except for (A7.18) when the window is too high. So in real cases these illuminance errors seem to be practically unimportant unless the sky luminance pattern is far from uniform.
- If the width angle is very wide, e.g., $\varphi_0 = 50 - 90°$, the error increases after (A7.14) and the relative percentage error after Fig. A7.4 rises in accordance with the width angle, with the maximum of 100% error if $\varepsilon_0 = 20°$ and $\varphi_0 = 50°$, which seems to be an enormous error.
- If the separation of ε_0 and φ_0 influences is realized after (A7.18), the exaggerated errors can occur under higher apertures but small width angles as documented in Fig. A7.4c.

Separation of the influences of ε_0 and φ_0 angles is possible after (A7.14) with relatively small errors when both these angles are small. With respect to the advantages of the Waldram diagrams – a very clear and illustrative "design and check" oriented tools criticized by Moon and Spencer (1946) cannot be justified because the approximation errors in the case of very wide windows are compensated by the droop-line reductions following the lune vanishing points. A comparison of the results obtained with graphical diagrams or protractors is given in the appendix in Chap. 8. In fact the droop-line tool was used earlier by Molesworth (1902) for graphical estimation of obstructions, while Pleier (1907) measured obstractions by solid angles.

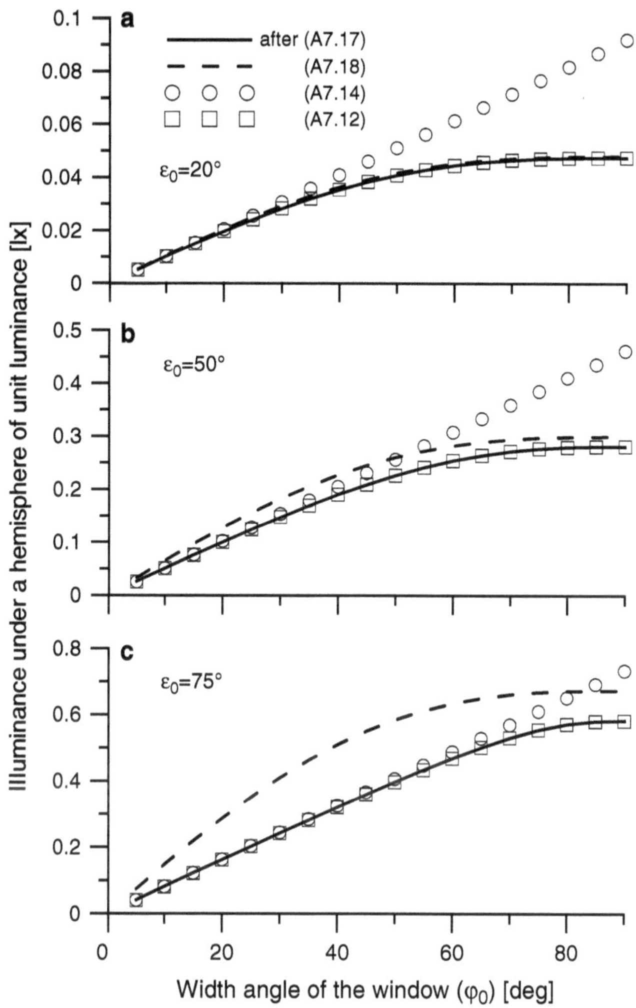

Fig. A7.3 Comparison of the relative illuminances obtained by approximate formulae and by numerical integration

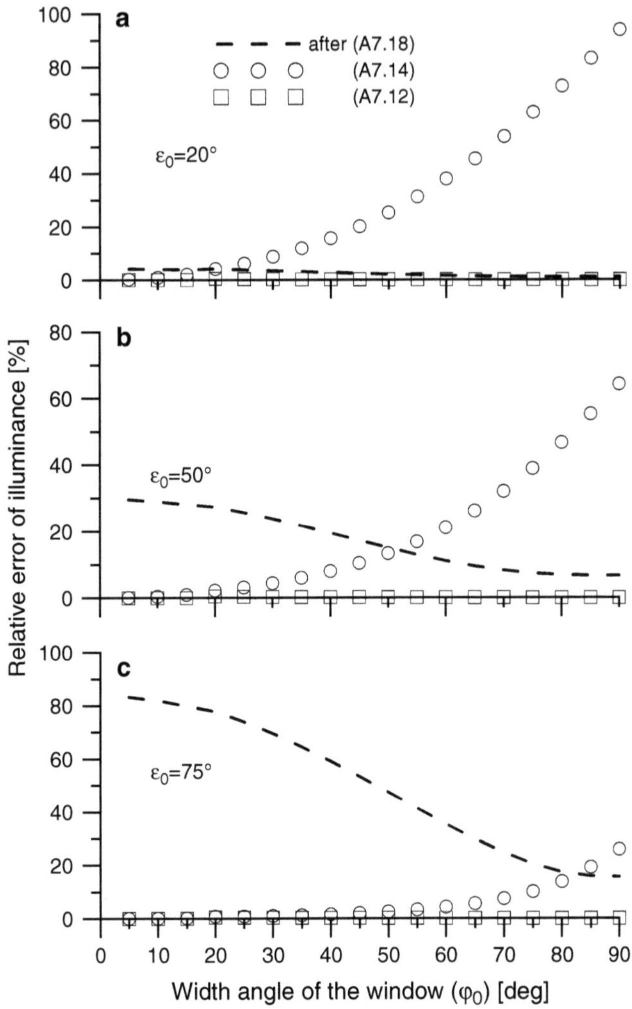

Fig. A7.4 Relative errors connected with approximations

References

Beer, A.: Grundriss des photometrischen Calcüles. Vieweg Verlag, Braunschweig (1854)

Burchard, A.: Die natürliche Beleuchtung der Strasse. Zentralblatt der Bauverwaltung, 7–8, 38–39 (1919a)

Burchard, A.: Die natürliche Beleuchtung des Hofes. Zentralblatt der Bauverwaltung, 100, 597–600 (1919b)

CIE – Commission Internationale de l'Éclairage: Natural Daylight, Official recommendation. Compte Rendu CIE 13th Session, **2**, part 3.2, 2–4 (1955)

CIE – Commission Internationale de l'Éclairage: Spatial distribution of daylight – CIE Standard General Sky. CIE Standard S 011/E:2003, CB CIE Vienna (2003)

Cox, H.: The law and science of ancient lights. H. Sweet, Law Publ., London (1871)

Du, J. and Sharples, S.: Daylight in atrium buildings: Geometric shape and vertical sky components. Lighting Res. Technol., **42**, 4, 385–397 (2010)

ISO – International Standardisation Organisation: Spatial distribution of daylight – CIE Standard General Sky. ISO Standard 15409:2004 (2004)

Kastendeuch, P.P. and Najjar, G.: Simulation and validation of radiative transfers in urbanised areas. Solar Energy, **83**, 3, 333–341 (2009)

Kittler, R., Ondrejička, Š., Tiňo, J.: Základy výpočtových metód osvetlenia z plošných zdrojov. (Basis of illumination calculation methods from planar sources) Private Res. Report, Institute of Construction and Architecture, Slovak Academy of Sciences, Bratislava (1962)

Kittler, R. and Mikler, J.: Zaklady vyuzıvania slnecneho ziarenia. (In Slovak. Basis of the utilization of solar radiation.) Veda Publ., Bratislava (1986)

Kittler, R. and Pulpitlova, J.: Basis of the utilization of daylight. (In Slovak. Zaklady vyuzıvania prırodneho svetla). VEDA Bratislava (1988)

Kittler, R.: Comment 2 on Matusiak, B., Aschehoug, Ø., 2002, Light. Res. and Technol., 34, 2, 145–147 (2002)

Lambert, J.H.: Photometria sive de mensura et gradibus luminis, colorum et umbrae. Augsburg (1760), German translation by Anding, E.: Lamberts Fotometrie. Klett Publ., Leipzig (1892), English translation with introduction notes by DiLaura D.L.: Photometry, or On the measure and gradation of light, colors and shade. Publ. IESNA, N.Y. (2001)

Littlefair, P.: Daylight prediction in atrium buildings. Solar Energy, **73**, 2, 105–109 (2002)

Markus, T.A. and Morris, E.N.: Buildings, climate and energy. Pitman Publ. Ltd., London (1980)

Matusiak, B., Aschehoug, Ø., Littlefair, P.: Daylighting strategies for an infinity long atrium: An experimental evaluation. Light. Res. and Technol., **31**, 1, 23–34 (1999)

Matusiak, B. and Aschehoug, Ø.: Algorithms for calculation of daylight factors in streets. Light. Res. and Technol., **34**, 2, 135–147 (2002)

Pianykh, O.S., Tyler, J.M., Waggenspack, W.N.: Improved Monte Carlo form factor integration. Computer Graphics, **23**, 6, 723–734 (1998)

Seshadri, T.N.: Equations of sky component with a CIE standard overcast sky. Proc. Indian Acad. Sci., **57A**, 233–242. (1960)

Sharples, S. and Lash, D.: Reflectance distributions and vertical daylight illuminances in atria. Lighting res. Technol., **36**, 1, 45–57 (2004)

Swarbrick, J.: Easements of light. Vol. I. Modern methods of computing compensation. Batsford Ltd., London, Wykeham Press, Manchester (1930)

Tregenza, P. and Sharples, S.: Daylight algoritms. Report ETSU, University of Sheffield (1993)

Waldram, P.J. and Waldram, J.M.: Window design and the measurement and predetermination of daylight illumination. Illum. Engineer, **16**, 4–5, 90–122 (1923)

Waldram, P.J.: A measuring diagram for daylight illumination. Batsford, London (1950, 1980)

Walsh, J.W.T.: The Science of Daylight. Macdonald Publ., London (1961)

Zahn, J.: Oculus Artificialis Telediptricus sive Telescopium. Würzburg (1686, 1702)

References to Appendix

Danilyuk, A.M.: Raschot yestestvennogo osveshcheniya pomeshcheniy. (In Rusian. Calculation of natural illumination of rooms.) Gos. Izdat. Stroyit. Liter., Moscow, Leningrad (1941)

Dufton, A.F.: The computation of Daylight Factors in factory design. Journ. Sci. Instr., **17**, 9, 226–227 (1940)

Herman, R.A.: A treatise on geometrical optics. Cambridge University Press, Cambridge (1900)

Higbie, H.H.: Prediction of daylight from vertical windows. Transac. IES, N.Y. **20**, 5, 433–437 (1925)

Jones, B.: The mathematical theory of finite surface light sources. Transac. I.E.S., **4**, 216–239 (1909)

Jones, B.: On finite surface light sources, Transac. I.E.S., **5**, 281–299 (1910)

Kittler, R., Kocifaj, M., Darula, S.: The 250th anniversary of daylight science: Looking back and looking forward. Light. Res. and Technol., **42**, 4, 479–486 (2010)

Lambert, J.H.: Photometria sive de mensura et gradibus luminis, colorum et umbrae. Klett Publ., Augsburg (1760), German translation by Anding, E.: Lamberts Fotometrie. Klett Publ., Leipzig (1892), English translation by DiLaura, D.L.: Photometry, or On the measure and gradation of light, colors and shade. Publ. IESNA, N.Y. (2001)

Molesworth, H.B.: Obstruction to light. Spon and Co., London (1902)

Moon, P.: Scientific basis of illuminating engineering. Dover Publ. Inc., New York (1936, 1961)

Moon, P., Spencer, D.E.: Light distribution from rectangular sources. Journ. Fraklin Inst., **241**, 3, 195–227 (1946)

Pleier, F.: Measurement of the stereo-angle of skylight. Proc. Congrés Internat. D´Hygiene Scolaire. **I**, 362 (1907)

Stevenson, A.C.: On the mathematical and graphical determination of direct Daylight Factors. Proc. Intern. Illumination Congress. Reprinted in an extract by Swarbrick, J. (1933), in Appendix I., p.146-160 (1931)

Stevenson, A.C.: The photographic and visual determination of direct Daylight Factors. Journ.Sci. Instruments, **9**, 3, 96. Reprinted by Swarbrick, J. (1933), in Appendix II., p.168-180 (1932)

Stevenson, A.C.: The construction of Daylight Factor grilles. In: Swarbrick, J. (1933), Chapter VI, 99–121 (1933a)

Stevenson, A.C.: On direct Daylight Factor and its estimation. In: Swarbrick (1933), Chapter VII, 122–145 (1933b)

Waldram, P.J.: A measuring diagram for daylight illumination. Batsford, London (1980)

Wiener, C.: Lehrbuch der darstellenden Geometrie, Vol. I. B.G. Taubner Verlag, Leipzig (1884)

Yamauti, Z.: Geometrical calculation of illumination. Researches of the Electrot. Lab., Tokyo, 148 (1924)

Chapter 8
Analytical Calculation Methods and Tools for the Design of Unglazed Apertures

8.1 Historical Achievements in Calculating the Daylight Geometry of Rectangular Unglazed Apertures

The first studies by Lambert in geometry and photometry were oriented on practical problems of determining skylight illuminance through rectangular apertures without window frames and glazing. His abstract and simplifying concepts of a fictitious sky hemisphere with unity uniform luminance led him to the importance of the solid angle and its mathematical expressions. However, theoretical photometry also involved the interrelation of basic terms of luminance, luminous flux, and illuminance linking the primary light sources with secondly illuminated surfaces via solid angles. Lambert realized that the planar source luminance might not be either uniform or unitary as often as thought in the case of the sky vault and that within elements of solid angles are propagated elemental luminous fluxes falling onto an arbitrary inclined element of an illuminated plane at some inclination angle; therefore, after Fig. 8.1

$$\mathrm{d}^2E_2 = \frac{\mathrm{d}^2\Phi_1}{\mathrm{d}S_2} = L_{\mathrm{d}S1}\,\mathrm{d}\omega_{S1}\,\cos i = L_{\mathrm{d}S1}\,\mathrm{d}\omega_{\mathrm{p}}\,(\mathrm{lx}), \tag{8.1}$$

where $\mathrm{d}E_2$ is the elemental illuminance on the element of the illuminated plane area $\mathrm{d}S_2$ and $\mathrm{d}\Phi_1$ is the elemental luminous flux caused by an elemental source of luminance $\mathrm{d}L_{S1}$ passing from a light source element $\mathrm{d}S_1$ onto an element of the illuminated plane area $\mathrm{d}S_2$ via the solid angle $\mathrm{d}\omega_{S1}$, which is the elemental solid angle taken from the illuminated element:

$$\mathrm{d}\omega_{S1} = \frac{\mathrm{d}S_1\,\cos\psi}{l^2}\,(\mathrm{sr}) \tag{8.2}$$

where $\mathrm{d}S_1$ is the elemental source area in a direction seen from the illuminated plane, e.g., the element on the sky hemisphere at distance l from the illuminated

R. Kittler et al., *Daylight Science and Daylighting Technology*,
DOI 10.1007/978-1-4419-8816-4_8,

Fig. 8.1 Mutual position of the planar source and illuminated plane

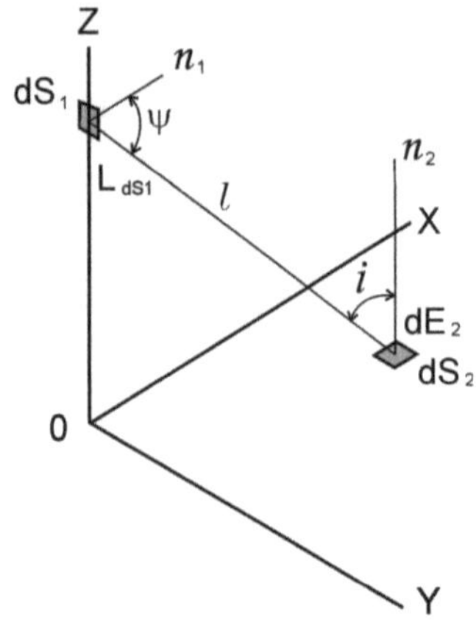

element. In the case of the sky this distance is questionable and cannot be measured; therefore, Lambert assumed a virtual hemisphere of unity radius with the illuminated element placed at its center.

The incidence angle i is equal to the space angle between the direction of the incoming flux and the normal of the illuminated plane which reduces the solid angle $d\omega_{S1}$ to its projected size $d\omega_p$, whereas for an arbitrarily positioned sloped illuminated plane

$$\cos i = \cos \beta \cos Z + \sin \beta \sin Z \cos|A_e - A_n| , \tag{8.3}$$

where β is the slope of the illuminated plane taken from the horizon, Z is the zenith angle of the sky element, A_e is the azimuth angle of the sky element, and A_n is the meridian azimuth angle of the illuminated plane normal.

Similarly, as in the case of sunbeam $\cos i$ in (3.25), (8.3) is simplified when the illuminated plane is horizontal or an arbitrarily oriented horizontal plane; then, $\cos i = \cos Z = \sin \varepsilon$, when ε is the elevation angle from the horizon, i.e., $\varepsilon + Z = \pi/2$.

A more analytical interrelation based on Lambert's assumptions was later published by Beer (1854), known as the Lambert–Beer formula for illuminance,

$$d^2E_2 = L_{dS1} \frac{dS_1 \cos \psi \cos i}{l^2} \text{ (lx)}, \tag{8.4}$$

where again after (8.2) it is simplified to the luminance of the source element multiplied by its projected solid angle as in (8.1).

Sometimes (8.4) is cited without the luminance multiplicand and then it seems that the illuminance is dependent only on the geometry, i.e., the solid angle and its projection, and the assumed unity luminance is forgotten.

When a rectangular infinitely high vertical aperture with its bottom on the horizon is imagined, then its solid angle resembles a vertical lune on the hemisphere

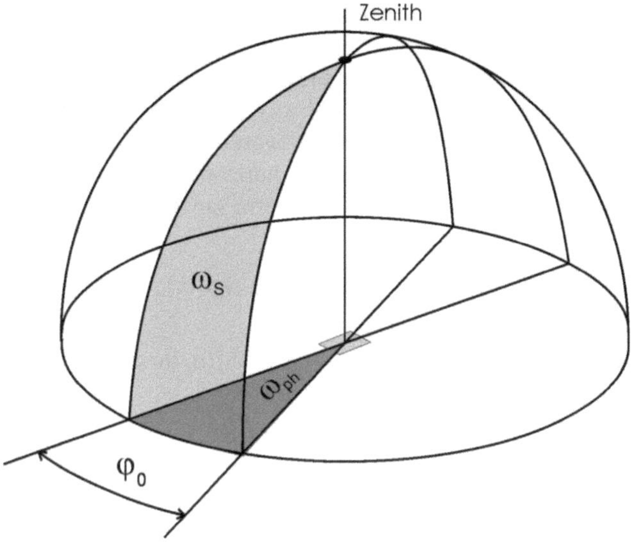

Fig. 8.2 The projection of the half-lune solid angle onto the horizontal plane

with its upper vanishing point at the zenith. If the hemisphere has unity radius and the overall uniform sky luminance is unity, then the area of such a half-lune is equal to the angle on hemisphere base $\mathrm{d}\varphi_0$; thus, the solid angle from the hemisphere center is (Fig. A3.3)

$$\mathrm{d}\omega_S = \mathrm{d}\varphi_0 \,(\mathrm{sr}) \tag{8.5}$$

and the whole hemisphere has a solid angle with φ_0 equal to 2π, or $\omega_S = 2\pi$.

For an unobstructed exterior sky lune, its projected solid angle onto the horizontal plane with its vertical normal (i.e., $\cos i = \cos Z = \sin \varepsilon$) is the area of the horizontal circle sector shown in Fig. 8.2,

$$\omega_p = \frac{\phi_0 \pi}{2\pi} = \frac{\phi_0}{2} \,[\mathrm{rad}] \tag{8.6}$$

whereas for the whole hemisphere its projected area forms the whole circular area in the horizon plane under a unity-radius hemisphere, i.e., $\omega_\mathrm{p} = \pi$.

When the window aperture is not infinitely high, the area of the upper spherical triangle has to be subtracted and in this sense Wiener (1884) interpreted the Lambertian angular expression in terms of descriptive geometry. Although Beer (1854) explained Lambert's ideas and basic principles in detail, several authors, e.g., Cox (1871), Mohrmann (1885), and Mentz (1887), did not succeed in the determination of interior illuminance in precise photometric terms. Herman (1900) was probably the first to define these trigonometric functions by the coordinates of a rectangular, vertical window situated at the bottom-right corner at the center of an x, y, and z coordinate system, taking its width as y, its height as z, and its distance from the illuminated horizontal element of the working plane as x. Thus, if the sky luminance is uniformly equal to unity, the horizontal illuminance on the interior working plane after Fig. A7.2 is in angular form same as in (A7.8):

$$E_{\mathrm{i}} = \frac{1}{2}(\varphi_0 - \varphi_1 \cos \varepsilon_0)\ (\mathrm{lx}), \tag{8.7}$$

where expressed in rectangular co-ordinates (A7.10), these are:

$$\varphi_0 = \arctan\frac{y}{x}\ (\mathrm{rad}), \tag{8.8}$$

$$\varphi_1 = \arctan\frac{y}{\sqrt{x^2 + z^2}}\ (\mathrm{rad}), \tag{8.9}$$

$$\cos \varepsilon_0 = \frac{x}{\sqrt{x^2 + z^2}}. \tag{8.10}$$

All these relations were influenced by the simplifications introduced by Lambert and later researchers as shown in the appendix in this chapter.

There were still several serious problems for future research:

- The general calculation procedure of differently oriented and sloped windows or zenith top lights for the illumination of vertical or tilted surfaces especially in industrial buildings
- The specification of different sky luminance distributions to be applied in aperture solid angles sometimes also linked with sunlight illumination by parallel sunbeams
- Complications and flux reductions by window frames and sun-shading devices (e.g., curtains, Venetian blinds, or operable louvers, and light shelves)
- The transmission of glazing materials and reflection of interior surfaces preferentially reducing, scattering, or redirecting the sunlight and skylight flux distribution into interior spaces
- Specific problems of daylight transport into deep interior spaces or windowless interiors (e.g., light wells and hollow light guides)

Attention was paid in the twentieth century to these and several other tasks, especially when linked with new and more precise measuring instrumentation, better knowledge of photometry and skills in mathematics and computing, and, not but at least, through international cooperation and standardization.

8.2 Calculation Methods Valid for Horizontal Illuminance from Vertical Rectangular Apertures

The simple traditional Lambertian principles to predict and evaluate skylight penetrating unglazed apertures with uniform unity SF(1:1) luminance and reaching an element of the horizontal working plane enabled application of "relative" criteria as opposed to criteria normalized by Lambert's absolute exterior horizontal illuminance level equal to π. With special regard to vertical windows, Basset (1909) and Higbie (1925) also studied their prediction in relative terms using a system of coordinates instead of angles. The criterion, previously called the sky factor, was then called the daylight factor and defined as the total horizontal interior illuminance E_{vi} without interreflections divided by the total exterior illuminance E_{vh} under the absolutely unobstructed overcast sky with uniform unity luminance preliminarily:

$$\mathrm{DF}(1:1) = \frac{E_{\mathrm{vi}}}{E_{\mathrm{vh}}}. \tag{8.11}$$

Under standardized conditions such a ratio either as an absolute value or as a percentage could be calculated when both illuminances were measured in real interiors or in their scale models. However, owing to the multiple and complex interreflection of the luminous flux entering the interior, for calculation reasons the daylight factor had to be subdivided into two components, i.e., the sky component and the externally reflected component as well as the internally reflected component. In theoretical studies with the uniform sky model and free unglazed openings in the wall, the sky component is called the sky factor (Hopkinson et al. 1966), which under the uniform unity sky with no gradation differences is denoted as SF$(1:1)$ assuming $E_{\mathrm{vh}} = \pi$.

Thus, using (8.7) and inserting the coordinates of (8.8)–(8.10), one obtains

$$\mathrm{SF}(1:1) = \frac{1}{2\pi}\left(\mathrm{arctg}\frac{y}{x} - \frac{x}{\sqrt{x^2+z^2}}\,\mathrm{arctg}\frac{y}{\sqrt{x^2+z^2}}\right) \tag{8.12}$$

or in the angular interpretation after (8.7) divided by $E_{\mathrm{vh}} = \pi$, one obtains

$$\mathrm{SF}(1:1) = \frac{1}{2\pi}(\varphi_0 - \varphi_1 \cos \varepsilon_0). \tag{8.13}$$

These formulae were tested and repeated in many references prior to the publication of the book by Moon (1936, 1961) and the article by Moon and Spencer (1946), where, besides the horizontal illuminance, also vertical illuminances in directions X, Y, and Z are derived for a vertical rectangular window without a sill.

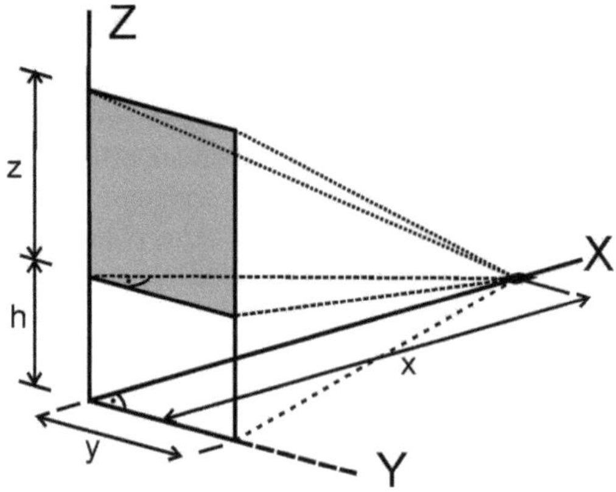

Fig. 8.3 Coordinates used in (8.12)

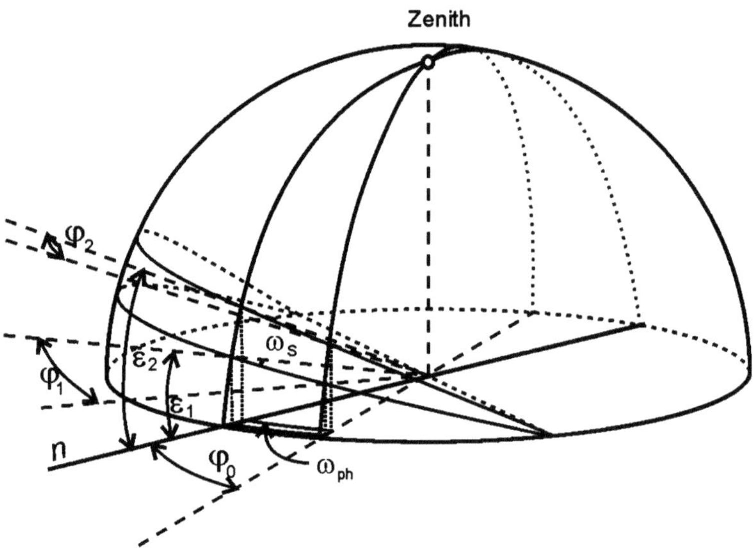

Fig. 8.4 Reduction of the solid and projected solid angle by the sill

For a vertical unglazed window opening with sill height h (or elevation ε_1) above the level of the working plane, this sill has to be taken into account after (Fig. 8.3); thus,

$$\mathrm{SF}(1:1) = \frac{1}{2\pi}\left(\frac{x}{\sqrt{x^2+h^2}}\,\mathrm{arctg}\frac{y}{\sqrt{x^2+h^2}} - \frac{x}{\sqrt{x^2+z^2}}\,\mathrm{arctg}\frac{y}{\sqrt{x^2+z^2}}\right), \quad (8.14)$$

or in the angular form (Fig. 8.4)

$$\mathrm{SF}(1:1) = \frac{1}{2\pi}(\varphi_1 \cos \varepsilon_1 - \varphi_2 \cos \varepsilon_2), \quad (8.15)$$

where both φ_1 and φ_2 are distorted owing to the "horizontal lune" and represented also by φ_0 in plan projection. Thus,

$\varphi_1 = \arctan(\cos\,\varepsilon_1\,\tan\,\varphi_0)$ and $\varphi_2 = \arctan(\cos\,\varepsilon_2\,\tan\,\varphi_0)$.

Instead of Lambert's approach and concept assuming a "vertical half-lune" to represent the endlessly high vertical aperture illuminating a planar element on a horizontal surface, a different approach is to start from the assumption of a "horizontal lune" covering the window sill and window head of an endless horizontal opening. Such integration was probably first done by Stevenson (1931) and later Seymour (1939) who was cited by Waldram (1946) and Hopkinson et al. (1966) as using the relation

$$\mathrm{SF}(1:1) = \frac{1}{\pi}\int_0^{\varepsilon}\int_0^{\varphi} \sin\,\varepsilon\,\cos\,\varepsilon\,\mathrm{d}\varepsilon\,\mathrm{d}\varphi, \tag{8.16}$$

which after integration yields

$$\mathrm{SF}(1:1) = \frac{\varphi}{2\pi}\sin^2\varepsilon = \frac{\varphi}{4\pi}(1-\cos\,2\varepsilon) \tag{8.17}$$

because $\sin^2\varepsilon = \frac{1}{2}(1-\cos\,2\varepsilon)$.

The advantage of (8.17) was in the separation of "plan" angles φ and "sectional" elevation angles ε commonly used in England since 1930 following trials to utilize photographic pinhole cameras to enable projection of the window with arbitrary obstructions on a cylindrical film (Beckett and Dufton 1932) and even earlier to measure angular grills using Swarbrick's phototheodolite (Swarbrick 1929, 1933).

Although Waldram (1946) questioned its approximation, his diagram was also based on (8.17), appreciating the fact that the influences of the width angles in plan projection and the elevation angles are separated and can be taken from architectural documentation or using his droop lines.

On the basis of the concept of a sky vault with uniform luminance, several graphical calculation tools were proposed and used in practice to evaluate SF(1 : 1) in interiors with unglazed windows of different size and placement and additionally transmittance losses and interreflection gains were approximately taken into account.

To obtain a user-friendly graphical tool some authors subdivided the sky hemisphere into a number of small areas with the same projected solid angles, thus defining SF(1 : 1) values by summing up their numbers.

After 1955 the luminance gradation 1:3 or 1:2 on overcast skies had to be taken into account. The integration within the aperture solid angle was also done and very complex equations expressing SF(1 : 3) or SF(1 : 2) were derived (Kittler et al. 1962a) but are now applicable only for tiresome computer programming. The findings of this study were partly published in Kittler and Ondrejička (1962b, 1963a, b, c) as well as in the discussion report of International Commission on Illumination (CIE) Expert Committee E-3.2 on daylight (Kittler 1964).

8.3 Graphical Tools for Unglazed Window Design and the Distribution of Skylight in Interiors

The simplest graphical tools for approximate calculation of sky factors or sky solid angles and their projections onto the illuminated planes were originally based on the assumption of a sky hemisphere of uniform and unity luminance.

Architects and builders are used to designing urban arrangements and buildings in orthogonal projections of proposed building shapes and spaces:

- In plans, i.e., whole buildings or their particular storeys projected vertically onto horizontal planes to represent the placement and width of windows and door connections
- In vertical sections of the designed building to specify vertical dimensions of walls, windows, and roofs and to reproduce staircase connections to floors, the basement, and the roof

It was assumed in daylighting that the sky hemisphere of uniform luminance could also be represented in plan projection by the base circle with an arrow indicating the orientation to the north or south and in section by a half circle over the horizontal line representing the local horizon. Whereas the plan projection for a circular plan of the sun path in true solar time has a definite orientation to the cardinal points, the sky hemisphere is often turned around to comply with the plan and section of the buildings or with the normal plane of the windows. Owing to the simultaneous influence of sunlight and skylight under sunny situations, the true orientation, i.e., azimuth angles for the sun position and house front orientation, has to be taken into account.

The first trial to introduce a practical graphical tool was probably made by the German architect Mentz (1887). He separated the sectional (elevation) part of the window solid angle and the plan (window width) part. He also indicated that the "lune" reduction of elevation angles followed a droop line, later used by Waldram (1946). In addition, elevation scaling on sectional protractors was applied in many protractors by Danilyuk (1931) and Dufton (1940) for SF(1 : 1) estimation.

The concept and derivation of the angular chart construction was explained in a book by Danilyuk (1941) as follows:

- The unit sky hemisphere with uniform luminance is taken as the only source of daylight, with sunlight being absent and no interreflections outdoors or indoors
- The Wiener (1884) formula (8.13) in its general form is suitable for the numerical integration and valid for a polygonal (e.g., triangular) aperture as

$$\mathrm{SF}(1:1) = \frac{1}{2\pi} \sum_{k=1}^{k=m} \varphi_k \cos \varepsilon_k, \tag{8.18}$$

where φ_k is the number of k angle radians under which the polynomial side (frame) of the aperture is seen from the illuminated element and ε_k is the number of k angles in radians under which the polynomial side is vertically stretched over the plane of the illuminated element.

In the case of a triangular aperture, only three sides have to be considered and in the case of a rectangular oblong aperture, four sides or the effects of two triangles forming the oblong aperture can be summed.

To find the sectional division of the angular chart, an endless aperture to both sides (a whole horizontal lune) to be determined has assuming that 100 partial lunes will cover the 90° vertical protractor. If on a horizontal planar element from an endlessly wide vertical aperture $\mathrm{SF}^{\infty}(1:1)$ is to be calculated, then the simplification $\varepsilon_3 = \varepsilon_4 = \pi/2$ with the cosines equal to zero and $\varphi_1 = \varphi_2 = \pi$ simplifies (8.18) to

$$\mathrm{SF}^{\infty}(1:1) = \frac{1}{2}(\cos\,\varepsilon_1 + \cos\,\varepsilon_2) = \frac{1}{2}(\sin\,\zeta_1 - \sin\,\zeta_2) = n_{\mathrm{I}}, \tag{8.19}$$

where the cosines of the vertical elevation angles ε_1 and ε_2 can be expressed using the zenith, i.e., incident, angles for the horizontal illuminated plane as $\cos\,\varepsilon_1 = \sin\,\zeta_1$ and $\cos\,\varepsilon_2 = -\sin\,\zeta_2$. This sectional distribution and number of particular lunes, each giving the same illuminance on a horizontal element, are counted within the window section as n_{I}.

In this way the separation of the sky patches from the drawing of the aperture section and plan was accomplished and then the reduction of the aperture solid angle due to the restricted window width can be calculated by the approximation component

$$n_{\mathrm{II}} = \left.\frac{\varphi + \sin\,\varphi\,\cos\,\varphi}{\pi}\right|_{\varphi_1}^{\varphi_2}, \tag{8.20}$$

where n_{II} is the number of patches within the sideway frame borderlines indicating the window width angle from the illuminated element.

Thus, the final result is given by summarizing the sky patches within the window section and multiplying these by the number of sky patches seen through the window width angle in plan projection:

$$\mathrm{SF}(1:1) = \frac{n_{\mathrm{I}}n_{\mathrm{II}}}{10{,}000} = \left.\frac{\sin\,\zeta}{2}\right|_{\zeta_1}^{\zeta_2}\left.\left(\frac{\varphi + \sin\,\varphi\,\cos\,\varphi}{\pi}\right)\right|_{\varphi_1}^{\varphi_2}. \tag{8.21}$$

So one takes only the number of patches of the sky seen through the aperture as part of the total number of 10,000 patches contained in both angular charts $\mathrm{SF}(1:1)$ can be an absolute value or divided by 100 in percent, and is easily determined. The approximation given in (8.21) is checked in Fig. 8.5. It is evident that (8.21) covers the exact results very nicely except when the apertures are very high or the illuminated element is very close to the window.

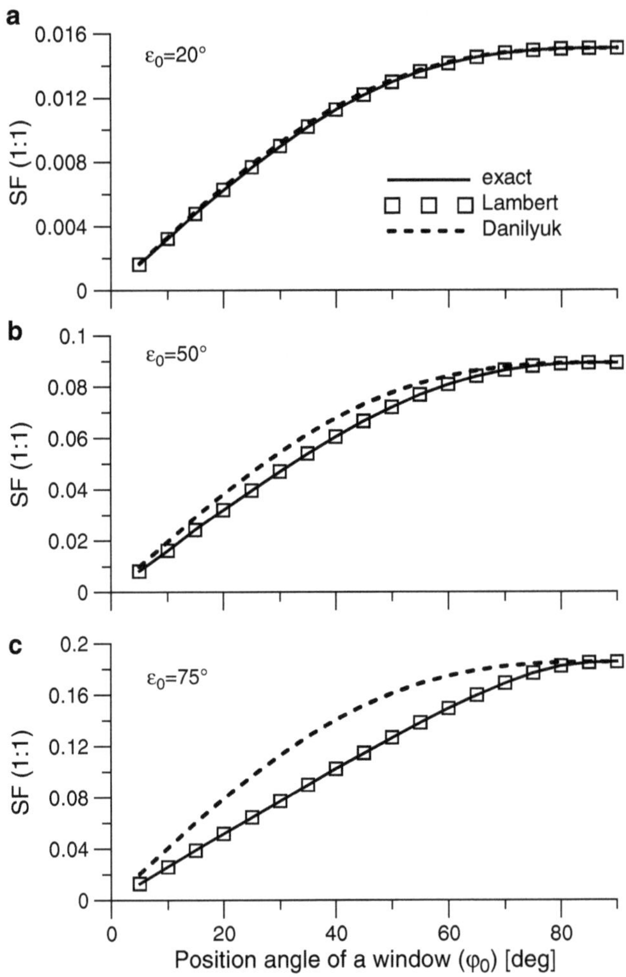

Fig. 8.5 Approximations of Danilyuk (8.21) and Lambert in comparison with exact SF(1 : 1) values

The original concept expressed in (8.21) also indicates the possible use of φ_k and ε_k or ζ_k angles, and thus also the use of vertical and horizontal protractors with special scaling.

The vertical scaling assumes an endlessly wide aperture with the same height angular proportions as a real particular window or vice versa. The interpretative grid for SF(1 : 1) determination follows the vertical strips with scaling after (8.17).

As indicated in Fig. 8.5, Danilyuk's approximation is quite good and only small errors can be expected if the window is very high or if the window is very near the illuminated element of the working plane (e.g., if ε_0 is around 75° and the window width φ_0 is between 20° and 60°). Danilyuk's angular charts can be used for arbitrarily sloped apertures also with a sill, as indicated on Fig. 8.6.

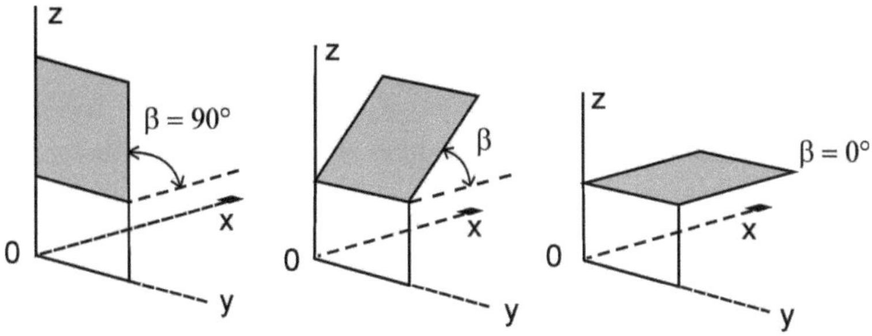

Fig. 8.6 Range of apertures with different slopes for applying Danilyuk's charts

A similar scaling was used first by Waldram and Waldram (1923) and also by Beckett and Dufton (1932) as well as for the construction of the Building Research Station (BRS) protractors (Dufton 1940, 1946), which became quite popular especially in the Commonwealth countries. Their use and some results are indicated in an example in the appendix in this chapter.

As mentioned already in the appendix in Chap. 2, in lawsuits the main litigation concerned the obstruction of window solid angles, the judge being easily persuaded with a very illustrative tool, the graphical representation of the Waldram diagram (Waldram and Waldram 1923, 1932, 1950). This diagram was based on approximation (8.17), which caused some errors (see the appendix in Chap. 7). However, these errors were compensated and corrected by the droop-line system. Later, in addition to including glass transmission characteristics, these diagrams were also innovated (Waldram 1936) and additionally adapted to take into account the CIE overcast sky gradation (1:3) (Walsh 1961; Hopkinson et al. 1966).

After the adoption of the CIE (1955), the overcast sky with luminance gradation 1:3, a very effective approximate correction was found in the relation between the 1:1 and the 1:3 overcast sky luminance standards at the mutually equal point at the elevation angle of 42°. Thus, the correction factor $k_\varepsilon(1:3)$ can be determined after

$$k_\varepsilon(1:3) = \frac{L_\varepsilon}{L_{42^\circ}} = \frac{1 + 2\sin\varepsilon}{1 + 2\sin(42^\circ)} = \frac{1 + 2\sin\varepsilon}{2.33826}. \tag{8.22}$$

Similarly, a correction to an overcast sky with luminance gradation 1:2 can be made using $k_\varepsilon(1:2)$.

These corrections are in accordance with the average elevation angle of the window reducing low solid angles or increasing high aperture angles in the $k_\varepsilon(1:3)$ range 0.43–1.28, and for $k_\varepsilon(1:2)$ the range is 0.6–1.2. These corrections were inserted in the Danilyuk charts by Kittler and Kittlerova (1968, 1975), and other older protractors as well in the equal-area diagram of Waldram which is documented in books by Walsh (1961) and Hopkinson et al. (1966).

8.4 Predicting Skylight from Unglazed Inclined and Horizontal Rectangular Apertures

The general formulae for unglazed apertures were derived in a research report (Kittler et al. 1962a) on the basis of the premise that when the illuminance of a rectangular inclined aperture with slope β placed on a vertical sill (Fig. 8.7) can be determined, then the resulting formulae can be simplified for a vertical window, i.e., $\beta = 90°$, or any slope until finally the horizontal aperture with $\beta = 0°$ is reached. General formulae after classical integration were derived for unglazed apertures, with simplifying cases shown in Kittler and Ondrejička (1962b) valid for coordinates. Formulae were also derived by Levin and Higbie (1926) for a sloped aperture without a sill. However, as previously discussed, these formulae were incorrect and were corrected by Bychinsky (1933) in his thesis, cited also by Higbie (1934). All these older calculation methods respected the Lambert unity luminance sky.

All older formulae for either a vertical window or a sloped opening followed the approximation of Lambert taking the lower horizontal edge of the window in the horizontal plane of the illuminated element either outdoors or indoors. The innovation of placing the opening on a sill (as shown in Fig. 8.6) enabled the derivation of a general complex formula with possible simplifications:

- When the aperture source is inclined from the horizontal plane either at its inclination angle $\beta = 90°$, which is equal also to the incidence angle, or at the angle between the source plane and its normal from the zenith $\zeta = 90°$, then the general formula leads to the same form as that for the vertical aperture.
- When the aperture is in a horizontal position, then $\beta = \zeta = 0°$ leads also to simplifications in the general formula which is valid for a horizontal aperture as indicated in the last case in Fig. 8.6.

Further applications of a general complex formula for forward or rearward inclined planar sources show such simplified formulae for vertical as well as horizontal apertures in the case of the Lambert unit sky (published also in Kittler and Ondrejička 1963a).

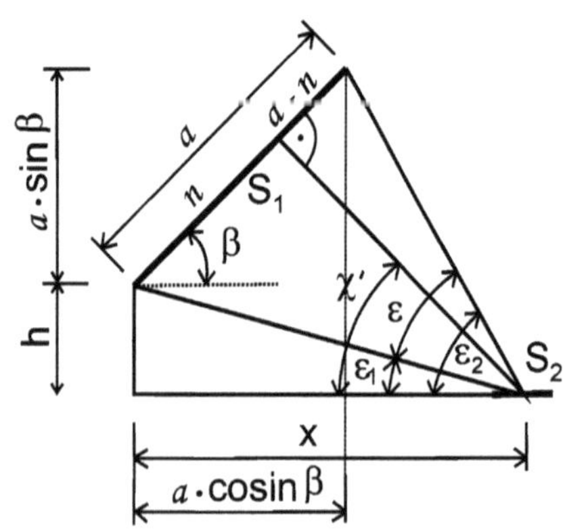

Fig. 8.7 Coordinates and angles defining the sloped source S_1

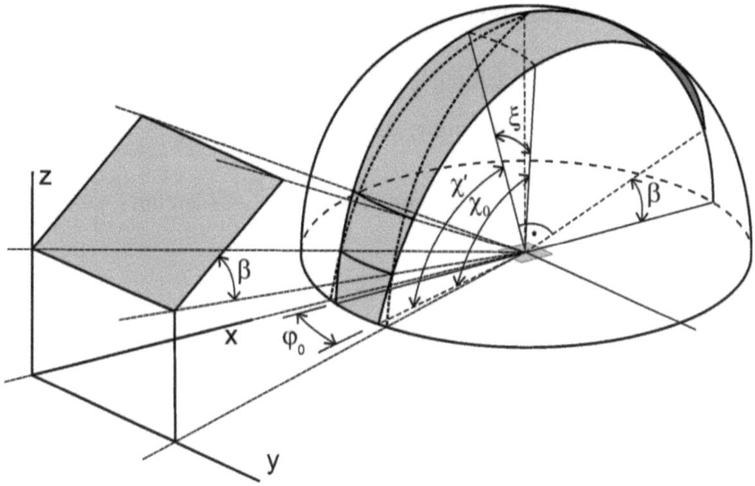

Fig. 8.8 The image of the solid angle for a sloped aperture on a sill

Such general formulae also include in the integration the CIE overcast sky gradation and were derived by Kittler and Ondrejička (1963b) in angular form and also using coordinates (Kittler and Ondrejička 1963) following the CIE (1955) standard gradation 1:3 after (5.17) assuming the darker ground reflectance and after (5.18) for a snow-covered terrain with gradation 1:2 (Kittler and Ondrejička 1964).

As the detailed integration was described elsewhere (Kittler and Ondrejička 1963b; Kittler and Kittlerova 1968, 1975) only the main steps following the derivation and the results in angular form after the angles are shown here in Figs. 8.7 and 8.8.

It is evident that the inclined lune predetermines the solid angle of the aperture with slope angle β, which is characterized by its width angle ξ:

$$\tan\,\xi = \frac{\cos\,\varepsilon_1\,\tan\,\varphi_0}{\sin(\beta + \varepsilon_1)}. \tag{8.23}$$

The full extent of the sloped lune is given by its width angle.

The basic relation to be integrated after the CIE (1955) overcast sky gradation

$$\mathrm{SF}(1:3) = \frac{L_{vz}}{\pi}\int\limits_{S_1}\left(\frac{1 + 2\,\cos\,Z}{3}\right)\frac{\cos\,\psi\,\cos\,Z}{l^2}\,\mathrm{d}S_2. \tag{8.24}$$

This integration includes the basic proportion of a certain level of the SF(1 : 1) component and the additional component expressing the gradation component SF(GC) in the relation

$$\mathrm{SF}(1:3) = \frac{3}{7}\mathrm{SF}(1:1) + \frac{2}{7}\mathrm{SF}(\mathrm{GC}). \tag{8.25}$$

After the classical integration, the final components of the general formula are

$$\mathrm{SF}(1:1) = \frac{1}{2\pi}\left\{\begin{array}{l} \cos\varepsilon_1 \arctan[\tan\xi \sin(\beta+\varepsilon_1)] \\ -\cos\varepsilon_2 \arctan[\tan\xi \sin(\beta+\varepsilon_2)] \\ +\dfrac{\tan\xi\cos\beta}{\sqrt{1+\tan^2\xi}}\left[\arctan\dfrac{\cot(\beta+\varepsilon_1)}{\sqrt{1+\tan^2\xi}} - \arctan\dfrac{\cot(\beta+\varepsilon_2)}{\sqrt{1+\tan^2\xi}}\right] \end{array}\right\} \tag{8.26}$$

and SF(GC) is even more complex:

$$\mathrm{SF}(\mathrm{GC}) = \frac{1}{\pi}\left\{\begin{array}{l} \arcsin\dfrac{\tan\xi\cos(\beta+\varepsilon_1)}{\sqrt{1+\tan^2\xi}} - \arcsin\dfrac{\tan\xi\cos(\beta+\varepsilon_2)}{\sqrt{1+\tan^2\xi}} \\ +\dfrac{\cos 2\beta - \tan^2\xi\sin^2\beta}{\tan\xi(1+\tan^2\xi)}\left[\begin{array}{l}\dfrac{\cos(\beta+\varepsilon_2)}{\sqrt{1+\tan^2\xi\sin^2(\beta+\varepsilon_2)}} - \\ -\dfrac{\cos(\beta+\varepsilon_1)}{\sqrt{1+\tan^2\xi\sin^2(\beta+\varepsilon_1)}}\end{array}\right] \\ -\dfrac{\cos 2\beta}{\tan\xi}\left[\begin{array}{l}\cos(\beta+\varepsilon_2)\sqrt{1+\tan^2\sin^2(\beta+\varepsilon_2)} - \\ -\cos(\beta+\varepsilon_1)\sqrt{1+\tan^2\sin^2(\beta+\varepsilon_1)}\end{array}\right] \\ -\tan\xi\sin 2\beta\left[\begin{array}{l}\dfrac{\sin^3(\beta+\varepsilon_2)}{\sqrt{1+\tan^2\xi\sin^2(\beta+\varepsilon_2)}} - \\ -\dfrac{\sin^3(\beta+\varepsilon_1)}{\sqrt{1+\tan^2\xi\sin^2(\beta+\varepsilon_1)}}\end{array}\right] \end{array}\right\} \tag{8.27}$$

A private research report by Kittler et al. (1962a) also contains the integration using actual dimensions, which make the classical integration even more complicated as:

- The aperture slope β is crucial because it changes the normal plane of the aperture (Fig. 8.8), which passes the illuminated element dS_2, and in this normal plane is the widest lune angle of the aperture solid angle.
- The oblong related to the aperture slope also determines a frame of the sill at distance n from the aperture normal as well as the upper aperture head $a - n$.

- The vertical sill with height h determines the elevation of the sloped window above the horizontal illuminated plane which also contains the arbitrary element dS_2; thus, the distance of an arbitrary aperture element from it is

$$l = \sqrt{(x - a \cos \beta)^2 + (h + a \sin \beta)^2 + y^2} \text{ (m)}. \tag{8.28}$$

- Any aperture element to the vertical normal in the horizontal illuminated element dS_1 has the angle

$$\cos Z = \frac{h + a \sin \beta}{l}. \tag{8.29}$$

The resulting formulae in dimensional form can be found in Kittler and Kittlerova (1968, 1975).

Only computer calculation programs can handle such complex general formulae but owing to the absence of practical projects with unglazed apertures probably no one has tried to solve concrete cases for streets, squares, or atria. However, a graphical method for predicting daylight illuminance from apertures of arbitrary shape and slope was published by Danilyuk (1935) assuming uniform sky luminance and unglazed openings. The later CIE overcast sky standard with graded luminance pattern (CIE 1955) demanded new research work as documented in this chapter and the CIE clear sky standard (CIE 1973) stimulated even more difficult development of new original calculation methods (Ondrejička 1973) and graphical tools (Kojić 1963).

Although in the first half of the twentieth century manual or rather primitive calculation methods with the help of graphical tools were quite common, it became clear that computer programs with simple user-friendly possibilities could cope with many complicated and tiresome procedures due to several sky patterns, e.g., applying the ISO (2004) or CIE (2003) general sky types. In fact there is now such a user-friendly computer tool for vertical windows (Roy et al. 2007). Energy-saving regulations and incentives have introduced further calculation difficulties, propelling solution by computers, especially because of complex mathematically defined interrelations and because the algorithms for the science behind the technology must be correct. Moreover, new requirements, shading device design or novel light transport systems (various anidolic and tubular light guides), supplementary daylight and artificial lighting systems and controls with dimming or switching regulations, etc. brought further computer program challenges and solutions.

8.5 Partial Conclusions

The fundamental calculation methods and formulae for vertical windows were derived by Lambert in the eighteenth century, and the simplification of the unity uniform sky luminance and the assumption of a free unglazed aperture were adopted then. However, respecting these assumptions, the historical circumstances, and means available at that time, his formulae are very accurate, as shown in the appendix in Chap. 7 even though no objective illuminance measurements could test them.

Many research results were obtained during the twentieth century, linked with new and more accurate measuring equipment, better knowledge of mathematics and computing, and international cooperation and standardization, etc.

However, future daylight research faces at least three serious major problems:

- To find and standardize more sophisticated criteria for calculation methods avoiding the obsolete sky factor or daylight factor system of evaluating or predicting daylighting of interiors. Possibly the best solution would be to adopt absolute values for luminance and illuminance criteria comparable to interior levels for artificial lighting together with year-round energy-saving computer simulation and measurement strategies.
- To adopt besides the relative luminance standard skies their standardized luminance distributions in absolute luminance values in accordance with the sky standardization for design purposes.
- To develop sophisticated computer programs and tools with possible application not only for rectangular windows, but also for round and arbitrarily shaped and sloped apertures and novel systems such as hollow light guides, light shelves, and shaded openings with louvers, jalousies, etc.

Appendix 8

Comparison of Graphical Tools for Daylight Prediction and Their Accuracy for Unglazed Apertures Under Uniform Skies

Although in the age of computers the graphical tools seem to be quite obsolete, some traditional assessment routines in practice often still use them. Especially, architects and builders are used to orthogonal projections in the design and drawings of architectural and building objects:

- In urban situations, where the exact orientation of each façade and window to cardinal points is described
- In plans of different storeys with the layout of different rooms as well as the placement of windows and roof lights
- In vertical sections of buildings where the vertical placement of windows and roof lights is given with designed dimensions of window sills, heights, and heads as well as the dimensions of hollow light guides

Also the yearly changes of the sun path and the luminance maps of the sky dome can be reproduced in different projection systems using the representation of the fictitious hemispheres or virtual domes. Many graphical tools such as diagrams, overlays, and protractors were produced and used by students or practitioners to predict and assess the possible sun insolation, to achieve sunlight and skylight illumination of interiors, or to avoid solar overheating gains or control sun glare.

When for the building documentation the orthogonal projection is used either on the horizontal plane (for plans) or on the vertical plane (for sections) these projections and drawings representing the actual design seem to be the most convenient for daylight graphical tools too. Therefore, the most often used design tools should be checked both for their accuracy and simplicity in application for practitioners.

At the Zurich CIE General Session where the CIE (1955) overcast sky standard with 1:3 gradation was adopted, the CIE Scope Committee requested the W-3.2 Daylight Committee "to decide on one or a small number of methods of daylight calculations which the CIE can recommend to the various interested professions." The aim was:

- To select a method applicable to any position in a room of any size and configuration with any kind of fenestration
- To find a favored method of relatively high accuracy capable of dealing with sky patches of any shape
- To prefer a method of day-to-day use which could be less accurate but readily understood and applicable at a design stage when only scale drawings are available
- To document also methods of lower accuracy but that are usable before scale drawings are decided on.

This proved a difficult task indeed. The CIE W-3.2 British member Peter Petherbridge (1959) prepared a draft and presented an excellent review of currently used graphical tools for calculating SF(1:1) and SF(1:3) in four categories using:

1. Graphical charts
2. Tabular techniques
3. Diagram techniques
4. Protractor tools

Although many foreign methods were included in the summary, logically British methods were preferred, i.e., Waldram (1950) diagrams and BRS protractors (Dufton 1946), but unfortunately without the comparison of accuracy and practical advantages. Owing to some opposition and further considerations, a simpler Australian method was proposed and a revised draft was published (CIE 1970) with a summary of 58 other methods briefly annotated and referenced. So the criteria of accuracy were totally abandoned and the calculation basis and assumptions behind the Australian minimum daylight factor diagrams including exterior and interior interreflections were vaguely stated in several correction factors or hidden altogether. So any comparison in terms of SF(1 : 1), SF(1 : 3) for unglazed apertures or accuracy compared with basic formulae to test the CIE (1970) diagram method proved impossible.

The sky factor SF(1 : 1) is, by the CIE (1939) definition, a geometrical quantity independent of the sky luminance distribution (assuming the Lambert unit uniform sky hemisphere) or glass transmission losses (assuming a free unglazed opening) and therefore can be calculated with a desired degree of precision. Of course, it has to be noted that applying the Lambert concept of relative terms based on the projection of the sky hemisphere solid angle onto the horizontal plane equal to π, SF(1 : 1) is inevitably linked as a ratio to the horizontal illuminance from the unobstructed sky outdoors E_h. Thus, it expresses also the indoor horizontal illuminance level E_i in lux in a black interior illuminated only by sky luminance equal to 1 cd/m^2 and

$$\frac{E_i}{E_h} = \frac{E_i}{\pi} = \mathrm{SF}(1:1). \tag{A8.1}$$

Currently the SF(1 : 1) criterion seems to be quite unreal but the uniform sky exists worldwide under certain weather situations and was adopted as sky type 5 in the set of standard skies (ISO 2004). Thus, if an absolute sky luminance L_{vz} in candelas per square meter is known, then

$$E_i = \mathrm{SF}(1:1)\pi L_{vz}\,(\mathrm{lx}). \tag{A8.2}$$

The relatively simple determination of the SF(1 : 1) value, either in angular or in coordinate form, enabled several tabular or graphical techniques to be developed as practical tools for the prediction of skylight in actual rooms.

After the comparison of the trigonometric basis of the aperture solid angle and its projection for the simple position of a rectangular vertical window in the appendix in Chap. 7, there is a possibility to evaluate also the graphical tools in the same style. It is evident that all graphical methods and techniques are based on the original Lambert SF(1 : 1) formula in angular form after (8.13) or (8.15) or after inserted aperture coordinates are applied in (8.12) or (8.14) results can be considered in the first class of accuracy. Suspect accuracy is associated with techniques and tools based on the simplified (8.17) if the "lune" vanishing perspective of rectangular apertures is not respected or additionally taken into account.

Although the first SF(1 : 1) diagrams after Higbie (1934) formulae were included in the basic illuminating engineering book of Moon (1936) and earlier a protractor method had been suggested (Higbie et al. 1930), even earlier more versatile graphical charts based on equal-area projection by Waldram and Waldram (1923, 1932) and Waldram (1946, 1950) were used in the UK. Later protractors by Dutton (1940, 1946), better known as BRS protractors, became preferred there, whereas in Russia and eastern Europe similar older nomograms were applied by Danilyuk (1931, 1934, 1935, 1941). Therefore, the last three tools will serve for the example and SF(1 : 1) accuracy test at first assuming for all tools the same uniform sky luminance and the same 3 m × 3 m unglazed window opening.

To compare SF(1 : 1) values obtained by different formulae or graphical tools, a simplified example of a window without a sill, in fact assuming the horizontal working plane coincides with the sill height, is used. The illuminated horizontal elements are placed on a perpendicular line from the 3 m × 3 m window corner at a

Table A8.1 Comparison of SF(1 : 1) values as percentages calculated after different formulae or graphical tools assuming unity sky luminance and a 3 m × 3 m unglazed vertical aperture

Window dimensions			SF(1:1) (%) after				
			Formula		Diagram		
ε_0 (°)	y (m)	x (m)	Exact	Lambert	Waldram	Danilyuk	BRS
70	3	1.092	15.33	15.338	15.150	15.62	15.700
50	3	2.517	7.19	7.200	6.951	7.40	7.574
40	3	3.610	4.14	4.150	4.076	4.20	4.260
30	3	5.196	1.94	1.943	1.994	2.01	2.135

BRS Building Research Station

distance from it so that their elevation angles $\varepsilon_0 = 70°,\ 50°,\ 40°$, and 30°, respectively. With these assumptions and configuration parameters, two formulae from the appendix in Chap. 7 can be used to calculate SF(1 : 1) values directly, i.e., the exact ones and their very good Lambert approximations are included in Table A8.1. Although angularly defined positions for critical illuminated horizontal elements were chosen, these can be easily determined also by the distances from the 3-m-high window by a simple tangent relation as

$$x = \frac{z}{\tan\,\varepsilon_0} = z\,\cot\,\varepsilon_0 \quad \text{or} \quad x = \frac{z}{\tan\,\varepsilon_0} = z\,\cot\,\varepsilon_0\ (\mathrm{m}), \tag{A8.3}$$

where z is the aperture height, i.e., $z = 3$ m and x is the distance from the window determined by the angular elevation of the window head ε_0. This distance is also given in Table A8.1 for each elevation angle.

In the original Waldram diagram, the droop lines for 70°, 50°, 40°, and 30° vertically and horizontally were chosen to calculate and sum up their area in each case to divide it by the area of the whole diagram. Similarly, after the summation of all relevant areas in the droop-line diagram, also corrections for $\varepsilon_0 = 70°,\ 50°$, and 40° were found and these were divided by the area of the whole diagram. Thus, the final SF(1 : 1) results as a percentage were calculated and are included in Table A8.1.

Of course, the details and calculation procedures using the original angular charts of Danilyuk (1931, 1934, 1935, 1941) are now out of date, so only the better known BRS (1944) protractor is shown in Fig. A8.1 as an example indicating how to predict the value of SF(1 : 1) from the vertical apertures at point B. Similarly, such overlays can be used also at points A, C, or D, respectively, or further protractors for different glazed and sloped apertures. Both Danilyuk and BRS protractors were later approximately corrected for SF(1 : 3) calculations, and Kittler's protractors (Kittler and Kittlerova 1968, 1975) were determined for SF(1 : 3)(TG) and SF(1 : 2)(TG) as well as for SF(1 : 3)(WG) and SF(1 : 2)(WG) for glazed apertures, in Chapter 9. An example of a protractor for a vertical window overlaid on the room section is shown in Fig. A8.2.

From Table A8.1 it is evident that all the results follow the same tendency of decreasing SF(1 : 1) values from the window opening to the rear places with only

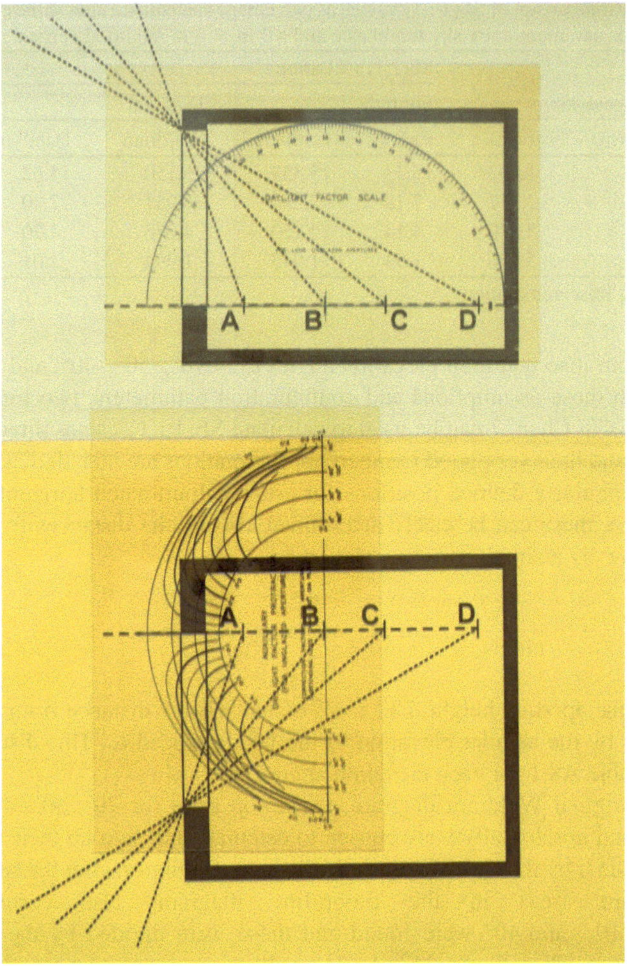

Fig. A8.1 Building Research Station protractor overlaid on sectional and plan drawing of a room

minor errors. Whereas Danilyuk and BRS protractors show a positive error, the Waldram diagram indicates a rather conservative minor decrease compared with the exact values or Lambert approximations except for the place furthest from the window. However, all the above-mentioned graphical tools take into account only dense overcast skies. The vertical windows are the most common rectangular apertures and correspond to the sky luminance raster of meridians; thus, a method of arbitrary meridians published by Kittler and Darula (2006) enabled a user-friendly computer program to be developed (Roy et al. 2007). This computer tool enables one to use all 15 sky types adopted by the ISO (2004) standard. Of course, this program can also calculate SF(1 : 3) values for unglazed vertical apertures if the input of normal transmittance $\tau_n = 1$ or any other transmittance, but no directional transmittance except the normal transmittance, is taken into account.

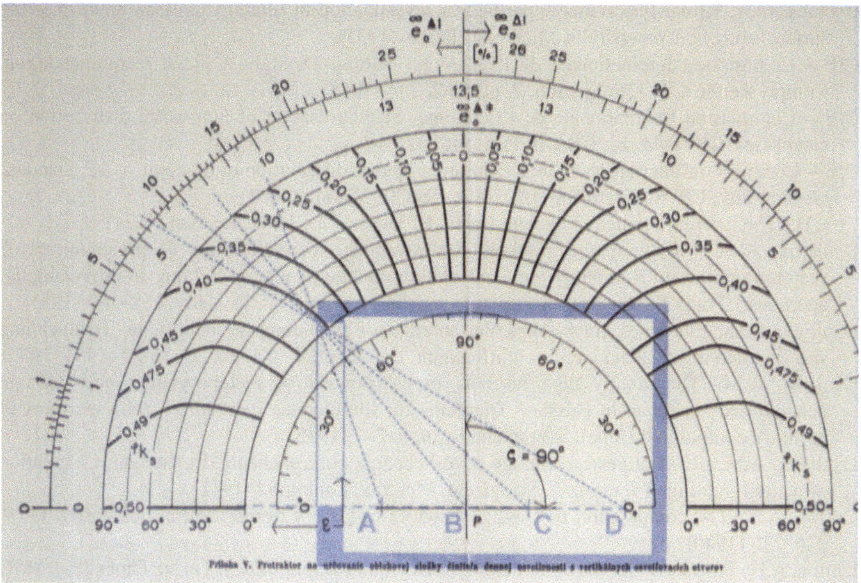

Fig. A8.2 Kittler's protractor overlaid on the section

Most existing graphical tools even in their corrected versions for SF(1 : 3) are relatively out of date. Some recent computer programs e.g., (Kota and Haberl 2009) are available for calculations. However, often their source information, assumptions, and software structures are unpublished. Some are restricted only for the determination of DF (1:3) values under overcast conditions as they include interreflections in rooms. Unfortunately, many user-friendly computer tools are oriented to prove that some prescribed criteria or standards were obeyed in building designs. So, users usually forget or do not care on what basis, formulae, or their simplifications the computer tool was developed, but they trust it because it gives apparently favorable results. However, not even the comparison of results produced by few computer software programs can help (Ubbelohde and Humann 1998). Recently the basic DF(1 : 3) criterion has been criticized because of the unrepresentative ratio system disregarding the lifetime performance of buildings in annual daylight climate as well as because of the rigidity respecting only overcast winter conditions in temperate zones while ignoring regions where extensive sun-shading and air-conditioning is of prime importance.

References

Basset, J. Jr.: The mathematical theory of finite surface light sources. Trans. IES, N.Y., 4, 4 (1909)

Beckett, H.E. and Dufton, A.F.: A photographic method of determining Daylight Factors and periods of insolation. Journ. Sci. Instruments, **9**, 5, 158–164 (1932)

Beer, A.: Grundriss des Photometrischen Calcüles. F. Vieweg Verlag, Braunschweig (1854)

Bychinsky, W.A.: Analytical and experimental investigation of illumination from interreflecting surface sources. University of Michigan, Thesis (1933)
CIE – Commission Internationale de l'Éclairage: Natural Daylight, Official recommendation. Compte Rendu CIE 13th Session, 2, part 3.2, 2–4 (1955)
CIE – Commission Internationale de l'Éclairage: Standardisation of luminance distribution on clear skies. CIE Publ. 22. CB CIE Paris (1973)
CIE – Commission Internationale de l'Éclairage: Spatial distribution of daylight – CIE Standard General Sky. CIE Standard S 011/E:2003, CB CIE Vienna (2003)
Cox, H.: The law and science of ancient lights. H. Sweet, Law Publ., London (1871)
Danilyuk, A.M.: Novyy metod raschota yestestvennogo osveshcheniya ot pryamougolnych svetoproyomov. (In Russian: A new calculation method of natural lighting from rectangular apertures.) Trudy II. Vsesoyuznoy svetotechnicheskoy konf. IV. Doklady, 168–180 (1931)
Daniljuk, A.M.: Ein neues graphisches Verfahren zur Berechnung der natürlichen Beleuchtung von Räumen mit rechteckigen Lichtöffnungen. Das Licht, **4**, 241–243, **5**, 16-20 (1934, 1935)
Daniljuk, A.M.: Diagrammy dlya raschota osveshchenosti ot svetoproyomov proizvolnogo ochertaniya i naklona.(In Russian: Diagrams for illuminance calculation from apertures of arbitrary contour and slope). Svetotechnika, **6**, 6, 7–9 (1935)
Danilyuk, A.M.: Raschot yestestvennogo osveshcheniya pomeshcheniy.(In Russian: Calculation of natural lighting of interiors.), Stroyizdat, Moscow, Leningrad (1941)
Dufton, A.F.: The computation of Daylight Factors in factory design. Journ. Sci. Instr., **17**, 9, 226–227 (1940)
Dufton, A.F.: Protractors for the computation of Daylight Factors. B.R.S. Techn. Paper 28, HMSO (1946)
Herman, R.A.: A treatise on geometrical optics. Cambridge University Press, Cambridge (1900)
Higbie, H.H.: Prediction of daylight from vertical windows. Transac. IES, N.Y. **20**, 5, 433–437 (1925)
Higbie, H.H.: Lighting calculations. Wiley Publ., New York (1934)
Hopkinson, R.G., Petherbridge, P., Longmore, J.: Daylighting. Heinemann, London (1966)
ISO – International Standardisation Organisation: Spatial distribution of daylight – CIE Standard General Sky. ISO Standard 15409:2004 (2004)
Kittler, R., Ondrejička, Š., Tiňo, J.: Základy výpočtových metód osvetlenia z plošných zdrojov. (Basis of illumination calculation methods from planar sources. In Slovak.) Private Res. Report, Institute of Construction and Architecture, Slovak Academy of Sciences, Bratislava (1962a)
Kittler, R. and Ondrejička, Š.: Raschot osveshcheniya ot naklonnogo ploskogo istochnika sveta. (In Russian: Computation of illuminance from a sloped planar light source.) Svetotekhnika, **8**, 9, 11–13 (1962b)
Kittler, R. and Ondrejička, Š.: Allgemeine Formel für die Ermittlung der Beleuchtungsstärke von zerstreut strahlenden Rechteckflächen. Lichttechnik, **15**, 9, 457–461 (1963a)
Kittler, R. and Ondrejička, Š.: Exact determination of the Daylight (Sky Component) from Rectangular Sloping window aperture with a "CIE Overcast Sky". Monograph of the IES, No.6B. IES Publ., London (1963b)
Kittler, R. and Ondrejička, Š.: Utočnennyi sposob rascheta yestestvennogo osveshcheniya. (More specified way of daylight illumination computation. In Rusian.) Izvestiya Akad. Strojit. i architekt. SSSR, **5**, 1, 93–97 (1963c)
Kittler, R. and Ondrejička, Š.: Stanovenie oblohovej zložky súčiniteľa dennej osvetlenosti pre vertikálne obdĺžnuikové osvetľovacie otvory. (In Slovak: Determination of sky-component for vertical rectangular daylighting apertures.) Stavebnícky časopis, **12**, 5, 261–278 (1964)
Kittler, R.: Discussion du Rapport du Comité E-3.2. Compte Rendu Quinziéme Session, Vol.B, 382–384. Publ.CIE No 11 B. Bureau Central de la CIE, Paris (1964)
Kittler, R. and Kittlerova L.: Návrh a hodnotenie denného osvetlenia. (In Slovak: Designing and evaluating of daylighting), Alfa Publ., Bratislava (1968, 1975)

Kojič, B.: The graphical method for the determination of interior daylighting under clear sky conditions. Bull. De Acad. Serbe des Sciences et des Arts, **30**, Sci. Techn., 6 (1963)
Levin, A. and Higbie, H.H.: Prediction of daylighting from sloping windows. Transac. IES, N.Y. **21**, 3, 273–278 (1926)
Mentz, R.: Berechnung der Tagesbeleutung innerer Räume und Messtäbe dazu. Deutsche Bauzeitung, **21**, 257–264 (1887)
Mohrmann, K.: Über die Tagesbeleuchtung innerer Räume. Berlin (1885)
Moon, P.: Scientific basis of illuminating engineering. Dover Publ. Inc., New York (1936, 1961)
Moon, P. and Spencer, D.E.: Light distribution from rectangular sources. Journ. Franklin Inst., **241**, 3, 195–227 (1946)
Ondrejička, Š.: Pole osvetľovacieho otvoru pri jasnej oblohe. (In Slovak: Field of the daylight aperture under the cloudless sky.) Stavebnícky časopis, **21**, 3–5, 357–366 (1973)
Roy, G.G., Kittler, R., Darula, S.: An implementation of the Method of Aperture Meridians for the ISO/CIE Standard General Sky. Light. Res. and Technol., **39**, 3, 253–264 (2007)
Stevenson, A.C.: On the mathematical and graphical determination of direct Daylight Factors. Proc. Intern. Illumination Congress. Reprinted in an extract by Swarbrick (1933), in Appendix I., p.146-160 (1931)
Seymour, W.D.: The heating, ventilation and lighting in school buildings. Oxford University Press, Oxford (1939)
Swarbrick, J.: The Swarbrick Daylight Factor Theodolite. British Patent No. 356.822 (1929)
Swarbrick, J.: Easements of light. Vol.II. Batsford Ltd., London, Wykeham Press, Manchester (1933)
Waldram, P.J. and Waldram, J.M.: Penetration of daylight and sunlight into buildings. Illum. Engineer, **16**, 90–95 and Illum. Res. Techn. Paper 7, HMSO, London (1923, 1932)
Waldram, P.J.: The transmission of light through window glass: A new form of daylight measuring diagram for glazed openings. Journal RIBA, **43**, 1072–1077 (1936)
Waldram, P.J.: The approximate measurement of daylight. Journ. Roy. Inst. of British Architects, **53**, 7, 290–294 (1946)
Waldram, P.J.: A measuring diagram for daylight illumination. Batsford Ltd., London (1950)
Walsh, J.W.T.: The science of daylight. MacDonald, London (1961)
Wiener, C.: Lehrbuch der darstellenden Geometrie. Band I. B.G. Taubner Verlag, Leipzig (1884)

References to Appendix

B.R.S. – Building Research Station : Daylight Factor Protractors. H.M.S.O. (1944)
CIE – Commission Internationale de l'Éclairage: Proc. CIE General Session, V.3, p.12 (1939)
CIE – Commission Internationale de l'Éclairage: Natural Daylight, Official recommendation. Compte Rendu CIE 13th Session, 2, part 3.2, 2–4 (1955)
CIE – Commission Internationale de l'Éclairage: Daylight. Technical Report, Publ. No.CIE 16, Central Bureau CIE, Vienna (1970)
Danilyuk, A.M.: Novyy metod raschota yestestvennogo osveshcheniya ot pryamougolnych svetoproyomov. (In Russian: A new calculation method of natural lighting from rectangular apertures.) Trudy II. Vsesoyuznoy svetotechnicheskoy konf. IV. Doklady, 168–180 (1931)
Daniljuk, A.M.: Ein neues graphisches Verfahren zur Berechnung der natürlichen Beleuchtung von Räumen mit rechteckigen Lichtöffnungen. Das Licht, 4, 241–243, 5, 16–20 (1934, 1935)
Daniljuk, A.M.: Diagrammy dlya raschota osveshchenosti ot svetoproyomov proizvolnogo ochertaniya i naklona.(In Russian: Diagrams for illuminance calculation from apertures of arbitrary contour and slope). Svetotechnika, **6**, 6, 7–9 (1935)
Danilyuk, A.M.: Raschot yestestvennogo osveshcheniya pomeshcheniy. (In Russian: Calculation of natural lighting of interiors.), Stroyizdat, Moscow, Leningrad (1941)

Dufton, A.F.: The computation of Daylight Factors in factory design. Journ. Sci. Instr., **17**, 9, 226–227 (1940)

Dufton, A.F.: Protractors for the computation of Daylight Factors. B.R.S. Techn. Paper 28, HMSO (1946)

Higbie, H.H., Turner-Szymanowski, W.: Calculation of daylight and indirect artificial lighting by protractor method. Transac. IES (N.Y.), **25**, 3, 213–261 (1930)

Higbie, H.H.: Lighting calculations. J.Wiley Inc., New York (1934)

ISO – International Standardisation Organisation: Spatial distribution of daylight – CIE Standard General Sky. ISO Standard 15409:2004 (2004)

Kittler, R. and Kittlerová, L.: 1968. Návrh a hodnotenie denného osvetlenia. (In Slovak: The design and evaluation of daylight). Slovak Publ. of Techn. Liter., Bratislava (1968, 1975)

Kittler, R. and Darula, S.: The method of aperture meridians : A simple calculation tool for applying the ISO/CIE Standard General Sky. Light. Res. and Technol., **38**, 2, 109–119 (2006)

Kota, S. and Haberl, J.S.: Historical Survey of daylighting calculations methods and their use in energy performance simulations. Proc. Intern. Conf. for enhanced building operations, ESL-IC-09-11-07, Austin (2009)

Moon, P.: The scientific basis of illuminating engineering. McGraw-Hill Inc., New York, London (1936. 1961)

Petherbridge, P.: C.I.E. draft guide to natural daylight calculation, part I – Overcast sky conditions. Private document of CIE W-3.2 Committee (1959)

Roy, G.G., Kittler R., Darula S.: An implementation of the Method of Aperture Meridians for the ISO/CIE Standard General Sky. Light. Res. and Technol., **39**, 3, 253–264 (2007)

Ubbelohde, M. S. and Humann, C.: Comparative evaluation of four daylighting software programs.Proc. ACEE Summer Study on energy efficiency in Buildings (1998)

Waldram, P.J. and Waldram, J.M.: Penetration of daylight and sunlight into buildings. Illum. Engineer, 16, 90–95 and Illum. Res. Techn. Paper 7, HMSO, London (1923, 1932)

Waldram, P.J.: The approximate measurement of daylight. Journ. Roy. Inst. of British Architects, **53**, 7, 290–294 (1946)

Waldram, P.J.: A measuring diagram for daylight illumination. Batsford Ltd., London (1950)

Chapter 9
Daylight Methods and Tools to Design Glazed Windows and Skylights

9.1 Light Transmission through Glazing Materials

The general trend in recent building has been to use glass much more than in the past because of its lower price and modern technology enabling larger windows in durable frames. Historically, these technology advancements of the Industrial Revolution were demonstrated by the cast iron and glass building of the Crystal Palace in London housing the 1851 Great Exhibition. This represented the starting era of glass architecture including the novel modular design, panel prefabricated construction and also documenting problems of sunshine overheating and of thermal insulation and water vapor condensation as well as questions of visual privacy.

The quality of the various types of glazing materials is important if full benefit is to be taken. For daylighting purposes, transparent glass is usually used. The two main types are known as sheet glass and polished plate, both with very good transmittance when clean: for normal beams the normal transmittance is roughly 90% and for diffuse skylight approximately 85% of incident light is transmitted. Partly translucent glasses such as rough-cast glass or rough-cast wired glass for industrial buildings may be reinforced with either rectangular or honeycomb wire nets and have lower transmittance, about or even lower than 80% is normal.

It is rarely realized that a glazed aperture is a selective glass filter which transmits, reflects, and absorbs incident flux preferentially. This means that any luminance seen by an arbitrary glass element within the whole outer half-space will be reduced in accordance with its direction in relation to the glass normal. Of course, the normal to surface transmittance is maximal; therefore, the orientation and slope of glazing (directional transmittance) has to be taken into account. As a consequence, all vertical glazed windows will transmit most effectively the horizon luminance, which is normally facing the aperture, whereas top lights with horizontal glazing will maximally transmit zenith sky luminance.

The incidence angle taken from the glazing normal is the influencing parameter which determines the directional transmittance τ_{ig}, which can be normalized by the maximum transmittance in the normal direction τ_n, i.e., τ_{ig}/τ_n.

R. Kittler et al., *Daylight Science and Daylighting Technology*,
DOI 10.1007/978-1-4419-8816-4_9, © Springer Science+Business Media, LLC 2012

Light passing through the glass is also refracted, but this factor is usually ignored as it does not influence the view outside. Skylight luminance and sunbeam transmitted flux reaching from the upper hemisphere are predominantly directed downward into interiors. However, as soon as the glazing has a broken surface, is ornamented, or is diffuse, the light scattering and redirection can cause serious conditions of glare in spite of the fact that the amount of transmitted light is actually reduced. A considerable, and negative, difference on visual comfort is to be expected when materials penetrable by light are either translucent or partly transparent (hammered or ornamental glass) or fully diffusing, (e.g., acid-etched, single-sided or double-sided sandblasted glass). Light transmittance τ, reflectance ρ, and absorption α losses follow a three-component relation as

$$\frac{\Phi_\rho}{\Phi_\mathrm{t}} + \frac{\Phi_\alpha}{\Phi_\mathrm{t}} + \frac{\Phi_\tau}{\Phi_\mathrm{t}} = \rho + \alpha + \tau = 1, \tag{9.1}$$

where Φ_ρ is the luminous flux reflected, Φ_α is the luminous flux absorbed, Φ_τ is the luminous flux transmitted, and Φ_t is the total luminous flux falling on the glass surface.

Absorption losses are visually unnoticed; therefore, (9.1) can be reduced to only two components:

$$\frac{\Phi_\rho}{\Phi_\mathrm{t}} + \frac{\Phi_\tau}{\Phi_\mathrm{t}} = \rho + \tau = 1 \quad \text{or} \quad \tau = 1 - \rho\,. \tag{9.2}$$

Transparent materials such as glass and Plexiglas are characterized by normal transmittance, i.e., maximal beam flow in a direction perpendicular to the material surface, which for clean clear glass of 2–3-mm thickness is roughly $\tau_\mathrm{n} = 0.9$–0.92. With respect to the beam incidence angle, the path through the penetrated material thickness is elongated; thus, transmittance is reduced.

Laboratory measurements of glass transmittance of different commercial glasses (Taylor and Grieveson 1926; Waldram 1929) demonstrated other influences which have to be considered, for instance, dirtiness, tarnish, spectral tint, surface irregularities, and the effects of a difference in the light source characteristics were stressed, i.e., whether transmitted from parallel beams (sunlight) or diffused light (skylight) (Walsh 1926). Different glass types were measured routinely (D.S.I.R 1932, 1936) for both directional and diffuse transmittance characteristics, but surprisingly no applications of daylight calculation tools were introduced except by Waldram (1936), who realized that some critics rightly considered his diagrams only to be accurate for unglazed openings and tried to correct them. He suggested the directional reduction for oblique transmittance should follow the curve shown in Waldram (1936). The transmittance is reduced with different beam incidence angles taken from the glass normal, at first only slightly but when the incidence angle is over 40°, the reduction is considerable.

The original theoretical study based on refraction by Holmes (1947) separately determined the two fluxes in (9.2) for the clear untarnished glass surfaces. In the seventeenth century the refraction of light within different transparent layers had been found to result in gradual small changes in the direction of the transferring light beam, especially noticeable at the surface of the layers owing to the difference in the light speed in two differently dense environments. Snell in Holland discovered the law of refraction in 1618 and later Descartes in France determined its formulation in terms of sine angles for oblique light beams.

As the absorbed luminous flux in clear glass is almost absent, the direct interrelation of reflection and transmission is evidently $\tau = 1 - \rho$ and is dependent on the refractive index of glass ($n = 1.5$) for a single air–glass surface under the normal flux incidence

$$\rho_{n1} = \frac{(n-1)^2}{(n+1)^2} = \frac{(1.5-1^2)}{(1.5+1^2)} = 0.04\,, \tag{9.3}$$

$$\tau_{n1} = \frac{4n}{(n+1)^2} = \frac{6}{(1.5+1^2)} = 0.96\,, \tag{9.4}$$

but for two glass surfaces the reflection factor ρ_{n2} is less than twice ρ_{n1}, i.e.,

$$\rho_{n2} = \frac{2\rho_{n1}}{1+\rho_{n1}} = \frac{0.08}{1.04} = 0.077 \tag{9.5}$$

and

$$\tau_{n2} = \frac{\tau_{n1}}{2-\tau_{n1}} = \frac{0.96}{1.04} = 0.923. \tag{9.6}$$

However, if the luminous flux is falling on the glass surface under oblique incidence, then generally

$$n = 1.5 = \frac{\sin i}{\sin\psi}\,, \tag{9.7}$$

where i is the beam incidence angle on the air side and ψ is the angle of the transmitted beam behind the glass surface taken from its normal, and which after (9.7) can be represented as $\sin\psi = \frac{\sin i}{1.5}$.

Thus, a more complicated refraction distribution between the reflection and transmission processes takes place, resulting in the transmittance component relations for two surfaces after the classical Fresnel formula:

$$\tau_\psi = 1 - \rho_\psi = 1 - \frac{1}{2}\left[\frac{\sin(i-\psi)}{\sin(i+\psi)}\right]^2 + \frac{1}{2}\left[\frac{\tan(i-\psi)}{\tan(i+\psi)}\right]^2. \tag{9.8}$$

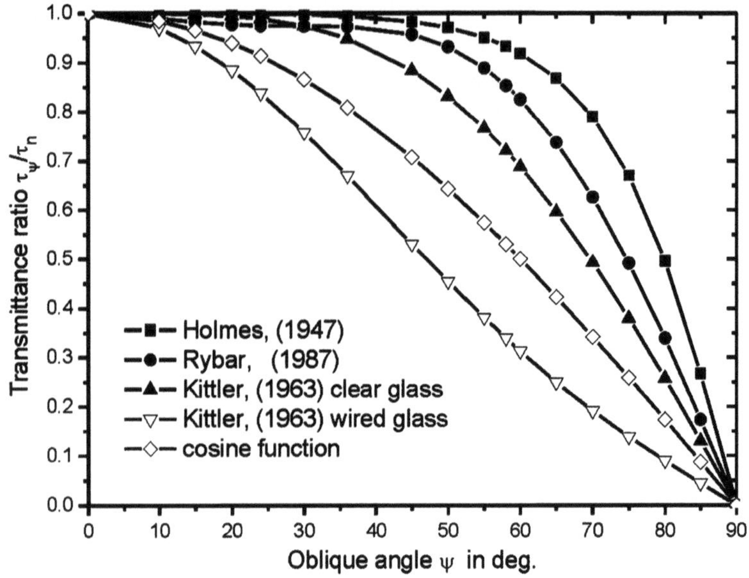

Fig. 9.1 Comparison of different oblique transmittances defined by simple formulae

Holmes (1947) produced tables showing the reduction of glass transmittance for oblique beam angles; these are reproduced in Fig. 9.1.

This relation was later expressed by simpler best-fit formulae:

- By Rivero (1958)

$$\frac{\tau_\psi}{\tau_n} = 1.018\tau_r \left(\cos\psi + \sin^3\psi \, \cos\psi\right), \tag{9.9}$$

where the reference $\tau_r = 0.8$, i.e., $1.018\tau_r = 0.8144$.

- By Kittler (1963) in two simple positive and negative variations of the ψ dependence, which enabled simplified integrations, i.e.,

  – For plain clear glass,

$$\frac{\tau_\psi}{\tau_n} = \cos\psi \left(1 + \frac{1}{2}\sin^2\psi\right). \tag{9.10}$$

  – For rough-cast wired glass

$$\frac{\tau_\psi}{\tau_n} = \cos\psi \left(1 - \frac{1}{2}\sin^2\psi\right). \tag{9.11}$$

- By Rybár (1987), who tried to express the dirtiness of sheet glass by its directional influences in the relation

$$\frac{\tau_\psi}{\tau_n} = \cos\psi\left(1+\sin^3\psi\right) + e^{-(\psi/27)} + e^{-(\psi/26)}, \qquad (9.12)$$

which can be simplified by disregarding the small exponential influences to

$$\frac{\tau_\psi}{\tau_n} = \cos\psi\left(1+\sin^3\psi\right). \qquad (9.13)$$

Equation (9.13) resembles well the Rivero equation (9.9) except for his reference constant 0.8144.

The simpler best-fit formulae are all reproduced in Fig. 9.1, where Rybar's and Rivero's curves can be taken as identical, and Kittler's curve after (9.10) includes some dirt effects rising with the oblique angles. The wired glasses because of the mesh behave differently (open marks in Fig. 9.1), whereas because of the quality of the rough-cast glass and the mesh density the transmittance drop is in the region from around the cosine function, as determined by Randall and Martin (1930), to that determined by Kittler (1963) for thicker rough-cast glasses used in industrial buildings.

These losses are valid for one pane of glass but for several glasses with the same thickness and transparency the same directional losses occur, lowered by their multiplied normal transmittance, i.e., for double-glazed windows ${\tau_n}^2 = 0.92^2 = 0.846$ and for triple glazing ${\tau_n}^3 = 0.92^3 = 0.7787$.

In addition to sky luminance changes, directionally changing transmittance characteristics of glazing had to be introduced into the integration formula (A7.13) within the solid angle elements together with the luminance changes. Therefore, only the simplest transmittance formulae (9.10) and (9.11) allowed the classical integration procedure to be performed.

Sometimes diffuse transmittance is applied as a single value either averaged for a half-sphere sky component or also specified showing its variations with incidence angle. In general, the diffuse transmittance τ_d is defined by an integral:

$$\tau_d = \tau_n \int_0^\pi \sin\psi\,\cos\psi\left(\frac{\tau_\psi}{\tau_n}\right)d\psi. \qquad (9.14)$$

The net transmittance of a particular aperture or skylight depends on the multiple layering, shape, i.e., whether flat or domed, the presence of shades or louvers, etc.

In the case of domed skylights with a decreasing thickness of the sheet at the center, IESNA (2000) recommends modification of the dome transmittance τ_{DM} with respect to flat-sheet transmittance τ_{FS}:

$$\tau_{DM} = 1.25\,\tau_{FS}\left(1.18 - 0.416\,\tau_{FS}\right). \qquad (9.15)$$

Contemporary skylight cupolas are also double-domed and often with a transparent dome over a translucent one. Then the overall transmittance of the whole unit with interreflection between them after Kreider and Kreith (1982) is

$$\tau = \frac{\tau_1 \tau_2}{1 - \rho_1 \rho_2}, \tag{9.16}$$

where τ_1 and τ_2 are transmittances and ρ_1 and ρ_2 are reflectances of the first and second domes, respectively.

9.2 Calculation Methods Valid for Vertical Glazed Windows Illuminating Horizontal Planes

The same integration procedure for unglazed rectangular apertures in (8.25) and (8.26) can be followed, but including also the directional characteristics of glazing. To avoid difficulties with an arbitrary placing of the window aperture with respect to the horizontal illuminated plane, the simplest case in Fig. 9.2 is theoretically solved in mathematical terms.

Then the basic relation to be integrated after CIE (1955) overcast sky gradation for transparent glazing (TG) with ordinary window glass (either single or double panes) is the sky factor:

$$\mathrm{SF}(1:3)(\mathrm{TG}) = \frac{3\tau_n}{7\pi} \int\limits_{S_1} (1 + 2\cos Z) \frac{\cos^2\psi \cos Z}{l^2} \left(1 + \frac{1}{2}\sin^2\psi\right) \mathrm{d}S_1, \tag{9.17}$$

which is transformed for the rough-cast wired glass (WG) used in industrial buildings to

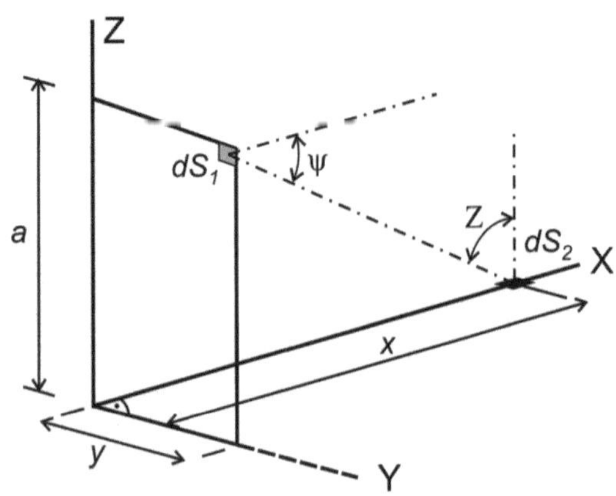

Fig. 9.2 A rectangular window without the sill

$$\mathrm{SF}(1:3)(\mathrm{WG}) = \frac{3\tau_\mathrm{n}}{7\pi} \int\limits_{S_1} (1 + 2\cos Z) \frac{\cos^2\psi \cos Z}{l^2} \left(1 - \frac{1}{2}\sin^2\psi\right) \mathrm{d}S_1. \quad (9.18)$$

These integrals as well as similar ones for the sky luminance gradation SF(1:2) (TG) and SF(1 : 2)(WG) were treated classically for an arbitrarily sloped rectangular aperture on a vertical sill by Kittler and Ondrejička (1963, 1966), with results published by Kittler and Ondrejička (1963) and in Kittler and Kittlerova (1968, 1975, pp. 148–152) with derived graphs and protractors (pp. 154–161 and appendices). Because of new computer possibilities, the numerical calculations were programmed and processed and afterward summarized in tables for a chosen set of two window parameters:

1. For ratios $L = a/x$ and $N = y/x$
2. For width parameter $\check{S} = y/a$ and for window-point distance parameter $D = x/a$

In both cases a is the window height dimension for window without a sill, y is the window width, and x is the distance of the illuminated horizontal element from the window.

In the case of vertical windows, four tables and four diagrams resulted for SF (1:3) (TG) and SF(1 : 3)(WG) as well as for SF(1 : 2)(TG) and SF(1 : 2)(WG), and were published in Kittler and Ondrejička (1964) as well as later in Kittler and Kittlerova (1968, 1975).

The simple placement of the vertical rectangular opening with its right downward corner in the zero center of the coordinate system has several advantages:

- The window distance from the illuminated horizontal element is taken from the zero point as x, the window width is y, and its height from the horizontal plane is $z = a$.
- The overcast sky luminance gradation is defined from the horizontal level that can be imagined superimposed on the illuminated horizontal plane.
- The directional transmittance is continually effective and increased from the normal to the window opening, i.e., from the right downward corner upward and sideways.
- Because of this placement, the simplest and shortest calculation formula is expected in spite of both sky gradation and directional transmittance relations being taken into account.
- Because only the simple overcast gradations 1:3 and 1:2 defined by $\cos Z$ functions and the simple directional transmittance relations for transparent glass with only $\pm\, 1/2\sin^2\psi$ difference in (9.17) and (9.18) were considered, a simplified classical integration can be applied.
- Although the resulting formula determines only the special placement of the opening, it can be used several times in the case of arbitrary window positions because the superposition relations in Fig. 9.3 exist.
- Because of several similar partial subcomponents in the solution, the final formulae for SF(1 : 3)(TG), SF(1 : 3)(WG), SF(1 : 2)(TG), and SF(1 : 2) (WG) can be expressed for computer programs in the following forms in percent:

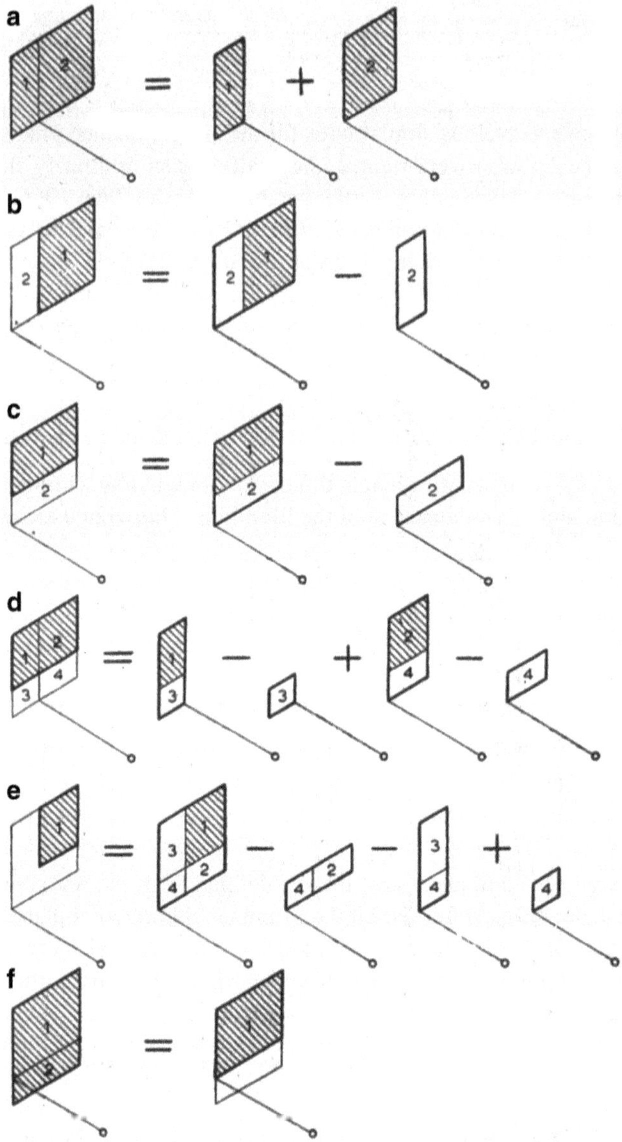

Fig. 9.3 Superposition relations for rectangular vertical windows

$$\mathrm{SF}(1:3)(\mathrm{TG}) = 0.28421\tau_{\mathrm{n}}[\mathrm{W1} + \mathrm{X1} - M - \mathrm{Y1} - \mathrm{Z1}]\,(\%), \tag{9.19}$$

$$\mathrm{SF}(1:3)(\mathrm{WG}) = 0.28421\tau_{\mathrm{n}}[\mathrm{W2} + \mathrm{X2} + M + \mathrm{Y2} - \mathrm{Z2}]\,(\%), \tag{9.20}$$

$$\mathrm{SF}(1:2)(\mathrm{TG}) = 0.19894\tau_{\mathrm{n}}[\mathrm{W1} + \mathrm{X3} - M - 2\mathrm{Y1} - \mathrm{Z1}]\,(\%), \tag{9.21}$$

$$\mathrm{SF}(1:2)(\mathrm{WG}) = 0.19894\tau_{\mathrm{n}}[\mathrm{W2} + \mathrm{X4} + M + 2\mathrm{Y2} - \mathrm{Z2}]\,(\%). \tag{9.22}$$

Of course, all these subcomponents are to be programmed in the form of relative coordinates, e.g., in ratios $L = a/x$ and $N = y/x$, so

$$\mathrm{W1} = \frac{L^3(13 + 16L^2)}{(1 + L^2)^{5/2}} \arctan\frac{N}{\sqrt{1 + L^2}}, \tag{9.23}$$

$$\mathrm{W2} = \frac{L^3(11 + 8L^2)}{(1 + L^2)^{5/2}} \arctan\frac{N}{\sqrt{1 + L^2}}, \tag{9.24}$$

$$\mathrm{X1} = \frac{N}{(1 + N^2)^{3/2}}\left[(15 + 16N^2)\arctan\frac{L}{\sqrt{1 + N^2}} + 1.6(12 + 13N^2)\right], \tag{9.25}$$

$$\mathrm{X2} = \frac{N}{(1 + N^2)^{3/2}}\left[(9 + 8N^2)\arctan\frac{L}{\sqrt{1 + N^2}} + 1.6(8 + 7N^2)\right], \tag{9.26}$$

$$\mathrm{X3} = \frac{N}{(1 + N^2)^{3/2}}\left[(15 + 16N^2)\arctan\frac{L}{\sqrt{1 + N^2}} + 3.2\,(12 + 13N^2)\right], \tag{9.27}$$

$$\mathrm{X4} = \frac{N}{(1 + N^2)^{3/2}}\left[(9 + 8N^2)\arctan\frac{L}{\sqrt{1 + N^2}} + 3.2\,(8 + 7N^2)\right], \tag{9.28}$$

$$M = \frac{N}{L(1 + N^2)}, \tag{9.29}$$

$$\mathrm{Y1} = \frac{N}{(1 + L^2)^2}\left[1.6\frac{N^2 - 3(1 + L^2 + N^2)(4 + 5L^2)}{(1 + L^2 + N^2)^{3/2}}\right], \tag{9.30}$$

$$\mathrm{Y2} = \frac{N}{(1 + L^2)^2}\left[1.6\frac{N^2 + 3(1 + L^2 + N^2)(8 + 5L^2)}{(1 + L^2 + N^2)^{3/2}}\right], \tag{9.31}$$

$$\mathrm{Z1} = \frac{(1 + L^2 + N^2)(1 - 12L^2 - 18L^4) - 2L^2N^2}{L(1 + L^2 + N^2)^2}, \tag{9.32}$$

$$\mathrm{Z2} = \frac{(1+L^2+N^2)(1+12L^2+6L^4)-2L^2N^2}{L(1+L^2+N^2)^2}. \tag{9.33}$$

In the case of an endlessly high aperture in one direction from the zero point, i.e., if $L \to \infty$, e.g.,

$$\mathrm{SF}(1:3)(\mathrm{TG}) = 0.28421\tau_\mathrm{n}\left\langle \frac{N}{(1+N^2)^{3/2}}\left[\begin{matrix}(15+16N^2)\frac{\pi}{2}+\\ +1.6(12+13N^2)\end{matrix}\right]\right\rangle (\%), \tag{9.34}$$

and, in the extreme case of the whole quarter of the sky hemisphere, $N \to \infty$ and $L \to \infty$, and a further simplification results in

$$\mathrm{SF}(1:3)(\mathrm{TG}) = 0.28421\tau_\mathrm{n}(8\pi+20.8) = 13.055\tau_\mathrm{n}\,(\%), \tag{9.35}$$

$$\mathrm{SF}(1:3)(\mathrm{WG}) = 0.28421\tau_\mathrm{n}(4\pi+11.2) = 9.755\tau_\mathrm{n}\,(\%), \tag{9.36}$$

$$\mathrm{SF}(1:2)(\mathrm{TG}) = 0.19894\tau_\mathrm{n}(8\pi+41.6) = 13.275\tau_\mathrm{n}\,(\%), \tag{9.37}$$

$$\mathrm{SF}(1:2)(\mathrm{WG}) = 0.19894\tau_\mathrm{n}(4\pi+22.4) = 6.956\tau_\mathrm{n}\,(\%). \tag{9.38}$$

This means that in the first case the virtual vertical glazed aperture with normal transmittance $\tau_\mathrm{n} = 1$ will cover the whole quarter of the sky hemisphere with luminance gradation 1:3, but yields the SF(1 : 3)(TG) value of 13.054% i.e., under the whole sky yields 52.22%. Of course, any real normal transmittance of the actual glazing will reduce these values accordingly. Similar tables were given by Hopkinson et al. (1958) and Hopkinson et al. (1966), where in Table 5.1 using the ideal Holmes directional transmittance curve SF(1 : 3)(TG) = 15% is shown and thus under the whole sky the value is 60%.

In actual cases, the solution of the general (9.19)–(9.22) with the ratios $L = a/x$ and $N = y/x$ or $\check{S} = Y/a$ and $D = x/a$ as well as using all parametric components W1, W2, X1, X2, M, Y1, Y2, Z1, and Z2 after (9.23)–(9.33) was a time-consuming and difficult computation task for even a computer. Therefore, Puškaš (1979) tried to approximate Danilyuk's (8.21) by applying the elevation and aperture width angle for the window center only, thus simplifying the formulae for use of a Texas Instruments calculator, using the redenoted vertical window ratios $d = L = a/x$ and $s = N = y/x$ and an additional constant:

$$k = 5.44 - 1.2\,\exp\left(-\frac{1+s}{d}\right). \tag{9.39}$$

So, only four simple angular parameters were needed for simple calculators:

$$m = \cos^2 \frac{\arc \tan d}{2}, \tag{9.40}$$

$$n = \sin \arc \cos m, \tag{9.41}$$

$$p = \sin \arc\ \tan(s\ m), \tag{9.42}$$

$$r = \cos \arc \sin \frac{p}{1.6}. \tag{9.43}$$

With these simplified parameter programs, all sky components of the daylight factor could be determined:

$$\mathrm{SF}(1:3)(\mathrm{TG}) = \tau_\mathrm{n}\, k\, m\, p\, r(0.5 + 3nr)\left(3 - m^2r^2\right)(1 - m)\left(5 - p^2\right)\ (\%), \tag{9.44}$$

$$\mathrm{SF}(1:3)(\mathrm{WG}) = -\tau_\mathrm{n}\, k\, m\, p\, r(0.5 + 3nr)\left(-1.1 - m^2r^2\right)(1 - m)\left(5 - p^2\right)\ (\%), \tag{9.45}$$

$$\mathrm{SF}(1:2)(\mathrm{TG}) = 0.71\tau_\mathrm{n}\, k\, m\, p\, r(1 + 3nr)\left(3 - m^2r^2\right)(1 - m)\left(5 - p^2\right)\ (\%), \tag{9.46}$$

$$\mathrm{SF}(1:2)(\mathrm{WG}) = -0.71\tau_\mathrm{n}\, k\, m\, p\, r(1 + 3nr)\left(-1.1 - m^2r^2\right)(1 - m)\left(5 - p^2\right)\ (\%). \tag{9.47}$$

9.3 Application of Graphical Tools for Window Design

Window design or evaluations of skylight behind glazed openings using the methods mentioned in Chap. 8 are often corrected only by the normal glass transmittance without taking into account the losses caused by the directional reductions of transmittance due to higher or wider incidence angles $i = \psi$. In the USA Randall and Martin (1930) measured the glazed window luminance decrease with the window width and with a different tendency when clear and hammered/ wire rough glass with the same normal transmittance was used. In their fenestra method and tables for daylighting predetermination, they proposed and used instead of the theoretical directional transmittance after Fresnel (9.8) a lower reduction tendency following the ψ angles from the normal which was close to that in (9.10), whereas for wired rough glazing their curve closely followed the $\cos\psi$ reduction (Kittler 1964). In the same article (Randall and Martin 1930) a proposal was made for correcting the glass transmittance due to collected dirt on glazing which is dependent on the glass slope and the washing frequency in months.

The first graphical tool corrected for the directional glass transmittance was published by Waldram (1936), who proposed the correction of the droop-line system either by reducing the droop-line net both vertically and horizontally or so as to "confine the necessary correction to vertical dimensions only." It is evident that the first correction corresponds to the spherical $i = \psi$ influence, but would alter the regular azimuth scale of the original Waldram diagram and thus would destroy the easy insetting of window and obstruction contours. As Waldram (1936) probably realized also the more important directional reduction of transmittance of vertical rays from the sky penetrating a vertical window, he recommended the second way of correcting the droop-line system.

A notable tool was luminous intensity azimuthal multiplier (LIAM) diagrams (MacGowan 1973), which incorporated precise glass transmission correction, droop lines, and respected the proposed (then later standardized) International Commission on Illumination (CIE) clear sky azimuthal luminance distributions. These diagrams were constructed by computer summation over a multiple of 32,400 elemental arrays. However, their publication was restricted because they were research tools, never designed for practice, only to show graphically to the interested CIE TC4.2 group the influence of the sky luminance distribution, glazing type, and orientation on daylighting prediction for the new clear sky distributions with indicatrices by Kittler and Gusev and the tropical cosec distribution and, the glass transmittance as measured and supplied by Petherbridge and by Kittler's and Rivero's formulae.

9.4 Possibilities to Predict Skylight from Inclined Rectangular Openings

The general formulae for SF(1 : 3)(TG), SF(1 : 3)(WG) in (9.17) and (9.18) assuming CIE overcast sky luminance gradation 1:3 as well as those for SF(1 : 2) × (TG) and SF(1 : 2)(WG) for overcast sky luminance gradation 1:2 when the ground is covered with snow are valid also for sloped apertures illuminating a horizontal plane or element placed on the lowest line of the sloped aperture in Fig. 9.4. In contrast to the vertical window position (Fig. 9.2), the inclination of the aperture area presents further complications as mentioned already in Chap. 8, but

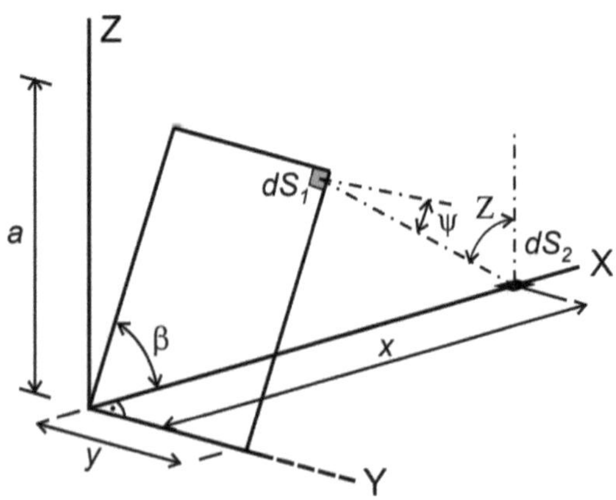

Fig. 9.4 The inclination of a rectangular aperture causes changes in influencing angles

additionally the indirect transmittance has to be taken into account when the angle from the window glass normal is changed in accordance with the glazing slope:

$$\cos\psi = \frac{x \sin\beta}{l}, \tag{9.48}$$

where

$$l = \sqrt{(x - a\cos\beta)^2 + (a\sin\beta)^2 + y^2}\ (\mathrm{m}). \tag{9.49}$$

However, the superposition relations are valid for the sloped aperture too, and therefore the sloped aperture placed at the zero point of the axis system by its right corner presents the simplest and basic case. Assuming the normal transmittance $\tau_n = 1$ and the window width parameter $\check{S} = y/a$ with the window-point distance parameter $D = x/a$, the resulting formulae are more complex because of the window slope β but are similar to (9.19)–(9.22):

$$\mathrm{SF}(1:3)(\mathrm{TG}) = 0.28421\tau_n[\mathrm{R1} + \mathrm{S1} + \mathrm{T1} + \mathrm{U1} + \mathrm{V1}]\ (\%), \tag{9.50}$$

$$\mathrm{SF}(1:3)(\mathrm{WG}) = 0.28421\tau_n[\mathrm{R2} + \mathrm{S2} + \mathrm{T2} + \mathrm{U2} + \mathrm{V2}]\ (\%), \tag{9.51}$$

$$\mathrm{SF}(1:2)(\mathrm{TG}) = 0.19894\tau_n[\mathrm{R1} + \mathrm{S1}/2 + \mathrm{T1}/2 + \mathrm{U1}/2 + \mathrm{V1}]\ (\%), \tag{9.52}$$

$$\mathrm{SF}(1:2)(\mathrm{WG}) = 0.19894\tau_n[\mathrm{R2} + \mathrm{S2}/2 + \mathrm{T2}/2 + \mathrm{U2}/2 + \mathrm{V2}]\ (\%). \tag{9.53}$$

So, for any arbitrary slope β further simplified parameters were used:

$$B = \frac{\check{S}}{\sqrt{1 + D^2 + \check{S} - 2D\cos\beta}}, \tag{9.54}$$

$$C = \frac{\check{S}}{\sqrt{D^2 + \check{S},}} \tag{9.55}$$

$$F = 1 - D\cos\beta, \tag{9.56}$$

$$G = 1 + D^2 - 2D\cos\beta, \tag{9.57}$$

$$Q = \frac{\check{S}}{\sqrt{D^2\sin^2\beta + \check{S}^2}}. \tag{9.58}$$

The resulting components cited by Kittler and Kittlerová (1968, 1975) are in the final solutions applied subcomponents as follows:

$R1 = R01(\cos\beta/15)$ and $R2 = R02(\cos\beta/15)$, $S1 = S01(\cos^2\beta/48)$ and $S2 = S02(\cos^2\beta/48)$, $T1 = T01(\sin^2\beta/48)$ and $T2 = T02(\sin^2\beta/48)$, $U1 = U01((D^3\sin^2\beta\sin^3\beta)/8)$ and $U2 = U02((D^3\sin^2\beta\sin^3\beta)/8)$, $V1 = V01(D^2\sin^3\beta/30)$ and $V2 = V02(D^2\sin^3\beta/30)$. Then,

$$\begin{aligned} R01 = \; & 6\left[\arctan\frac{BF}{D\sin\beta} + \arctan(C\cot\beta)\right] \\ & + Q^4 D\sin\beta\left[\frac{1}{BF} + \frac{1}{CD\cos\beta}\right] \\ & - \frac{BD^3\sin^3\beta}{2}\left(\frac{15G^2 - 16D^2G\sin^2\beta + 2D^4\sin^4\beta}{F^3G^2}\right) \\ & - \frac{C\tan^3\beta(15 - 16\sin^2\beta + 2\sin^4\beta)}{2} \\ & + \frac{Q^2D\sin\beta}{2}\left(\frac{10G - 9D^2\sin^2\beta}{BF^2} + \frac{10 - 9\sin^2\beta}{CD\cos^3\beta}\right) \\ & + \frac{D^5\sin^5\beta}{2\check{S}^2}\left(\frac{B^3}{FG} + \frac{C^3}{D^3\cos\beta}\right), \end{aligned} \tag{9.59}$$

$$\begin{aligned} R02 = \; & 4\left[\arctan\frac{BF}{D\sin\beta} + \arctan(C\cot\beta)\right] - Q^4D\sin\beta\left[\frac{1}{BF} + \frac{1}{CD\cos\beta}\right] \\ & - \frac{BD^3\sin^3\beta}{2}\left(\frac{5G^2 - 4D^2G\sin^2\beta - 2D^4\sin^4\beta}{F^3G^2}\right) \\ & - \frac{C\tan^3\beta(5 - 4\sin^2\beta - 2\sin^4\beta)}{2} \\ & + \frac{Q^2D\sin\beta}{2}\left(\frac{10G - 11D^2\sin^2\beta}{BF^2} + \frac{10 - 11\sin^2\beta}{CD\cos^3\beta}\right) \\ & - \frac{D^5\sin^5\beta}{2\check{S}^2}\left(\frac{B^3}{FG} + \frac{C^3}{D^3\cos\beta}\right), \end{aligned} \tag{9.60}$$

$$\begin{aligned} S01 = \; & \frac{F(28G^2 + 14D^2G\sin^2\beta - 3D^4\sin^4\beta)}{G^{5/2}}\arctan\frac{\check{S}}{\sqrt{G}} \\ & + \cos\beta(28 + 14\sin^2\beta - 3\sin^4\beta)\arctan\frac{\check{S}}{\sqrt{G}} \\ & + Q\left[28 + \frac{D^2\sin^2\beta(11Q^2 + 3Q^4)}{\check{S}^2}\right]\left[\arctan\frac{QF}{\check{S}} + \arctan\frac{QD\cos\beta}{\check{S}}\right] \\ & + \frac{Q^2D^2\sin^2\beta}{\check{S}}\left[\frac{14G - 13D^2\sin^2\beta}{F^3} + \frac{14 - 13\sin^2\beta}{D\cos^3\beta}\right] \\ & - \frac{D^4\sin^4\beta}{\check{S}}\left[\frac{B^2(18G^2 - 20D^2G\sin^2\beta + 3D^4\sin^4\beta)}{F^3G^2} + \frac{C^2(18 - 20\sin^2\beta + 3\sin^4\beta)}{D^3\cos^3\beta}\right] \\ & + \frac{2D^6\sin^6\beta}{\check{S}^3}\left[\frac{B^4}{FG} + \frac{C^4}{D^3\cos\beta}\right] - \frac{3Q^4D^3\sin^4\beta}{F\check{S}^3\cos\beta}, \end{aligned} \tag{9.61}$$

$$
\begin{aligned}
\mathrm{S02} = {} & \frac{F\left(20G^2 + 10D^2G\sin^2\beta + 3D^4\sin^4\beta\right)}{G^{5/2}} \arctan\frac{\check{S}}{\sqrt{G}} \\
& + \cos\beta\left(20 + 10\sin^2\beta + 3\sin^4\beta\right) \arctan\frac{\check{S}}{\sqrt{D}} \\
& + \frac{Q}{\check{S}}\left[20\,\check{S}^4 + 10\,\check{S}^2 Q^2D^2\sin^2\beta + 3Q^4D^4\sin^4\beta\right]\left[\arctan\frac{QF}{\check{S}} + \arctan\frac{QD\cos\beta}{\check{S}}\right] \\
& + \frac{Q^2D^2\sin^2\beta}{\check{S}}\left[\frac{10G - 11D^2\sin^2\beta}{F^3} + \frac{10 - 11\sin^2\beta}{D\cos^3\beta}\right] \\
& - \frac{D^4\sin^4\beta}{\check{S}}\left[\frac{B^2\left(6G^2 - 4D^2G\sin^2\beta - 3D^4\sin^4\beta\right)}{F^3G^2} + \frac{C^2\left(6 - 4\sin^2\beta - 3\sin^4\beta\right)}{D^3\cos^3\beta}\right] \\
& - \frac{2D^6\sin^6\beta}{\check{S}^3}\left[\frac{B^4}{FG} + \frac{C^4}{D^3\cos\beta}\right] + \frac{3Q^4D^3\sin^4\beta}{F\check{S}^3\cos\beta},
\end{aligned}
\tag{9.62}
$$

$$
\begin{aligned}
\mathrm{T01} = {} & \frac{F\left(16G^2 - 19D^2G\sin^2\beta + 3D^4\sin^4\beta\right)}{G^{5/2}} \arctan\frac{\check{S}}{\sqrt{G}} \\
& + \cos\beta\left(16 - 19\sin^2\beta + 3\sin^4\beta\right) \arctan\frac{\check{S}}{D} \\
& + Q\left(15 + Q^2\right)\left[\arctan\frac{FQ}{\check{S}} + \arctan\frac{QD\cos\beta}{\check{S}}\right] \\
& - \frac{Q^2D\sin^2\beta}{F\check{S}\cos\beta} \\
& - \frac{D^2\sin^2\beta}{\check{S}}\left[\frac{B^2\left(18G^2 - 22D^2G\sin^2\beta + 3D^4\sin^4\beta\right)}{FG^2} + \frac{C^2\left(18 - 22\sin^2\beta + 3\sin^4\beta\right)}{D\cos\beta}\right] \\
& + \frac{2D^4\sin^4\beta}{\check{S}^3}\left[\frac{B^4F}{G} + \frac{C^4\cos\beta}{D}\right],
\end{aligned}
\tag{9.63}
$$

$$
\begin{aligned}
\mathrm{T02} = {} & \frac{F\left(8G^2 - 5D^2G\sin^2\beta - 3D^4\sin^4\beta\right)}{G^{5/2}} \arctan\frac{\check{S}}{\sqrt{G}} \\
& + \cos\beta\left(8 - 5\sin^2\beta - 3\sin^4\beta\right) \arctan\frac{\check{S}}{D} \\
& + Q\left(8 + \frac{Q^2D^2\sin^2\beta}{\check{S}^2}\right)\left[\arctan\frac{FQ}{\check{S}} + \arctan\frac{QD\cos\beta}{\check{S}}\right] \\
& + \frac{Q^2D\sin^2\beta}{F\check{S}\cos\beta} - \frac{2D^4\sin^4\beta}{\check{S}^3}\left[\frac{B^4F}{G} + \frac{C^4\cos\beta}{D}\right] \\
& - \frac{D^2\sin^2\beta}{\check{S}}\left[\frac{B^2\left(6G^2 - 2D^2G\sin^2\beta - 3D^4\sin^4\beta\right)}{FG^2} + \frac{C^2\left(6 - 2\sin^2\beta - 3\sin^4\beta\right)}{D\cos\beta}\right],
\end{aligned}
\tag{9.64}
$$

$$\begin{aligned}\mathrm{U01} =& \frac{D^2\sin^2\beta - 6G}{2G^{5/2}}\arctan\frac{\check{S}}{\sqrt{G}} - \frac{\sin^2\beta - 6}{2D^3}\arctan\frac{\check{S}}{D}\\ &+\frac{1}{2\check{S}}\left[\frac{B^2\left(D^2\sin^2\beta - 6G\right)}{G^2} - \frac{C^2\left(\sin^2\beta - 6\right)}{D^2}\right]\\ &+\frac{D^2\sin^2\beta}{3\check{S}^3}\left[\frac{B^4}{G} - \frac{C^4}{D^2}\right],\end{aligned} \tag{9.65}$$

$$\begin{aligned}\mathrm{U02} =& \frac{2+\sin^2\beta}{2D^3}\arctan\frac{\check{S}}{D} - \frac{2G + D^2\sin^2\beta}{2G^{5/2}}\arctan\frac{\check{S}}{\sqrt{G}}\\ &+\frac{1}{2\check{S}}\left[\frac{C^2\left(2+\sin^2\beta\right)}{D^2} - \frac{B^2\left(2G + D^2\sin^2\beta\right)}{G^2}\right]\\ &+\frac{D^2\sin^2\beta}{3\check{S}^3}\left[\frac{C^4}{D^2} - \frac{B^4}{G}\right],\end{aligned} \tag{9.66}$$

$$\begin{aligned}\mathrm{V01} =& \frac{B\left(2D^2\sin^2\beta - 15G\right)}{G^2} - \frac{C\left(2\sin^2\beta - 15\right)}{D^2}\\ &+\frac{D^2\sin^2\beta}{\check{S}^2}\left[\frac{B^3}{G} - \frac{C^3}{D^2}\right],\end{aligned} \tag{9.67}$$

$$\begin{aligned}\mathrm{V02} =& \frac{C\left(5+2\sin^2\beta\right)}{D^2} - \frac{B\left(5G + 2D^2\sin^2\beta\right)}{G^2}\\ &+\frac{D^2\sin^2\beta}{\check{S}^2}\left[\frac{C^3}{D^2} - \frac{B^3}{G}\right].\end{aligned} \tag{9.68}$$

These formulae were programmed and processed by a computer for apertures with slopes from 5° to 85° in 5° steps (archived at the Institute of Construction and Architecture, Slovak Academy of Sciences). Calculation graphs for apertures with wired glazing are included in Kittler and Kittlerova (1968, 1975) on a reduced scale for slopes of 75°, 60°, and 45°.

9.5 Light Propagation through Circular Apertures and Hollow Light Guides

Novel hollow guides were recently used to illuminate either deep interior without windows or those in the building core spaces (Aizenberg 2009; Darula et al. 2009). Some approximate calculation methods were proposed that usually applied flux propagation and reflection within tubes (Carter 2002; Jenkins et al. 2005). However, only few considered the changes of sky luminance patterns and direct sunlight rays entering the tube. In this respect, a sophisticated analytical solution called HOLIGILM (for "hollow light guide interior illumination method") was published by (Kocifaj et al. 2008) and was based on the backward ray tracing procedure with a user-friendly tool (Kocifaj and Kondracik 2009). This method for straight vertical tubes enables one to determine the luminance distribution of the transparent (Kocifaj 2009a) or translucent (Kocifaj 2009) glazing of the circular light source in the interior ceiling as well as the illuminance distribution on the horizontal working plane or on the floor. Examples of the resulting solution alternatives were demonstrated in several articles (Darula et al. 2010a) also considering light lube advantages in sunny tropical daylight climate (Darula et al. 2010b). Furthermore, the analytical method was extended for bended tubes (Kocifaj et al. 2010) with applications and examples (Darula et al. 2010c; Kocifaj 2010). Daylight efficiency measurements of light tubes were also made under the artificial sky (Darula et al. 2010d).

A review of special problems of light guides and combinations of light guides placed horizontally on the ceiling and shelves adjusted to catch skylight on the house front were considered (Beltran et al. 1997; McCluney 1998).

9.6 Daylighting Calculations with Computers

Architectural and lighting practice has increasingly adopted and advantageously used computers for the presentation of designs and their evaluations especially when computer graphics and tools were refined (Kota and Haberl 2009). Daylighting software was critically accepted (Ubbelohde and Humann 1998) and evaluated even for annual daylight simulations (Reinhart and Herkel 2000).

9.7 Partial Conclusions

Plate glass as a glazing material for daylighting apertures has very favorable properties. It does not allow rain and wind to penetrate indoors, and its very high light transmission enables sunlight beams and skylight even in oblique directions to enter the interior quite effectively. However, compared with unglazed openings, all glazed interiors suffer from some transmission losses and these losses have to be taken into account additionally to glazing losses caused by window frames, dirt or possible obstructions, furniture, or interior equipment. Usually in practical cases when evaluating daylighting, one expresses these losses by approximate reduction coefficients. The only exception is the directional transmittance of glass, which is traditionally included in calculations of the sky component as well as in graphical tools such as Waldram diagrams or protractors. Therefore, all calculation methods for glazed apertures including practical tools should accept these influences as a reality.

In contrast to Lambert's historical simplifications, all newer formulae have to incorporate more real conditions such as sky luminance distributions as well as directional glazing transmittance corrections. Advanced fenestration systems are increasingly being used to distribute sunlight and skylight purposefully into building cores and deep interior spaces. Distribution of visible light can be optimized to enhance daylighting and reduce electricity consumption and thermal loads. Novel daylighting systems such as anidolic ducts and hollow tube light guides are more efficient when their year-round performance is considered instead of their usefulness under overcast sky conditions. Some trials and examples applying the ISO (2004) standard sky luminance patterns using new computer possibilities, such as backward ray tracing, were shown using HOLIGILM for calculating the illuminance distribution from vertical and bent tubes.

With respect to energy-saving policies, the average annual effectiveness of daylighting systems is becoming more important than the minimum illuminance levels under overcast sky conditions because minima can be saturated more easily now by temporarily controlled and switched on artificial lighting systems. Thus, more fundamental research and theory resulting in new daylight criteria and sophisticated computer calculation methods as well as user-friendly calculation tools have to be expected.

Appendix 9

Comparison of Calculation Tables and Tools for the Sky Components from Vertical or Sloped Glazed Windows

With the mass production of relatively cheap glass during the Industrial Revolution in the second half of the nineteenth century, larger glazed windows became the main features of house fronts as well as in long-span buildings (e.g., exhibition halls and railway stations). The small light losses of roughly 10% of the light flux falling perpendicularly on the glazed surface were an advertisement slogan. Even in renowned books on building materials (e.g., Handisyde 1950, p. 284) the daylight transmission of glass was introduced by two sentences: "Ordinary polished plate or drawn sheet glass, when clean, transmits about 90% of daylight which falls upon it. In passing through the glass the light is slightly refracted, but this factor can usually be ignored." However, as shown in Fig. A9.1, only a few glazed window elements are penetrated by sky luminance normally, so the reduction of transmittance caused by arbitrary rays of nonnormal direction has to be considered.

The directional angle ψ is defined as the angle between the arbitrary directional ray and the normal to the glass plane. Thus, the maximum transmittance τ_n of the glazing is when $\psi = 0°$, i.e., for the shortest pass of the ray through the least glass thickness. So, the most effective element of any vertical, sloped, or horizontal aperture is that which is facing the illuminated place with the largest solid angle as well as the normal transmittance (Fig. A9.1). Furthermore, such normal glass elements (in red) are even more effective when momentarily facing the sun position or very high sky luminance patches owing to window orientation and time on a particular day. Every extended oblong aperture due to the pair of parallel frames has a solid angle gradually decreasing toward the lune vanishing points. Thus, together with the solid angle, also the directional transmittance τ_ψ toward the possible four vanishing points is decreased, so finally $\tau_\psi = 0$ as $\psi = 90°$, as indicated in Fig. A9.1.

Although the directional or oblique transmittance of glazing was differently taken into account, several tables were published for the sky factor or sky component either for vertical rectangular windows or for sloped apertures illuminating the element of the horizontal illuminated plane. For the different tables or diagrams two different ratios of window dimensions were used for the simplest position of the window with its lowest corner at its zero point of the coordinate system, i.e.:

- The window width is only one side to the normal taken as the width coordinate W, or L, or the coordinate axis direction y.
- The height of the window head above the horizontal plane of the illuminated element as H or the direction of the vertical axis z, or even arbitrarily as a.
- The distance of the illuminated element is D, or it can be taken as the prolongation of the axis x.

Various authors used different ratios in their formulae or tables, e.g.:

- In the comprehensive series of daylight tables published by Rivero (1958), ratios $H/D = z/x$ and $L/D = y/x$ are preferred.
- In Building Research Station simplified daylight tables by Hopkinson et al. (1958, 1966) the ratios $H/D = z/x$ and $W/D = L/D = y/x$ determine the window dimensions.
- Kittler and Ondrejička (1964, 1966) used both $L = a/x = z/x$ and $N = y/x$ as well as the width parameter $\check{S} = y/a = y/z$ and the point distance parameter $D = x/a = x/z$ (Kittler and Kittlerova 1968, 1975).

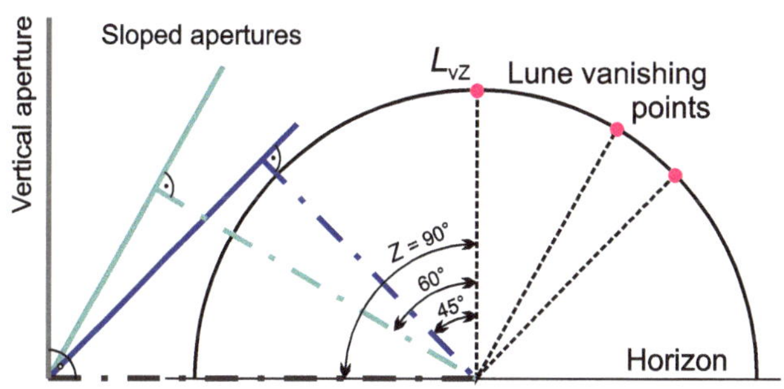

Fig. A9.1 Differently tilted apertures have also the glazing with different normal and directional transmission

- Puškaš (1979) denoted vertical window ratios as $d = L = a/x$ and $s = N = y/x$.

There are also several possibilities for how to derive the dependence of the sky component on the above parametric ratios. The renowned sky component tables published by the Building Research Station (Hopkinson et al. 1958) and reproduced also in the seminal book by Hopkinson et al. (1966) as Table 5.1 was deduced from summated values of the sky component for a uniform sky, then corrected for the CIE overcast sky (1:3) and glazing losses obtained from a large-scale Waldram (1929) diagram. To demonstrate the differences in the sky component values, the same ratios $H/D = z/x = L$ and $W/D = L/D = y/x = N$ were applied by Kittler and Ondrejička (1964, 1966).

To determine sky component results in the case of vertical windows, one can use either tables or the graphical tools mentioned in the appendix in Chap. 8, but in the twenty-first century more sophisticated computer programs are favored. These can simulate besides overcast sky conditions also the whole range of ISO/CIE sky types and instead of relative sky component values can respect the local sun paths and sky luminance patterns at any time in absolute luminance or illuminance values either outdoors and indoors (Darula and Kittler 2005). Such a program is based on the possibility to calculate absolute sky luminance along the window meridian using the Method for Aperture Meridians (MAM) (Kittler and Darula 2006) by applying the sophisticated software modeling of the hour and date sun position in an arbitrary geographical location, then simulating any of the 15 ISO/CIE sky patterns with absolute luminance and illuminance levels and with their representation within the window solid angle as well as interior sky illuminance in lux on the chosen element of the horizontal working plane (Roy et al. 2007). Such a user-friendly program called MAMmodeller is freely available in an online form at http://www.cadplan.com.au.

References

Aizenberg, J.B.: Hollow light guides. Znack Publ., Moscow (2009)

Beltran, L.O., Lee, E.S., Selkowitz, S.E.: Advanced optical daylight systems: Light shelves and light pipes. Journ. IES, 91–106, (1997)

Carter, D.J.: The measured and predicted performance of passive solar light pipe systems. Lighting res. technol., **34**, 1, 39–52 (2002)

CIE – Commission Internationale de l'E´ clairage: Natural Daylight, Official recommendation. Compte Rendu CIE 13th Session, **2**, part 3.2, 2–4 (1955)

CIE: Tubular light guide system. CIE Publ. 173:2006, Central Bureau, Vienna (2006)

Darula, S., Kittler, R., Kocifaj, M., Plch, J., Mohelnikova, J., Vajkay, F.: Osvětlování světlovody. (In Czech: Guide-light illumination). Grada, Prague (2009)

Darula, S., Kundracik, F., Kocifaj, M., Kittler, R.: Tubular Light Guides: Estimation on indoor illuminance levels. Luekos, **6**, 3, 241–252 (2010a)

Darula, S., Kittler, R., Kocifaj, M.: Luminous effectiveness of tubular light-guides in tropics. Applied Energy, **87**, 11, 3460–3466 (2010b)

Darula, S., Kocifaj, M., Kittler, R.: Illumination of interior spaces by bended hollow light guides: Application of the theoretical light propagation method. Solar Energy 84, 12, 2112–2119 (2010c)

Darula S., Rybár, P., Mohelnikova, J., Popeliš, M.: Measurement of tubular light guide efficiency under the artificial sky. Przeglad elektrotechn., (Polish Electrical Review), 86, 10, 177–180 (2010)

Department of Scientific and Industrial Research, (D.S.I.R.): Penetration of daylight and sunlight into buildings. Illum. Res. Techn. Paper 7 (1932)

Department of Scientific and Industrial Research, (D.S.I.R.): The transmission of light through window glasses. Illum. Res. Techn. Paper 18 (1936)

Holmes, J.G.: The reflection factor of glass. Transac. IES, **12**, 5, 108–114 (1947)

Hopkinson, R.G., Longmore, J., Graham, A.M.: Simplified daylight tables. Nat. Building Studies Special Report 26. Deptm. Sci. and Industr. Research, B.R.S., London (1958)

Hopkinson, R.G., Petherbridge, P., Longmore, J.: Daylighting. Heinemann, London (1966)

IESNA - Illuminating Engineering Society of North America: Lighting Handbook: Daylighting p.8.1-8.28, 9th Edition, IESNA, New York (2000)

ISO – International Standardisation Organisation: Spatial distribution of daylight – CIE Standard General Sky. ISO Standard 15409:2004 (2004)

Jenkins, D., Muneer, T., Kubie, J.: A design tool for predicting the performance of light pipes. Energy and Building, **37**, 5, 485–492 (2005)

Kittler, R.: Smerová charakteristika svetelnej priepustnosti priehľadných plochých stavebných skiel. (In Slovak: Directional characteristic of light transmittance of transparent planar building glass), Stavebnícky časopis, **11**, 1–2, 161–168 (1963)

Kittler, R. and Ondrejička, Š.: Nové výpočtové vzťahy pre určovanie oblohovej zložky dennej osvetlenosti zo skloneného pravoúhleho osvetľovacieho otvoru s priehľadným zasklením. (In Slovak: New calculation relations for the determination of the sky component from a sloped rectangular opening with transparent glazing.) Stavebnícky časopis, **11**, 8, 469–485 (1963)

Kittler, R.: Discussion du Rapport du Comité E-3.2. Compte Rendu Quinziéme Session, Vol.B, 382–384. Publ.CIE No 11 B. Bureau Central de la CIE, Paris (1964)

Kittler, R. and Ondrejička, Š.: Stanovenie oblohovej zložky súčiniteľa dennej osvetlenosti pre vertikálne obdĺžnikové osvetľovacie otvory. (In Slovak: Prediction of the sky component of the daylight factor for vertical rectangular illumination apertures.) Stavebnícky časopis, **12**, 5, 261–278 (1964)

Kittler, R. and Ondrejička, Š.: Graficheskiy raschot osveshchniya ot pasmurnogo neba cherez osteklennye svetoproyomy. (In Russian: Graphical prediction of illumination from the overcast sky through glazed apertures.) Svetotekhnika, **12**, 12, 10–12 (1966)

Kittler, R. and Kittlerová, L.: Návrh a hodnotenie denného osvetlenia. (In Slovak: Designing and evaluating of daylighting), Alfa Publ., Bratislava (1968, 1975)

Kocifaj, M, Darula, S., Kittler, R.: HOLIGILM: Hollow Light Guide Interior Illumination Method – an analytic calculation approach for cylindrical light-tubes. Solar Energy, **82**, 3, 247–259 (2008)

Kocifaj, M. and Kundracik, F.: HOLIGILM free software for illumination calculation. http://www.holigilm.info (2009)

Kocifaj, M. and Kundracik, F., Darula, S., Kittler, R.: Theoretical solution for light transmission of a bended hollow light guide. Solar Energy, **84**, 8, 1422–1432 (2010)

Kocifaj, M.: HOLIGILM-based simulations for bended light guides. Przeglad elektrotechn., (Polish Electrical Review), 86, 10, 218–221 (2010)

Kota S. and Haberl J.S.: Historical survey of daylighting calculation methods and their use in energy performance simulations. Proc. 9th Intern. Conf. for Enhanced Building operations, ESL-IC-09-11-07 (2009)

Kreider, J.F. and Kreith, F.: Solar heating and cooling: Active and passive design. 2nd Edition, McGraw-Hill, New York (1982)

MacGowan, D.: Creative Application of Digital Computers: Daylighting II: Condensed version. Department of Architecture, University of Washington, 98195, USA. In University of Adelaide Barr Smith Library Main collection Book. 729.28 M146c.C, 1–28 (1973)

McCluney, R.: Advanced fenestration and daylighting systems. Proc. Daylighting 1998 Conf., 317–324 (1998)
Puškáš, J.: Nová výpočtová metóda predurčovania oblohovej zložky činiteľa dennej osvetlenosti. (In Slovak: A new calculation method for the predetermination of the sky component of the daylighting). Zdravotní technika a vzduchotechnika, **22**, 3, 151–155 (1979)
Randall, W.C. and Martin, A.J.: Predetermination of daylighting by fenestra method. Transac. I.E. S. (N.Y.), (1930)
Reinhart, C.F. and Herkel, S.: The simulation of annual daylight illuminance distribution – a state-of-the-art comparison of six RADIANCE-based methods. Energy and Buildings, **32**, 167–187 (2000)
Rivero, R.: Illuminacion natural. Calculo del factor de dia directe du ventanas sin vidrios y con vidrios y para cielos uniformes y no uniformes. Instituto de la Construccion de Edificios, Montevideo. Translation by R.G. Hopkinson in B.R.S. Library Comm. 860 (1958)
Rybár, P.: Priestup svetla cez ploché stavebné sklo. (In Slovak: Light transmission through structural sheet glass). Světelná technika, **20**, 3, 37–40 (1987)
Taylor, A.K. and Grieveson, C.J.W.: The transmission factors of commercial window glasses. Illum. Res. Techn. Paper 2, Stationery Office, London (1926)
Ubbelohde M.S. and Humann C.: A comparative evaluation of daylighting software: SuperLite, Lumen Micro, Lightscape and Radiance. Proc. Daylighting '98 Conf., 97–104 (1998)
Waldram, J.M.: The precise measurement of optical transmission, reflection and absorption factors. Proc. CIE General Session, 1020–1046 (1929)
Waldram, P.J.: The transmission of light through window glass: A new form of daylight measuring diagram for glazed openings. Journal RIBA, **43**, 1072–77 (1936)
Walsh, J.W.T.: The reflection factors of a polished glass surface for diffused light. Appendix in Illum. Res. Techn. Paper 2, Stationery Office, London (1926)

References to Appendix

Darula, S. and Kittler, R.: New trends in daylight theory based on the new ISO/CIE Sky Standard: 3. Zenith luminance formula verified by measurement data under cloudless skies. Build. Res. Journ., **53**, 1, 9–31 (2005)
Handisyde, C.C.: Building materials, science and practice. The Architectural Press, London (1950)
Hopkinson, R.G., Longmore, J., Graham, A.M.: Simplified daylight tables. Nat. Building Studies Special Report 26. Deptm. Sci. and Industr. Research, B.R.S., London (1958)
Hopkinson, R.G., Petherbridge, P., Longmore, J.: Daylighting. Heinemann, London (1966)
Kittler, R. and Ondrejička, Š.: Nové výpočtové vzťahy pre určovanie oblohovej zložky dennej osvetlenosti zo skloneného pravoúhleho osvetľovacieho otvoru s priehľadným zasklením. (In Slovak: New calculation relations for the determination of the sky component from a sloped rectangular opening with transparent glazing.), Stavebnícky časopis, **11**, 8, 469–485 (1963)
Kittler, R. and Ondrejička Š.: Stanovenie oblohovej zložky súčiniteľa dennej osvetlenosti pre vertikálne obdĺžnikové osvetľovacie otvory. (In Slovak: Prediction of the sky component of the daylight factor for vertical rectangular illumination apertures.) Stavebnícky časopis, **12**, 5, 261–278 (1964)
Kittler, R. and Ondrejička, Š.: Graficheskiy raschot osveshchniya ot pasmurnogo neba cherez osteklennye svetoproyomy. (In Russian: Graphical prediction of illumination from the overcast sky through glazed apertures.) Svetotekhnika, **12**, 12, 10–12 (1966)
Kittler, R. and Kittlerova, L. : Návrh a hodnotenie denného osvetlenia. (In Slovak : Designing and evaluating of daylighting), Alfa Publ., Bratislava (1968, 1975)
Kittler, R. and Darula, S. : The method of aperture meridians: A simple calculation tool for applying the ISO/CIE Standard General Sky. Light. Res. and Technol., **38**, 2, 109–119 (2006)

Puškáš, J.: Nová výpočtová metóda predurčovania oblohovej zložky činiteľa dennej osvetlenosti. (In Slovak: A new calculation method for the predetermination of the sky component of the daylighting). Zdravotní technika a vzduchotechnika, **22**, 3, 151–155 (1979)

Rivero, R.: Illuminacion natural. Calculo del factor de dia directe du ventanas sin vidrios y con vidrios y para cielos uniformes y no uniformes. Instituto de la Construccion de Edificios, Montevideo. Translation by R.G. Hopkinson in B.R.S. Library Comm. 860 (1958)

Roy, G.G., Kittler, R., Darula, S.: An implementation of the Method of Aperture Meridians for the ISO/CIE Standard General Sky. Light. Res. and Technol., **39**, 3, 253–264 (2007)

Waldram, J.M.: The precise measurement of optical transmission, reflection and absoption factors. Proc. CIE General Session, 1020–1046 (1929)

Chapter 10
Modeling Daylight Distribution in Complex Architectural Spaces

10.1 Reflection, Absorption, and Transmission Properties of Materials and Surfaces

Everybody knows that white surfaces are the brightest and color surfaces reflect mostly that light which is in the wavelength range of the perceived color, that mirrors reflect directly (specularly), and that many matt paints diffuse or scatter and reflect light rays into many directions. Glass that transmits daylight into interiors and also enables view is a common, although specific transparent type with low reflectance. Obviously, the characteristics of reflecting and transmitting characteristics of media are important when interreflection has to be predicted or when a particular class of diffusing materials is to be utilized in interiors, in calculations, or in measurements.

The specular surfaces with their advantageous directly reflecting properties were enjoyed by Egyptians 5,000 years ago when polished bronze, silver, or gold plates were used as mirrors or for reflecting sunbeams in a required direction. Such an example is the well-known story of Archimedes' war experiment in 214 BC, where troops used warrior shields as "burning mirrors" to redirect concentrated sun rays toward enemy ships with devastating results. The same principle of sun ray concentration from many mirrors is used in current solar ovens or solar electricity power stations where not only sunlight but also infrared solar radiation is utilized.

The old rule of direct beam reflection – that the incidence and reflection angles taken from the mirror normal are the same – is frequently applied by boys worldwide. Currently, in daylight design regular reflection is utilized in hollow light guides with extremely high reflecting specular inner surfaces, or to redirect sunlight by light shelves. The clear mirror or silver foil reflectance ρ_M is very high in the reflected direction (roughly 0.9–0.93) and in the case of a favorable sun position the redirected sunlight reduction is under 10%. Therefore, horizontal shading shelves with the specular upper surface on southern house fronts can penetrate deep interiors considerably and enhance places far from the side

R. Kittler et al., *Daylight Science and Daylighting Technology*,
DOI 10.1007/978-1-4419-8816-4_10, © Springer Science+Business Media, LLC 2012

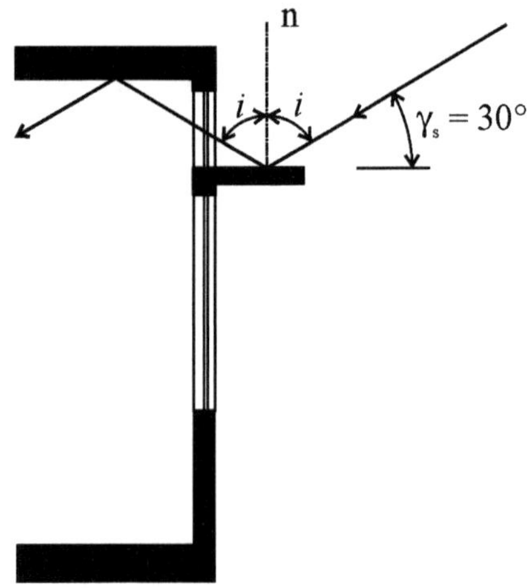

Fig. 10.1 Reflected beam from the shelf with a mirror upper surface

windows. Of course, the effectiveness is dependent on the unshaded sun position and slope of the shelves.

The reflectance factor ρ_M in the direction of reflection can be defined by two ratios:

$$\rho_M = \frac{\Phi_{M\rho}}{\Phi_t} = \frac{L_{M\rho}}{E_t}, \tag{10.1}$$

where $\Phi_{M\rho}$ is the luminous flux reflected by the mirror, Φ_t is total luminous flux falling on the surface, $L_{M\rho}$ is the luminance of the surface in the direction of mirror reflection, and E_t is the total illuminance of the surface.

In a simplified virtual case if only parallel sunbeams fall on the horizontal shelf mirror in Fig. 10.1 under solar altitude $\gamma_s = 30°$ and the sunbeams are normal to the house front, then under assumed turbidity conditions $T_v = 4$ (3.20) can be used to calculate the sunlight illuminance P_v with additional m and a_v values for $\gamma_s = 30°$ of $m = 1.996$ after (3.19) and after (3.18) $a_v = 0.0909$. Thus, the directional luminance of the upper shelf directed toward the interior ceiling will be

$$L_{M\rho} - P_v \rho_M \ (\text{cd/m}^2), \tag{10.2}$$

where

$$\begin{aligned} P_v &= 133{,}334 \exp(-a_v\, m\, T_v) \sin\gamma_s = 66{,}667 \exp(-0.72575) \\ &= 32{,}264.26 \ (\text{lx}), \end{aligned} \tag{10.3}$$

assuming a mirror surface with $\rho_M = 0.9$ and the transmittance of the double transparent glass in the direction of an angle $\psi = 90° - \gamma_s = 60°$. Thus, the directional transmittance $\tau_\psi = 0.1875$, and $\tau_g = \tau_n\, \tau_\psi = 0.81(0.1875) = 0.152$, i.e.,

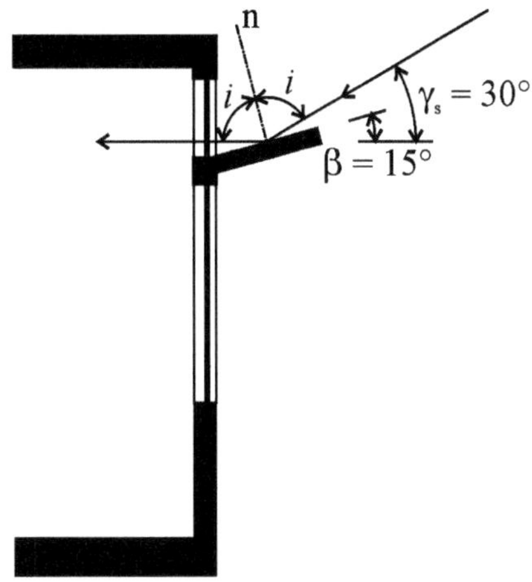

Fig. 10.2 Solar beam reflected from a 15° sloped shelf

$$L_{M\rho} = P_v \rho_M \tau_g = 32\,264.26\,(0.9)(0.152) = 4,413.75\ (\text{cd/m}^2). \tag{10.4}$$

However, in the case of the sloping shelf with a 15° tilt inwards in Fig. 10.2, the sunbeam incidence angle from the shelf normal is $i = 75°$; thus, the beam enters horizontally and in the normal direction of glazing, i.e., $\tau_g = \tau_n \tau_\psi = 0.81$. Thus, appropriate values have to be inset into (10.4) for a sloped shelf,

$$\begin{aligned} P_{v\angle} &= 133,334\,\exp(-a_v m T_v)\cos i = 133,334\,\exp\ (-0.72575)(0.25882) \\ &= 16,701.27\ (\text{lx}), \end{aligned} \tag{10.5}$$

i.e., owing to the reduced illuminance on the exterior surface but a higher transmittance through the glazing, the luminance reflected by the mirror shelf is

$$L_{M\rho} = P_v\,\rho_M\,\tau_g = 16,701.27(0.9)(0.81) = 1,217.23\ (\text{cd/m}^2).$$

In reality both shelves also get some luminance from the sky, especially from the sky close to the sun corona, and these additional luminances contribute within their solid angles from different shelf elements to the total brightness of the interior. When sunlight is absent owing to either cloudiness or the orientation of the house front with respect to the momentary sun position, then only diffuse sky luminance patterns are specularly projected into the interior. Only an interior having totally black surfaces will have no reflected daylight component, and only the sky component alone will provide interior daylighting.

The opposite extreme reflectance is that of a perfectly diffusing surface such as that produced by a magnesium oxide coating (reflectance ρ_D roughly 0.98, or 98%) (Preston 1929–1930), or even by a clean matt white paint (reflectance 0.9, or 90%) (Powell and Kellog 1926). The diffusing properties of such high reflectance cause by interreflectance almost an ideal light intensity distribution in all directions within the surface half space, resulting in a near constant luminance indicatrix.

The diffuse reflection is characterized by the fact that the total luminous flux whether in beam form or already diffused before falling on the surface is dispersed into the whole half space π after reflection from the surface; therefore, the diffuse reflectance

$$\rho_{\mathrm{D}} = \frac{\Phi_{\mathrm{D}\rho}}{\Phi_{\mathrm{t}}} = \frac{\pi L_{\mathrm{D}\rho}}{E_{\mathrm{t}}}, \tag{10.6}$$

where $\Phi_{\mathrm{D}\rho}$ is the luminous flux diffusely reflected, Φ_{t} is the total luminous flux falling on the surface, $L_{\mathrm{D}\rho}$ is the luminance of the surface in any direction of reflection, and E_{t} is the total illuminance of the surface.

In the example in Fig. 10.1 when the sunlight falls on, for instance, a matt dark-gray carpet or floor with reflectance $\rho_{\mathrm{D}} = 0.3$, the indoor parallel sunbeam illuminance after (10.3) or (10.5) will be the same but the floor luminance $L_{\mathrm{D}\rho}$ in any direction will be reduced by the diffuse reflectance of the carpet:

$$L_{\mathrm{M}\rho} = \frac{P_{\mathrm{v}}\, \rho_{\mathrm{M}} \tau_{\mathrm{g}} \rho_{\mathrm{D}}}{\pi} = \frac{4,413.75(0.3)}{\pi} = 421.48\ (\mathrm{cd/m^2}). \tag{10.7}$$

Sunlight and skylight in the exterior are similarly reflected from the ground surface owing to reflectance sometimes named the albedo. Many surfaces in nature reflect diffusely; however, calm water surfaces or flat ice reflect specularly, but most surfaces reflect both specular and diffuse components. Those with quasi-diffuse or irregular reflectance have to be measured and characterized in different angular directions, either by intensity or by luminance bodies with their sectional indicatrix curves.

Colorful objects or surfaces have spectrally selective reflectance which is determined by colorimetric measurement methods and representation means (Schanda et al. 2006).

Nontransparent materials and surfaces impenetrable by light can only reflect or absorb the luminous flux falling on them; thus,

$$\frac{\Phi_{\rho}}{\Phi_{\mathrm{t}}} + \frac{\Phi_{\alpha}}{\Phi_{\mathrm{t}}} = \rho + \alpha = 1, \tag{10.8}$$

i.e., the more light is reflected from a surface, the less light the surface absorbs.

10.2 Multiple Interreflection of Daylight in Interiors

Daylight penetrating through any apertures into interiors from outdoor sources (from the sun and sky or reflected from the ground external obstructions, house front surfaces, etc.) illuminates all interior surfaces, with multiple interreflection among them. The nature and quantities of these interreflections are influenced by:

- The illuminance or luminous fluxes reaching the illuminated plane first, e.g., surfaces facing upward such as floors or horizontal working planes are illuminated by the sources from the upper half space especially by sunlight and skylight, whereas downward-facing planes such as the ceiling are illuminated only by indoor reflected daylight or from upward-directed reflections from outer terrain and obstructing house front surfaces. However, vertical walls can receive both direct and reflected light flux.
- The reflecting properties of surfaces, which can be:
  - Owing to the surface structure either specular or diffuse or both specular and diffuse, and which can be represented by their space luminance body/solid and its sectional distribution curves called also reflection indicatrices, which specify the directional reflection into the whole half space of the illuminated planar surface element.
  - Due to the ability of the reflectance "quantity," expressed by the reflectance factor/coefficient ρ defining the ratio of the reflected flux to the flux received, with the range from maximally reflecting white surfaces through a scale of intermediate grays to absolutely nonreflecting black.
  - Due to spectral properties with selective reflectance of different colors, which can be defined by several colorimetric methods (Schanda et al. 2006).

In the interiors of buildings there are usually built-in light-obstructing or light-reflecting surfaces of low reflectance or dark textures and paints. However, these redistribute the incoming light flux in all directions, whereas the light intensity is reduced in accordance with the cosine angle taken from the surface normal. It is logical that, owing to multiple interreflection, a certain unifying luminance balance is achieved in every interior depending on the average reflectance in the "interior cavity," which traps the flux incoming through the aperture.

After the Edison-Swan Electric Company started in 1878, the mass production of cheap incandescent bulbs and the simultaneous electrification progress from 1883 resulted in a new era of artificial electric lighting of streets and interiors in the USA and Europe. So many older buildings were reconstructed and candle or gas lighting fixtures were replaced by new electric ones. This boom had some consequences also in lighting efficacy and performance studies especially applied to larger community halls and interior gathering spaces. In this respect Mascart (1888) was probably the first investigator who studied interior interreflections. He was curious and wondered about the increase of artificial light inside the representative rooms of the Palace of Versailles, especially the mirror-clad hall, and also in other Paris halls and theaters. His theoretical finding indicated that a light source placed in the interior emits the total luminous flux Φ_I, which is many times reflected in a row sequence of decreasing intensity, with the final balance summarized in Φ_ρ; thus,

$$\Phi_\rho = \Phi_I\left(1 + \rho + \rho^2 + \rho^3 + \cdots\right) = \Phi_I \frac{1}{1-\rho} \text{ (lm)}, \tag{10.9}$$

where ρ is the average reflectance factor of the inner cavity and the reflection increase r is

$$r = \frac{1}{1-\rho}. \tag{10.10}$$

Mascart also noticed that in an extreme case of a totally white interior with $\rho = 0.95$ the original flux could be increased 20 times in comparison with the same room with absolutely black surfaces.

The same idea was later explored by Ulbricht (1900) to discover that the most effective interreflection cavity was in the form of a hollow white sphere, called also a spherical integrator. (Ulbricht 1920; Taylor 1920). Although some errors were identified in Ulbricht's measurement method (McNicholas 1928; Taylor 1935), they were insufficient to prevent determination of the diffuse reflectance factor of any sample using an Ulbricht sphere (Moon 1936, 1941).

The total illuminance on the walls of the sphere with radius r due to an infinite number of interreflections is

$$E_\mathrm{i} = \frac{\Phi_\mathrm{I}}{4\pi\, r^2\,(1-\rho)}\ (\mathrm{lx}). \tag{10.11}$$

The approximate expression of the ideal interreflection with coefficient ρ in (10.9) and (10.11) or r defined in (10.10) is valid in the case of an absolutely closed cavity with an internal light source. If an interior is illuminated by the exterior light source through a window with area W, then this is an entrance for the incoming luminous flux, and also an exit for the interreflected flux escaping from the cavity. Hence, the interreflection coefficient is partly reduced (Meshkov 1957, p. 201):

$$r_0 = \frac{\Phi_{\mathrm{i}\rho}}{\Phi_\mathrm{ew}} = \frac{1}{1-\rho(1-(W/A))}, \tag{10.12}$$

where $\Phi_{\mathrm{i}\rho}/\Phi_\mathrm{ew}$ is the ratio of the interreflected luminous flux in the interior normalized by the luminous flux entering through the window, W is the window area, and A is the total area of all interior surfaces.

Thus, the interreflected luminous flux leaving the interior through the window is

$$\frac{\Phi_\mathrm{ix}}{\Phi_\mathrm{ew}} = \frac{\rho\,(W/A)}{1-\rho(1-(W/A))}. \tag{10.13}$$

A more or less evenly distributed light flux in daylight situations is experienced in vertical light wells if the uniform sky is modeled and reaches the aperture from all sides. Such model studies, illuminating an inner staircase, were conducted by Hannauer (1940, 1941), who discovered an exponential increase of SF(1:1) entering the light well; thus, the value of the daylight factor (DF) including interreflections DF(1:1) is

$$\mathrm{DF}(1:1) = \mathrm{SF}(1:1)^{(1-0.556\rho)}. \tag{10.14}$$

For practical use a very simple diagram for the calculation of DF(1 : 1) in side-lit rooms was proposed by Kittler (1956, 1957), who used model measurements by Pleijel (1949) to insert additional possibilities to take into account obstructions and reflection gains behind the no-sky line.

In side-lit interiors the position of windows because of skylight directed from the upper sky hemisphere toward the usually darker floor as well as the reflected daylight from the darker terrain upward to the ceiling presents a considerable reduction in the first and second reflection. Although Mascart's interest was in the interreflection increase of the interior flux supplied by luminaires, in daylighting theory the fact that outdoor flux penetrates through windows posed a rather different problem. So another system to calculate the interior reflection was first suggested by Arndt (1938), who respected the daylight situation in side-lit rooms:

- The mean reflectance factor for the room ρ_m weighted by the area of interior surfaces A_n has to be considered:

$$\rho_m = \frac{A_n \rho_n}{\sum A_n}. \tag{10.15}$$

- The internally reflected illuminance in any distant places from the windows E_{ind} can have its value increased in relation to the incoming direct illuminance E_{dir}:

$$E_{ind} = E_{dir} \frac{\rho_m}{1 - \rho_m}. \tag{10.16}$$

- The incoming skylight flux was already defined by the sky illuminance or the sky factor (SF) value, i.e., sky illuminance normalized to outdoor horizontal illuminance under an unobstructed sky, and was therefore separated from the internally reflected component (IRC); thus,

$$\mathrm{DF} = \mathrm{SF} + \mathrm{IRC} = \frac{E_{dir}}{E_{vh}} + \frac{E_{ind}}{E_{vh}}, \tag{10.17}$$

where E_{vh} is horizontal illuminance under an unobstructed sky.

Later Arndt (1955) defined the IRC value

$$\mathrm{IRC} = \frac{E_W}{E_{vh}} \frac{W \rho_m}{A(1 - \rho_m)}, \tag{10.18}$$

where W is the window area, later corrected to only the glazed area reduced by the transmission of daylight, A is the total room area, i.e., the area of the ceiling, floor, and walls, including the window, ρ_m is the mean reflectance factor of ceiling, walls, and floor, E_W is the vertical illuminance on the window, which defines the incoming flux as $\Phi_I = WE_W$, and E_{vh} is the horizontal illuminance from the unobstructed sky, which is normalized for the DF(1 : 1) value. Finally, Arndt also included in his formula for IRC(1 : 1) also the empirical coefficient K expressing the uniform $K(1:1)$ or International Commission on Illumination (CIE)-graded sky luminance $K(1:3)$ corrections, respectively, for different obstruction angles, including also the transmittance of the glazing. Thus,

$$\mathrm{IRC}(1:1) = \frac{W\rho_m}{A(1-\rho_m)} K(1:1)\ (\%) \tag{10.19}$$

or

$$\mathrm{IRC}(1:3) = \frac{W\rho_m}{A_t(1-\rho_m)} K(1:3)\ (\%). \tag{10.20}$$

Dresler (1954) tested Arndt's formulae by applying model measurement results obtained by Pleijel (1949) and found quite good agreement, whereas Hopkinson et al. (1954) tried to improve the incoming flux distribution expression as well as the interior reflection expression. The concept was to define the interreflections in a side-lit room by dividing the inner-acting surfaces with their reflectances excluding the window wall into the upper and downward part/cavity with slightly different initial light fluxes. Respecting the glazed window area W_g and the standard CIE overcast sky conditions, Hopkinson et al. (1954) proposed the formula

$$\mathrm{IRC}(1:3) = \frac{0.85\,W_g}{A\,(1-\rho_m)} (K_{BRS}\rho_d + 5\rho_u)\ (\%), \tag{10.21}$$

where K_{BRS} is a function of the sky luminance distribution and the obstruction angle, and ρ_d and ρ_u are reflectance factors of the upper and downward parts of the interior cut through the window center in two reflecting cavities.

The Building Research Station (BRS) also developed the BRS reflected light nomograms (Hopkinson et al. 1966) to help practitioners who favor graphical tools to determine IRC values, and in addition to those for side-lit rooms, the BRS also developed a nomogram valid for roof lights (Hopkinson et al. 1954).

However, these empirical formulae did not comprehensively determine the different influences on daylighting of the exterior space, especially when the side-lit windows face facades with or without windows, as well as the exterior terrain reflectances: therefore, after several model measurements Krochmann (1962) and Kittler (1964) tried to approximate the minimum as well as average IRC(1 : 3), respectively. Extensive model measurements under the box-type artificial sky determined approximately the externally reflected component ERC(1 : 3) impact with a free horizon under different terrain reflectances when the average value [ERC(1 : 3) + IRC(1 : 3)](AVG) was determined for glazed windows under the standard CIE overcast sky by Kittler and Kittlerová (1968, 1975):

$$[\mathrm{ERC}(1:3)+\mathrm{IRC}(1:3)](\mathrm{AVG}) = \frac{85\,{W_\mathrm{g}}^{0.7}}{A_\mathrm{t}\,(1-\rho_\mathrm{m})}\left[\begin{array}{r}0.785\rho_\mathrm{d}(1-\sin\varepsilon_\mathrm{o})^{1.5}+1.24\rho_\mathrm{d}\rho_\mathrm{F}(1+4\rho_\mathrm{T})\sin\varepsilon_\mathrm{o}\\ +1.475\rho_\mathrm{T}\rho_\mathrm{u}\cos\varepsilon_\mathrm{o}\end{array}\right](\%), \tag{10.22}$$

where additionally reflectance of the obstructing facades ρ_F and terrain ρ_T are included with respect to the obstructing angle elevation function of ε_o.

For the interior without any obstructions, (10.22) is simplified to

$$[\mathrm{ERC}(1:3)+\mathrm{IRC}(1:3)](\mathrm{AVG}) = \frac{85\,{W_\mathrm{g}}^{0.7}}{A_t\,(1-\rho_\mathrm{m})}(0.785\rho_\mathrm{d}+1.475\rho_\mathrm{T}\rho_\mathrm{u})\;(\%). \tag{10.23}$$

Similar approximation formulae were found for the minimum values of reflected components: [ERC(1 : 3) + IRC(1 : 3)](MIN), published also by Kittler and Kittlerova (1968, 1975), where the coefficient 0.785 is replaced by 0.5, whereas 1.24 and 1.475 are replaced by 1.

Owing to the multiple reflection in the outdoor space, some authors, e.g., Hopkinson et al. (1966, p. 222), determined also ERC(1:3) of the DF under either the uniform sky or the CIE overcast sky. The modification of (10.19) was proposed by Tregenza (1989a) in the case of large external obstructions.

10.3 Approximate Flux-Type Predictions of Interior Interreflection

Approximate calculation methods based on the zonal cavity abstraction of the interior, sometimes also called the lumen method as for electrical lighting, were originally developed in Germany (Frühling 1928) and were later taken as very practical in North America (Griffith et al. 1955; IES 1971).

Frühling's concept for side-lit rooms was based on assumptions of a constant distribution of the sky luminance, and then the average illuminance on the vertical window E_vw is expressed by a window factor f_w, which in its maximum for an unobstructed situation can be half of the horizontal exterior illuminance $E_\mathrm{h} = D_\mathrm{v}$, i.e.,

$$E_\mathrm{vw} = f_\mathrm{w} D_\mathrm{v}, \quad \text{where} f_\mathrm{w} \le 0.5\ (\mathrm{lx}). \tag{10.24}$$

Then the average needed or usable luminous flux on the working plane Φ_n is normalized by the incoming flux through the window Φ_W and their ratio is called the coefficient of utilization (η or CU), i.e.,

$$\eta = \mathrm{CU} = \frac{\Phi_\mathrm{n}}{\Phi_\mathrm{W}} = \frac{E_\mathrm{aih}\,A_\mathrm{f}}{E_\mathrm{W}\,A_\mathrm{W}}, \tag{10.25}$$

where E_aih is the average interior horizontal illuminance on the working plane, A_f is the floor area, E_W is the vertical illuminance at the window center, and A_W is the window area or its glazed area W.

So, either the horizontal indoor illumination level E_aih or the average DF (ADF) can be calculated with dependence on the ratio of window to floor areas:

$$E_\mathrm{aih} = f_\mathrm{w} D_\mathrm{v} \mathrm{CU} \frac{A_\mathrm{w}}{A_\mathrm{f}} \;(\mathrm{lx}), \tag{10.26}$$

or

$$\mathrm{ADF} = \frac{E_\mathrm{aih}}{D_\mathrm{v}} = \frac{A_\mathrm{W}}{A_\mathrm{f}} f_\mathrm{w}\,\mathrm{CU}\;(\%), \tag{10.27}$$

where the exterior horizontal illuminance D_v is valid only for the overcast sky conditions.

Frühling's flux method was soon criticized in the USA by Turner-Szymanowski (1931), who claimed that the essential deficiency of the average illumination is not sufficient for a correct design of windows, their dimensions, and placement.

Several trials were made to improve these deficiencies, e.g., Grifith et al. (1955) introduced a better net aperture transmittance, room dimensions, and distance from the window with the concluding indication that the flux method can be extended to take into account also interreflections. However, in a practical guide (Griffith 1958) the so-called daylight prediction formula had two split components for skylight and reflected daylight from the ground, resembling Frühling's method but instead of the floor area a numerical multiplying factor K_R was introduced to represent room dimensions and reflectances. Thus,

$$E_\mathrm{aih} = E_\mathrm{ex}\,A_\mathrm{wt}\,\mathrm{CU}\,K_\mathrm{R}\;(\mathrm{lx}), \tag{10.28}$$

where E_ex is the illuminance either from the sky or from the ground and A_wt is the window transmission area. The interior illuminance was predicted at three room depth points with maximum, middle, and minimum illuminance. Such a relatively simple basic formula of the lumen method is still recommended and included in the US handbook by IESNA (2000):

$$E_\mathrm{i} = E_\mathrm{ex}\mathrm{NT}\;\mathrm{CU}\;(\mathrm{lx}), \tag{10.29}$$

where E_i and E_ex are the interior and the exterior illuminance in lux at a prescribed point and NT is the net transmittance of the aperture. CU is the coefficient of utilization.

Generally, in the case of side-lit rooms, averaging of illuminances on the working plane where with the distance from the window these, without high interreflection, decrease rapidly toward the rear interior places results in average values that are close to reality only under many ideal conditions, e.g.,

- It assumes an empty rectangular room with shape close to a cubic or spherical enclosure.
- High reflectances of interior surfaces will even the interreflected component.
- The influence of exterior obstructions will reduce the overall illuminance level.
- Special attention has to be given to shading and light shelves.
- The first approximation is made by the assumption that the illuminance at the window center represents the average over all the window frame, whereas the second and further errors stem from the variable surface luminances.
- Relatively large sets of tables with CU have to be used to correct the daylight distribution in a room under different window dimension, room depth, glazing, shading, and overcast sky conditions.
- Exterior horizontal illuminance in lux as a basic value is usually determined under overcast conditions not respecting any variation due to window orientation.

Gradually, under the pressure of the energy crisis and calls for window design at the early stages of the architectural design process, the ADF with the flux-based interreflection theory attracted Lynes (1979) and Longmore (1975, 1978) to investigate and revisit the flux-type method in Europe too.

Lynes tried to specify better the possible window vertical illuminance with respect to obstructions, comparing the $(C\rho_{\mathrm{d}} + 5\rho_{\mathrm{u}})$ value in (10.21) with the half value of the zenith obstruction angle in degrees $\theta/2$ and found a possible approximation. He suggested expressing the flux entering the window as

$$\Phi_{\mathrm{W}} = \frac{W_{\mathrm{g}}\, D_{\mathrm{v}}\, \tau_{\mathrm{g}}}{100}\, \frac{\theta}{2}\ (\mathrm{lm}) \tag{10.30}$$

and the flux absorbed by interior surfaces is

$$\Phi_{\mathrm{ai}} = \mathrm{ADF}\frac{A_{\mathrm{t}} D_{\mathrm{v}}\,(1-\rho_{\mathrm{m}})}{100}\ (\mathrm{lm}). \tag{10.31}$$

Owing to the conservation law, equations (10.30) and (10.31) have to be equal; thus, the ADF is

$$\frac{E_{\mathrm{aih}}}{D_{\mathrm{v}}}\,100 = \mathrm{ADF} = \frac{W_{\mathrm{g}}\,\tau_{\mathrm{g}}}{A_{\mathrm{t}}\,(1-\rho_{\mathrm{m}})}\,\frac{\theta}{2}\ (\%) \tag{10.32}$$

and thus the glazed area of the window necessary to produce the required or standardized ADF value indoors can be determined from (10.32) when the critical ADF is known.

Lynes (1979) respected the basic interreflection formula in (10.20) and (10.27) with its influential parameters $W_g = A_g$, but $A_t(1 - \rho_m)$ expresses the interreflection from all interior surfaces taking into account their mean reflectance, whereas θ is the sky lune angle of the endless street from the window center and τ_g is the glazing transmittance. The ADF as a percentage in (10.32) seems to express only the interreflections without the influence of direct skylight, but the contrary is true as the average illuminance indoors is related to the exterior horizontal diffuse illuminance:

$$E_{aih} = \frac{D_v \mathrm{ADF}}{100} \ (\mathrm{lx}). \tag{10.33}$$

Lynes also determined the depth in a side-lit room d_L in which probably the required ADF would satisfy

$$d_L = \frac{2/(1 - \rho_m)}{(1/w_r) + (1/h_{whh})} \ (\mathrm{m}), \tag{10.34}$$

where w_r is the room width in meters and h_{whh} is the window head height in meters.

Although the Lynes concept has the flux-type basis in contrast to the American lumen or utilization methods, their CU is substituted by the rough interreflection expression. However, all rely on the fundamental assumption of the overcast sky conditions with sunlight absent and with unity uniform sky luminance, i.e., $G_v = D_v$ and $L_{v\chi} = L_{vz} = 1$.

Recently the rough flux method used by Longmore (1975) determining the questionable average ADF value was totally misappropriately applied to estimate ADF under ISO (2004) general skies by Li and Cheung (2006), who reformulated ADF:

$$ADF = \tau_g W_g \left(\frac{C}{A_d} + \frac{C\rho_d + D\rho_u}{A_t(1 - \rho_m)} \right) \tag{10.35}$$

where A_d is the area of the floor and walls below window center, parameters C and D express daylight flux incident on the window center from above and below the horizon respectively. Later Li et al. (2011) have given changeable C parameters for 15 sky types without any dependence on the window orientation to sun position. Thus real time-changeable influences of sky paterns is ignored in both C and D values.

It is an enormous pity that recently it seems that in the USA, Canada, and the UK window design in side-lit rooms will still be done with the standardization of a thumb-based rule and the traditional CIE overcast sky (Reihart and LoVerso 2010) in the IESNA handbook (2000) as well as in CIBSE guide (1999). It is curious that the inaccurate Lynes formula (10.32) is replaced by its "modified version," justified only by specific model measurements in a scale model of a classroom (Crisp and Littlefair 1984),

$$\mathrm{ADF} = \frac{A_g \, \tau_g \, \theta}{A_t \left(1 - \rho_m^2\right)} \, (\%), \tag{10.36}$$

which means that the original $\theta/2$ component expressing approximately the obstruction angle as a percentage is included in the denominator, where $\left(1 - \rho_m^2\right)$ should be equal to $2(1 - \rho_m)$. This distorts the obstruction influence and for different ρ_m does not fit. A further substitution is the introduction of the window-to-wall ratio $\mathrm{WWR} = A_g/A_t$, and after a daylight feasibility study an assumption was made that the daylighted zone should have a daylight feasibility factor (DFF) of at least 0.22 in Canada or 0.25 in the USA, which should be compared with the adjusted effective aperture (AEA):

$$\mathrm{AEA} = \mathrm{WWR} \, \tau_g \, \mathrm{OF} > \mathrm{DFF} \, (\%), \tag{10.37}$$

where OF is the obstruction factor, which is equal to the zenith angle of the obstructing opposite facade divided by its maximum angle for a free horizon, i.e., $\theta/90°$.

Such a simple criterion as the DFF hides many simplifying assumptions, e.g., the overcast sky influence on the ADF and the obstruction situation without terrain or house front reflectances with interior interreflection influences missing. Such a criterion ignores the daylight theory principles and disingenuously claims that window design and daylighting is perfectly respected in the design process. Furthermore, if the minimum DFF values could be accepted for critical window design in temperate countries, then all other climate zones could adopt suitable DFF standards in accordance with their daylight availabilities based on the unrealistic extreme CIE overcast sky assumption without respecting prevailing sky luminance patterns in the region. All interreflection methods are based often on time-stable flux input which is true for constant artificial lighting. The average criterion ADF ignores the fact of fluent and sometimes dynamic variations of teal daylight conditions.

In trying to overcome such deficiencies, Tregenza and Waters (1983) introduced the daylight coefficient (DC) concept. In contrast to the DF definition (the ratio of the horizontal illuminance indoors normalized by the horizontal illuminance outdoors under an unobstructed sky vault), the DC normalizes the total illuminance at a point indoors to the actual sky luminance within the sky patch solid angle seen through the aperture:

$$\mathrm{DC} = \frac{\Delta E_{ia\chi}}{L_{a\chi} \, \Delta\omega_{a\chi}}, \tag{10.38}$$

where $E_{ia\chi}$ is the direct illuminance produced by the sky luminance as well as the indirect illuminance by interreflection from outdoor and indoor surfaces in lux, $L_{a\chi}$ is the luminance of the sky patch at an azimuth-orientated angular distance from the sun position in candelas per square meter, and $\omega_{a\chi}$ is the sky patch solid angle in steradians.

The dependence of the total interior illuminance on actual sky luminance within the aperture solid angle takes into account various sky luminance distributions, and if the total interior illuminance has to be influenced also by interreflections, then the DC depends also on the geometry of the room, surrounding buildings, and all the reflectances of their surfaces reduced by the transmittance of window glazing.

When the final or total DC value is calculated it has to be summarized from three additive components (similar to SF, ERC, and IRC) in calculations using the flux-type method for the upper room cavity and the lower room cavity (similar to the system of (10.22), (10.23)):

$$\mathrm{DC} = \mathrm{DC}_{\mathrm{sky}} + \mathrm{DC}_{\mathrm{uc}} + \mathrm{DC}_{\mathrm{dc}}. \tag{10.39}$$

Owing to the ground-reflected skylight and upper cavity reflection, the DC was enhanced by (Tregenza and Waters 1983)

$$\mathrm{DC}_{\mathrm{uc}} = \frac{W_{\mathrm{g}}\,\tau_{\mathrm{g}}\,\rho_{\mathrm{T}}\,\rho_{\mathrm{u}}\,\sin\varepsilon_0}{2A_{\mathrm{t}}(1-\rho_{\mathrm{m}})}, \tag{10.40}$$

and the interreflected component from the downward-entering flux from the vertical window with incidence angle i dependent on the window wall orientation

$$\mathrm{DC}_{\mathrm{dc}} = \frac{W_{\mathrm{g}}\,\tau_{\mathrm{g}}\,\rho_{\mathrm{d}}\,\rho_{\mathrm{u}}\,\cos i}{A_{\mathrm{t}}(1-\rho_{\mathrm{m}})}. \tag{10.41}$$

The DC concept and criterion is for the sky component assuming the point-by-point method for measurements and criteria application, but in the case of DC calculation the interreflections included as additional components after (10.40) and (10.41) are average values subtracted from the overall flux distribution and might be quite imprecise.

Mardaljevic (1999) analyzed the possibilities to apply the DC concept to the computer summation in the compact matrix formulation [the daylight coefficient matrix (DCM) containing 5,010 points evenly distributed over the hemisphere] and tried the DC implementation in the Radiance calculation algorithms. Studying the problem of sky discretization by different sky patches and hemispheric nets, whether circular solid angles or rectangular source angles in Radiance, he formulated for the calculation of the direct $\mathrm{DC}_{\mathrm{sky}}$. In the component equation (10.39), he added also the solar component $\mathrm{DC}_{\mathrm{sun}}$, i.e.,

$$\mathrm{DC}_{\mathrm{sun}} = \frac{E_{\mathrm{sun}}}{\omega_{\mathrm{sun}}\,L_{\mathrm{sun}}}. \tag{10.42}$$

Finally, he recommended the DC-based daylighting analysis procedure including a system to predict time-varying illuminances and applied the Kew test reference year (TRY) 1984 as basic irradiance data to recalculate them to illuminances using the constant luminous efficacy of 120 lm/W. The main deficiency of the DC_{sky} system was the unavailability of sky luminance distributions except the Perez model (see the appendix in Chap. 5) and therefore Mardaljevic used questionable "sky model blends" with a mixing function for the Kew test reference year assuming 60% intermediate skies, 24% overcast skies, and 16% overcast–intermediate blends. However, the fictitious "intermediate sky" covers erroneously all clear sky and many cloudy sky conditions, so distorting the London daylight climate markedly. Therefore, all such manipulations and estimates are unfortunately based on local irradiance data with the aforementioned irregularities and rough conversions to daylight illuminance levels.

10.4 Interreflections from Rectangular Sources within Rectangular Planes

Windows as well as the room envelopes are formed and usually represented by rectangular planes or strips with relatively good diffusivity and almost uniform luminance. These facts attracted Lambert (1760), who assumed under such conditions the possibility to express the luminous flux distribution from a rectangular plane source to another plane illuminated in a similar manner as in the case when a window produces the illuminance on a plane element, but the double integration has to cover the whole illuminated plane, i.e., the integration is to be repeated four times. Thus, mathematically, starting from Lambert's classical basic formulae valid for an arbitrary planar element of the large-area source dS_1 with luminance L_{vs} illuminating an arbitrarily positioned planar element dS_2, causing its illuminance dE_2, see Fig. 8.1,

$$dE_2 = \frac{L_{vs} \cos \vartheta_1 \cos \vartheta_2 \, dS_1}{l^2} \text{ (lx)}, \tag{10.43}$$

but because also the luminous flux from the source element to the illuminated element is already defined as

$$d^2\Phi_{12} = dE_2 \, dS_2 = \frac{L_{vs} \cos \vartheta_1 \cos \vartheta_2 \, dS_1 \, dS_2}{l^2} \text{ (lm)}. \tag{10.44}$$

Thus, as $dS_1 = dy_S \, dz_S$ a double integration is needed to obtain E_2, which has to be again double-integrated within the illuminated plane area $dS_2 = dy_I \, dz_I$ to get Φ_2, i.e.,

$$E_2 = \frac{L_{vs}}{\pi} \int_{y_{S_1}}^{y_{S_2}} \int_{z_{S_1}}^{z_{S_2}} \frac{\cos \vartheta_1 \cos \vartheta_2}{l^2} dy_S \, dz_S \text{ (lx)}, \tag{10.45}$$

$$\Phi_{12} = \frac{L_{vs}}{\pi} \int_{ys_1}^{ys_2} \int_{zs_1}^{zs_2} \int_{xI_1}^{xI_2} \int_{yI_1}^{yI_2} \frac{\cos \vartheta_1 \cos \vartheta_2}{l^2} \, dy_S \, dz_S \, dx_I dy_I \text{ (lm)}. \qquad (10.46)$$

If the luminous flux from the planar light source reaches only a surface element or a point of the illuminated plane, its effect is equal to the solid angle of the source projected into the illuminated plane under the assumption of a unity and uniform source luminance. However, if the luminous flux from a planar source is spread over a plane of finite dimensions, then the reciprocity exists that the flux from the source to the illuminated plane is the same as in the opposite direction. This reciprocity theorem (Moon 1936, 1941, p. 335) is valid when both planes have the same luminance and perfectly diffusing reflectance. Then their interrelation is characterized by the so-called form factors or flux functions, recently also termed interchange configuration factors. Originally, Lambert (1760), German translation is in 1892 p. 71 found a repetition of the angular function of the angle υ and ω (Fig. A7.1) in a basic type of the flux function:

$$\Phi(\upsilon) = \frac{1}{2}\left[\upsilon \tan \upsilon + \frac{\log \cos \upsilon}{2} - \frac{\tan^2 \upsilon \log \sin \upsilon}{2}\right], \qquad (10.47)$$

$$\Phi(\omega) = \frac{1}{2}\left[\omega \tan \omega + \frac{\ln \cos \omega}{2} - \frac{\tan^2 \omega \ln \sin \omega}{2}\right]. \qquad (10.48)$$

Herman (1900) tested also these functions and later Yamauti (1926, 1929) presented them in a rewritten form using cotangent instead of tangent functions.

The theory of the photic field and its application in the illuminance vector enabled also derivation of many older formulae anew by Fok (1924a, b), with applications for rectangles by Khoroshilov (1938) and Sapozhnikov (1960).

$$\Phi_{12} = \frac{L_1}{2} \oint_{l_1} \oint_{l_2} \ln r d\vec{l_1} \, d\vec{l_2} \text{ (lm)}, \qquad (10.49)$$

with its application to the light vector integration for the specific interchanging form factor u_{ik} corresponding to the flux ratio on the i surface equal to its normalizing $\pi S_i L_i$ or the form factor e_{ik} expressed as the average illuminance factor by Hamilton and Morgan (1952):

$$u_{ik} = \frac{1}{2\pi S_i} \oint_{l_1} \oint_{l_2} \ln r d\vec{l_1} \, d\vec{l_2} \qquad (10.50)$$

or

$$e_{ik} = \frac{1}{2\pi} \oint_{l_1} \oint_{l_2} \ln r d\vec{l_1} \, d\vec{l_2}, \qquad (10.51)$$

where r as the average distance between the plane of the light source and the illuminated plane has to be taken in accordance with the pair of enframing lines of both planes and their integration ranges either in the case of parallel planes or in the case of planes in rectangular position Kittler and Tino (1963). Thus, all interior surfaces shared the entering luminous flux with respect to their mutual position expressed by their form factors as well as their average luminance and reflectance either in their total area or divided in strips. Then the overall interior interreflection can be represented by a set of equations formulated first by Yamauti (1926). The number of equations in the set representing the surfaces or their strips with the same number of unknown fluxes, with inset reflectance, luminance, and form factor values, respectively, presented a tedious and complex calculation procedure that was seldom performed (Moon 1941).

Each surface or its strip i of quasi-uniform luminance contains the originally accepted luminous flux Φ'_i and some additional parts of the reflected flux from all other interior surfaces Φ_i in relation to their participation in the whole interreflection coefficient γ_i as well as due to their reflectance ρ_i and their specific interchanging form factor u_{ik} in the complex system:

$$\Phi_1 = \gamma_1(\Phi'_1 + \rho_2\Phi_2\, u_{21} + \cdots + \rho_n\Phi_n\, u_{n1})\ (\mathrm{lm}), \tag{10.52}$$

$$\Phi_2 = \gamma_2(\rho_1\Phi_1\, u_{12} + \Phi'_2 + \cdots + \rho_n\Phi_n\, u_{n2})\ (\mathrm{lm}), \tag{10.53}$$

$$\Phi_n = \gamma_n(\rho_1\Phi_1\, u_{1n} + \rho_2\Phi_2\, u_{2n} + \cdots + \Phi'_n)\ (\mathrm{lm}). \tag{10.54}$$

These interreflection equations express the mutual position of reflecting planes, their reflectances, and interchanging fluxes in the spatial configuration on the resulting redistribution state.

Only the new computer advantages enabled the set of finite difference equations in matrix form to be formulated and response and excitation vectors either for computer calculations or for analog models to be identified (O'Brien and Howard 1958; O'Brien 1959). New ray-tracing methods can be used to calculate interreflections (Ward et al. 1988).

With respect to the use of luminous flux, luminance, or illuminance from larger planes, it has to be noted that their Lambertian uniform diffuse reflectance ρ_D and some ideal or average values are usually assumed and then, e.g.:

- The surface luminance

$$L_s = \frac{\rho_D E_{av}}{\pi}\ (\mathrm{cd/m^2}). \tag{10.55}$$

- The average illuminance on the whole illuminated plane

$$E_{av} = L_s\, \omega_{pav}\ (\mathrm{lx}). \tag{10.56}$$

- The luminous flux on the whole illuminated plane

$$\Phi = E_{av} S_{il} = L_s S_{il}\, \omega_{pav}\ (\mathrm{lm}). \tag{10.57}$$

In reality, some of the averages do not correspond perfectly with the real distribution of luminance or illuminance on interior surfaces because of either reflectance or illuminance differences in nearer or relatively farther mutually positioned planar elements. In many interiors also the presence of furniture, partitions, or shelves creates secondary obstructions or cavities with randomly placed reflecting objects, surfaces, or elements which usually cannot be taken into account. In such complex situations there is the possibility only to measure the interreflection in the scale models during the design process or in real buildings.

10.5 Daylight Measurements in Real Interiors

Since the time of the Roman Empire, a law has been adopted declaring the right to receive sufficient daylight in interiors, and the need to set specific simple rules or criteria has been established. When interiors are illuminated by windows facing either a street or an atrium, some obstructions of the horizon can be expected. In such critical cases, Vitruvius (1970), in 10 BC, in Chap. VI of book VI of *De Architectura* recommended a simple visual rule to find and evaluate the situation roughly in the interior location or working place with minimum daylight. When kneeling down, with eyes at table level, one should see a strip of the sky above the outside obstruction under the upper window frame. If the visible sky strip is "sufficient," the daylighting is good enough. Later, probably in medieval architecture because of smaller windows this rule eliminated the "sufficient strip" requirement and replaced it by the so-called no-sky line, behind which it was considered that the interior would not serve for visual work. These ideas were followed when theodolite and photographic tools were available in the twentieth century (Swarbrick 1929, 1933; Beckett and Dufton 1932).

Only the invention of light-sensitive materials such as selenium (Sale 1873, Smith 1873) enabled the development of the first illuminance meters, where circular selenium plates collected the small electrical current to be measured by galvanometers or microammeters. These photometric selenium heads were covered for safety reasons and had to be corrected to measure also the low beams in correspondence with the cosine law as proposed by Pleijel (1949), Buck (1949), or Pleijel and Longmore (1952), besides the correction for the $V(\lambda)$ spectral sensitivity by yellow–green filters (Preston 1946). Thus, two calibrated, spectrally corrected and cosine-corrected illuminance meters could be used directly to measure simultaneously daylight illuminance levels inside E_i and outdoors E_{ext} under an unobstructed standard overcast sky to determine the DF, i.e., $\mathrm{DF} = E_i/E_{ext}$ as the basic criterion of daylighting. Some specified DF meters were also proposed and manufactured, e.g., by Turner (1960, 1961) and by Petherbridge and Collins (1961). When selecting a suitable place for the outdoor meter, usually on the roof of the building, one has to check the standard conditions at least by measuring roughly sky luminances in the normal direction to the window plane. Later, selenium photocells were replaced by stabler and smaller silicon cells (MacGowan 1965).

In outdoor studies, sky luminance distributions and illuminance levels have to be checked and measured in small steps from the start and throughout the whole model measurement period, to ascertain acceptability of their compliance with CIE standards. In cases of some small change, then the correct exterior illuminance was applied when calculating the momentary DF(1 : 3) value. Of course, if the outdoor sky distribution is not close to the CIE overcast sky standard, then the model measurements have to be postponed.

Therefore, under real conditions daylight measurements have to be made when a standard or particular sky luminance pattern exists. In some countries such standard conditions are required, with measured tests to prove their actual existence. In other special precautions, e.g., a Czechoslovak standard (1967) required checks of luminance gradation for measurements under overcast skies, whereas for Great Britain other recommendations were published in Anon (1961). Naturally, it is important to perform interior illuminance measurements simultaneously with those outdoors under an unobstructed sky. Such readings are then compared to obtain either DF values or critical illuminance levels on the working plane, which should be above the required minimum standard for the indoor activity. In reality, the resulting measurements are usually DF values, including all relevant and actual influences together. These would include the sky luminance distributions within the window solid angles, exterior reflections, interior interreflections, various glazing losses, obstruction influences, etc. Such indoor illuminance should correspond to the assumed DC value, which should be normalized by the sky luminance within the window solid angle (Tregenza and Waters 1983, Mardaljevic 1999).

It has to be realized that any daylight measurements in real interiors or their models are documenting only momentary local conditions which can change anytime. Therefore, without simultaneous exterior measurements indoor results are suspect and worthless.

10.6 Measurements in Complex Architectural Models to Evaluate Daylighting During Design

As architects quite often design unusual and extraordinary buildings, they often contact daylight experts for advice. However, some of their designs are so complicated in space organization and aperture positioning or shape that neither available calculation methods and tools nor computer programs can cope with the distribution of daylight and sunlight including interreflections in irregular spaces and surfaces. In such cases, during the design stage there is only the possibility to test the design via model measurements either outdoors under standard conditions or under a calibrated artificial sky in a laboratory.

Models are the reduced simulations, in scale 1:10 or 1:20, of buildings which are to be built; therefore, relatively good approximation of sizes, window frames, and reduced actual glazing material as well as indoor surfaces of designed colors and reflectances reproduced either in actual paints or wallpaper are all required. The possibility to operate and replace the illuminance sensors at working plane height has to be enabled either via the model floor grooves or by opening one side wall, but all stray light during the measurements has to be blocked fully. Some experience with the models can be found in Musgrove and Petherbridge (1964). The larger the model scale and window details, the more accurate are the illuminance readings inside the model. Previously, selenium cells with a diameter 2.5 or 5 cm were used, which measure in a model scale of 1:10 the average illuminance on a 25- or 50-cm circular plane element. Further, because of the horizon scale error inherently induced by the model to artificial sky size ratio, especially in mirrored-box-type artificial skies and, the hysteresis malady in the selenium photocell, MacGowan (1965) recommended use of the much smaller and more accurate silicon sensors universally employed today.

Satisfactory photometric instrumentation enabled illuminance or luminance distributions to be measured either in real building interiors or in their models under actual sunlight and skylight conditions. Thus, all relevant influences, whether sky luminance patterns or interreflection effects, are inherent in the resulting levels and different architectural solutions or alternatives can be checked on models during the design process.

Realizing these facts, the daylight criterion systems of both DFs and DCs assume the exact existence of a particular sky luminance pattern that has to await model measurements in nature to be done. Furthermore, these sky conditions have to be documented to prove constant existence during the measurement period. In general, model experiments are very important in research on skylight redistribution in interiors especially owing to interreflections and serious hypothesis to identify relevant influences. Such tiresome and long series of measurements can be realized in stable laboratory conditions and results with proofs are usually archived in private research reports and the bulk are not publishable in full. In this sense, sometimes panel models are used (Longmore 1962) and several objectives, questions, and goals can be followed (Spitzglas et al. 1985).

10.7 Artificial Skies for Laboratory Model Measurements

Historically, many trials to model some sky parts were tested as shown in the summary of first artificial skies by Kittler (1959), where all simulated either the uniform or overcast sky luminance distribution realized on their reflective surfaces. Longmore (1962) described some problems and aims of model measurements under artificial skies.

To overcome the measurement difficulties associated with the uniqueness and great variability of sky luminance patterns occurring in nature, artificial sky models of standard sky distributions were constructed in laboratory facilities. These facilities provided stability under defined conditions for the measurement of daylighting for research under stable and standard calibrated conditions. The models were initially small boxes with either filament or fluorescent sources covered with diffusing glazing. These were placed directly over shafts, light wells, hollow light guides, or window openings, below or beyond which illuminance distributions were measured. Later, these sky modeling boxes were placed on mirrored-box-type model "exterior" spaces that formed the so-called box-type artificial skies of different plan dimensions and various heights (Pleijel 1949; Hopkinson and Longmore 1954).

Any type of artificial sky for laboratory model measurements suitable to evaluate daylight conditions has to have a relevant or standardized luminance pattern which has to be stable and calibrated. Several types of artificial skies exist with respect to their shape:

- Box-type skies with a square or octagonal plan
- Hemispherical or semiellipsoidal skies, i.e., circular in plan
- Vertical lune-type skies representing a cut section of a fictitious hemispherical sky

However, soon it was realized that the multipurpose hemispherical artificial skies were capable of reproducing outdoor sky conditions more realistically than the box-type skies. So either reflective white or translucent diffuse domes with 4–12-m diameters were built in various daylight research laboratories and stations (Longmore 1962; Selkowitz 1981; Navaab 1996). The creation of the luminance pattern in dome-type artificial skies is produced on white reflective surfaces where the luminance distribution is simulated by light sources which project from below and within the sky simulator. Although such artificial skies modeled mainly overcast or uniform sky patterns, some were also equipped with artificial sun parabolic reflectors to simulate sunlight. However, after the standard of a CIE clear sky had been adopted, one artificial sky could model even that luminance distribution, together with the artificial sun facility (Kittler 1974).

Artificial sky simulators were also built to use translucent or transparent transilluminated surfaces in domes or in mirrored boxes, with or without segmental dimmable luminaires. The former sky type was constructed in Cambridge (Croghan 1964). Some larger artificial skies, e.g., at the Institute of Building Physics in Moscow and at the Sekisui Building Institute in Nara, Japan (Okado et al. 1997), tried to simulate sky patterns by dimmable sources. However, their designed luminance patterns were overwhelmed by the luminance unevenness caused by the extreme luminance of sources (hot spots) and unwanted dim (stripes) of the sky's structural support elements. In combination, these design gaffs created insurmountable distribution problems and these sky simulators were never commissioned or calibrated. In the large box-type artificial skies or domes, the structure supporting the light-delivery translucent panels inherently interfered with their required luminance. Tregenza (1989b) proposed a new solution to that malady: to construct only sections or meridian segments of hemispherical skies as part of any sky hemisphere or luminance distribution that could be modeled. Thus, DCs could be measured with smaller and cheaper equipment. However, the accuracy of such a solution depends on the division of the sky and the number of its finite elements participating in achieving a particular sky luminance pattern. However, such a portioned dome artificial sky was built at the Lighting Research and Experimentation Center (CERSIL) in Turin, Italy, simulating the sky with 25 circular luminaires with an additional sun simulator producing parallel rays with a flux output of 27,600 lm (Lo Verso 2000; Filippi et al. 2001).

In general, all artificial skies are not capable of representing absolute luminance and illuminance levels that occur in reality. Laboratory measurements simulate only unique conditions modeled by the particular artificial sky. In this respect, measurements under outdoor conditions can show better the real daylight changes and levels, especially those with complex obstructions, shading, and interreflection situations either in the stage of design or during the real performance and maintenance of buildings.

10.8 Partial Conclusions

Although the sky illuminance calculations presented a difficult task for Renaissance scholars, most were elegantly solved by Lambert, but the interreflection and complicated redistribution of luminances and fluxes in cavity-like interiors presented a challenge even for Lambert. His assumption of a uniform sky luminance also partly agreed with the diffuse interior surfaces and the image of relatively even light flux exchanges among interior surfaces. Thus, that attracted him to determine mathematically the interchange relations or abilities of rectangular planes in different mutual positions within rectilinear spaces. However, in reality, every plane surface or even its patch with different luminance is balanced by interreflection in accordance with its position or solid angle to which the light flux from the interior aperture can reach. Lambert (1760) angularly defined the interchange (configuration) factor formulae, which were rediscovered by Yamauti (1929) but are still seldom applied in the complicated system of sets of flux formulae. These sets are even more complex when real sky luminance patterns and penetration of sunbeams are taken into account. Such daily changing exterior conditions and influences of reflected daylight from exterior surfaces additionally elevate the influx and interreflection level indoors.

The complex interreflection of sunlight and skylight within any interior presents a difficult task for interpretation in calculation programs. Practical approximation formulae were derived from scale model measurements verified partly in real interiors. Scale model measurements are based on analog simulations of different interiors following the design alternatives. These alternatives are tested before arriving at final decisions or are aimed at dealing with theoretical and research questions or analysis to identify the contribution of variables and illuminance components with special regard to interreflection.

Computational algorithms and programs treat the direct luminance patterns within the window solid angle separately and differently from the way they treat the interreflection between the surfaces of the enclosure. However, the photosensors in any real interior or in its scale model respond to the resulting "stabilized" situation of the redistribution of luminances within the architectural space or even with particular influence to the outdoor scene. In fact, with use of a scale model, direct skylight can also be identified in a black model with fluent changes of the outdoor sky luminance pattern as well as changes of obstructions and their reflectances. So, initial model measurements can apply to only the simplest forms and detail to provide information for parametric computer simulations or gross rather than detailed improvements. Photometric measurements and the analysis of such measurements can resolve discrepancies between measured and calculated parameters or criteria.

Unfortunately there is no user-friendly computer program for the calculation of precise interior interreflection due to multiple sky luminance changes seen through the window by all interior surfaces.

In usual practical cases, to avoid the tiresome and lengthy interreflection calculations, several approximate formulae or tools are currently applied. However, computer programs are awaited to solve such tasks in a more accurate way with user-friendly possibilities, to achieve ERC and IRC results comparable to the SC precision. Nevertheless, unusual interior spaces with colored surfaces present problems that cannot be solved by calculations. Therefore, only model measurements can yield the answers to questions about how and under which conditions daylight interreflections will brighten any or every part of complex enclosed spaces.

Appendix 10

Special Laboratory Possibilities to Test Complex Interreflections in Designed Architectural Spaces

Unique architectural solutions using exceptional space forms, shapes, color rendering, or unusual fenestration systems for daylighting often present taxing design scenarios. Such novel applications can create very complex interreflection situations with consequences that cannot be predetermined either by calculation methods or by computer programs. However, in a research or testing laboratory equipped with an artificial sky and sun and appropriate photometric facilities, scale model measurements can be made to reveal the effects of complex interreflections or light redistribution influences. Scale models can simulate alternative design scenarios at the early stages of the design process. These can model a black-clad interior in order to separate direct skylight or sunlight influences from those caused by interior interreflectance. With such scale models placed under a sophisticated artificial sky or heliodon, the influence of the sky and sun can be tested under conditions in a sequence of hypothetical modeling alternative solutions.

The traditional use of scale models for the analysis of daylight and sunlight in rooms, atria, or whole architectural complexes can also be similarly scale-modeled and usefully extended to nontraditional special or novel devices such as anidolic systems or hollow light tubes, diffusers, laser-cut panels, and special glazing materials. Side-lit interiors are typically characterized by illuminance decreasing with the distance from the window. Nevertheless, technology can alter the rate of decrease of such lighting. Devices have been designed to transport light during sun-shaded situations or both sunlight and skylight during sunny situations. However, the efficiency of light-guiding systems is different in "enhancement" and "redirection" situations. Edmonds (1993) measured indoor illuminance under various sky conditions in a model room with laser-cut panels which were transparent and deflected skylight deeper into the interior. Courret et al. (1996) built a scale model with anidolic zenithal openings and set it in a heliodon to test the direct and diffuse light penetration into the modeled interior. These measurements confirmed the belief that under various sky luminance distributions there are significant influences on the efficiency of these novel transport or enhancement daylighting devices.

Innovative approaches and special laboratory equipment for the measurement of advanced fenestration daylight delivery systems are now complemented by faster analysis of the optical properties of special glazing materials and the development of digital imaging techniques in the determination of the daylighting performance characteristics of fenestration. Special laboratory instruments have also advanced the state of this art. Goniophotometers were designed and constructed especially for complex fenestration systems (Andersen and de Boer 2006), and for sunlight-simulated penetration using scale models (Andersen et al. 2005). Special glazing, with spectrally and angularly selective coatings, was tested using a scanning and projection goniophotometer and a spectral video-goniophotometer (Gayeski and Andersen 2007), or such innovative daylight systems set under a heliodon tested the direct and diffuse light penetration into the interior (Littlefair 1990). All devices transport light during a sun-shaded situation or both sunlight and skylight during a sunny situation, and the efficiency of light-guiding systems is different in both of these situations. Darula et al. (2010) tested penetration of daylight through a hollow tube under an artificial sky adjusted to uniform and CIE overcast sky luminance distribution. These measurements confirmed the supposition that under various sky luminance patterns there are significant influences on the efficiency of these novel devices.

New light delivery, shading, enhancement, and/or directing optical elements of building fabric, shading systems, and concentrators present many problems which can be investigated and tested in daylight research laboratories prior to field pilot evaluation and real-world validation studies (Müller 1994).

References

Anon: The standardization of daylight measurements. Light and Lighting, **54**, 71, (1961)

Arndt, W.: Praktische Lichttechnik. Union, Berlin (p.112) (1938)

Arndt, W.: The calculation of reflected daylight. Compte Rendu CIE, Vol.2, TC 3.2, A/10 (1955)

Beckett, H.E. and Dufton, A.F.: A photographic method of determining Daylight Factors and periods of insolation. Journ. Sci. Instruments, **9**, 5, 158–164 (1932)

Buck, G.B.: The correction of light sensitive cells for angle of incidence and spectral quality of light. Illum. Eng., **44**, 293–302 (1949)

CIBSE – The Chartered Institution of Building Services Engineers: Daylighting and window design, Lighting guide 10, CIBSE, London (1999)

Crisp, V.H.C. and Littlefair, P.J.: Average daylight factor prediction. Proc. Nat. Lighting Conf., Cambridge, 234–243, CIBSE, London (1984)

Croghan, D.: Transilluminated domical artificial sky. Light and Lighting, **57**, 10, 290–293, (1964)

Czechoslovak standard: Měření denního osvětlení. (In Czech: Measurement of daylighting.): ČSN 36 0014, Office for Normalization and Measurements, Prague (1967)

Dresler, A.: The "Reflected Component" in daylighting design. Transac. Illum.Eng.Soc.(London), **19**, 2, 50–60 (1954)

Filippi, M., Aghemo, C., Pellegrino, A., Lo Verso, V.: Daylight design through the use of an artificial sky. Proc. 12th Intern. Conf. Light, 117–124, High Tatras, Slovakia (2001)

Fok, V.A.: Osveshchennost ot poverkhnostey proizvolnoy formy. (In Russian: Illumination from surfaces of arbitrary shape). Trudy Gos. Opt. Inst. **3**, 28, 1–12. Leningrad (1924a)

Fok, V.A.: Zur Berechnung der Beleuchtungsstärke. Zeitschr. für Physik, **28**, 2, 102–113 (1924b)

Frühling, H.G.: Die Beleuchtung von Innenräumen durch Tageslicht, ihre Messung und ihre Berechnung nach der Wirkungsgradmethode. Lichttechnische Helfe der DBG, Berlin (1928)

Griffith, J.W., Arner, W.J., Conover, E.W.: A modified lumen improved method of daylighting design. Illum. Eng., **50**, 3, 106 (1955)

Griffith, J.W.: Predicting daylight as interior illumination. Libbey-Owens-Ford Glass Co., Ohio (1958)

Hamilton, D.C. and Morgan, W.R.: Radiant Interchange Configuration Factors. National Advisory Committee for Aeronautics, Techn. Note NASA 2836, Washington (1952)

Hannauer, K.: Mathematisch-graphisches Verfahren zur Ermittlung der Vertikalbeleuchtungsstärke auf den Seitenwände eines Lichthofes oder bei Berücksichtigung der Gesamtreflexion der Seitenwände. Das Licht **10**, 12, 153–154, **11**, 1 19–20 (1940, 1941)

Herman, R.A.: A treatise on geometrical optics. Cambridge University Press, Cambridge (1900)

Hopkinson, R.G., Longmore, J.: Study of the interreflection of daylight using models and artificial skies. Nat. Build. Studies Res. Paper 24, DSIR, London (1954)

Hopkinson, R.G., Longmore, J., Petherbridge, P.: An empirical formula for the computation of the Indirect component of Daylight Factor. Transac. Illum. Eng. Soc.(London), **19**, 7, 201–219 (1954)

Hopkinson, R.G., Petherbridge, P., Longmore, J.: Daylighting. Heinemann, London (1966)

IES – Illuminating Engineering Society: The calculation of utilization factors. The BZ method. (1971)

IESNA – Illuminating Engineering Society of North America: The IESNA lighting handbook, reference and application. IESNA New York (2000)

Khoroshilov P.I.: Raschot svetovogo potoka ot pryamougolnika na pryamougolnik. (In Russian: Calculation of the light flux from a rectangle on a rectangle). Svetotekhnika, **6**, 2, 43–47 (1938)

Kittler, R.: Razvitye metodov raschota yestestvennogo osveshcheniya pomeščeniy s uchetom otrazhennogo sveta. Svetotekhnika, 2, 2, 18–21 (1956a). English translation by R.G. Hopkinson: Development of methods for the calculation of natural illumination in building, taking into account reflected light. Library Com. 780, Build. Res. Station, Watford (1957)

Kittler, R.: An historical review of methods and instrumental daylight research by means of models and artificial skies. P-59.8, 319–334, Vol. B, Proc. CIE Session, Brussels (1959)

Kittler, R.: Slnko a svetlo v architektúre. SNTL, Bratislava (1956b). English translation by E. Bok: Sun and light in architecture, Pilkington Broth.Ltd., St. Helens (1963)

Kittler, R. and Tino, J.: Einstrahlzahlen bei der Flächenstrahlung in einem rechtwinkligen Raum. Gesundheits-Ingenieur, **84**, 1, 15–19 (1963)

Kittler, R.: Ob empiricheskikh formulyach dlya opredeleniya otrazhenoy sostavlyayuchey yestestvennoy osveshchennosti pri bokovom osveshcheniyi pomeshcheniy. (In Russian: About empirical formulae for the determination of the reflected component of daylight illuminance in side-lit rooms). Svetotekhnika, **10**, 3, 18–21 (1964)

Kittler, R. and Kittlerová, L.: Návrh a hodnotenie denného osvetlenia. (In Slovak. Design and assessment of daylight illumination). Slovak Publ. Techn. Lit., Alfa, Bratislava (1968, 1975)

Kittler, R.: A new artificial 'Overcast and Clear Sky' with an artificial sun for daylight research. Lighting Res. Technol., **6**, 4, 227–229, (1974)

Krochmann, J.: Über die Bestimmung des Innenreflexionsanteils des Tageslichtquotienten. Lichttechnik, **14**, 3, 105–108 (1962)

Lambert, J.H.: Photometria sive de mensura et gradibus luminis, colorum et umbrae. Klett Publ., Augsburg (1760). German translation by E. Anding: Lamberts Fotometrie. Klett Publ., Leipzig (1892). English translation by D.L. DiLaura: Photometry, or on the measure and gradation of light, colors and shade. Publ.IES Amer., N.Y. (2001)

Li, D.H.W. and Cheung, G.H.W.: Average daylight factor for the 15 CIE standard skies. Lighting Res., Technol., **38**, 2, 137–152 (2006)

Li, D.H.W., Lam, T.N.T., Lam, J.C.: Average daylight factor under various skies and external environments. Proc. 5th Solaris Conf., 174–179, Univ., of Technol., Brno (2011)

Longmore, J.: The role of models and artificial skies in daylight design. Transac. IES., **27**, 3, 121–138, (1962)

Longmore, J.: Daylighting: a current view. Light and Lighting, **68**, 3, 113–119 (1975)

Longmore, J.: The engineering of daylight. Developments in lighting - 1, 169–192 (1978)

Lo Verso, V.: Progetto di un cielo a luminanza variabile: equazionirelative ai differenti modelli di ciello. (In Italian: Design of a sky with variable luminance: equations for the various models of the sky). Proc. CODEA, Nat. Energy and Environment Conf., Materna, Italy (2000)

Lynes, J.A.: A sequence for daylighting design. Light., Res., Technol., **11**, 2, 102–106 (1979)

MacGowan, D.: Miniaturisation in daylight prediction. Light and Lighting, **58**, 8, 256–258 (1965)

Mardaljevic, J.: Daylight simulation: Validation, Sky Models and Daylight Coefficients. PhD Thesis, De Montford Univesity, Leicester (1999)

Mascart, E.: Sur la mesure de léclairement. Bull. Soc. Int. des Électr., 5, 103 (1888)

Meshkov, V.V.: Osnovy svetotekhniki. (In Russian. Basis of illuminating engineering.) State Energy Publ., Moscow (1957)

Moon, P.: The scientific basis of illuminating engineering. McGraw-Hill Inc., Revised edition: Dover Publ., Inc., New York, London (1936, 1961)

Moon, P.: Interreflections in rooms. Journ. Opt. Soc. Amer., **31**, 734–382, (1941)

McNicholas, H.J.: Absolute methods in reflectometry. Bureau of Standards Journ. of Res., 1, 29 (1928)

Musgrove, J. and Petherbridge, P.: Lighting-model construction for appraisals of building interiors. Archit., Journ., **125**, 3232, 216 (1957)

Navaab M.: Scale model photometry techniques under simulated sky conditions. Journ. IES, **1**, 57–68 (1996)

O'Bien, P.F. and Howard, J.A.: Predetermination of luminances by finite difference equations. Proc. Nat. Techn. Conf. I.E.S., Toronto, (1958)

O'Bien, P.F.: An analogue computer for the predetermination of luminance patterns. Paper P-59.18, 235–247, Vol. B, Proc. CIE Session, Brussels (1959)

Okado M., Goto, K., Nakamura, H., Koga, Y., Fujii, S.: Development of a new artificial sky. Proc. Lux Pacifica, E63-66 (1997)

Petherbridge, P. and Collins, W.M.: The EEL – BRS Daylight photometer. Journ. Sci. Instr., **28**, 9, 375 (1961)

Pleijel, G.: Daylight investigation. Statens Kom. f. Byggnadsforskning, Rap.17, Stockholm (1949)
Pleijel, G. and Longmore, J.: A method of correcting the cosine error of selenium rectifier photocells. Journ. Sci. Instr., **29**, 5, 137 (1952)
Powell, G.L. and Kellog, W.: Reflecting properties of white interior paints of varying compositions. I.E.S. Transac. **21**, 70–78 (1926)
Preston J.S., The reflection factor of magnesium oxide. Opt. Soc. Transac., **31**, 776 (1929–1930)
Preston, J.S.: The specification of the spectral correction filter for photometry with emission photocells. Journ. Sci. Instr., **23**, 211–216 (1946)
Reinhart, C.F. and LoVerso, V.L.K.: A rules of thumb based design sequence for diffuse daylight. Light. Res. and Technol., **42**, 1, 7–32 (2010)
Sale, R.E.: The action of light on the electrical resistance of selenium. Proc. of the Royal Soc., **21**, 283–285 (1873)
Sapozhnikov, R.A.: Teoreticheskaya fotometriya. (In Russian: Theoretic photometry). Gos. Energet. Izdat. Moscow, Leningrad (1960)
Selkowitz, S.: A hemispherical sky simulator for daylight model studies. Proc. 6th Passive Solar Conf., 850–854, University of Delaware (1981)
Selkowitz, S., MacGowan, D., McSwain, B., Navvab, M.: A hemispherical Sky Simulator for daylighting studies: Design, construction and operation. Proc. IES Annual Techn. Conf., Toronto, (1981)
Schanda, J. et al.: Colorimetry: Understanding the CIE System. CIE Preprint Edition. CIE CB, Vienna (2006)
Smith, W.: Effect of light on selenium during the passage of an electrical current. Nature, 303 and Amer. Journ. of Science, **5**, 301 (1873)
Spitzglas M., Navvab, M., Kim, J.J., Selkowitz, S.: Scale model measurements for a daylighting photometric data base. Journ. IES, **15**, 1, 41–61 (1985)
Swarbrick, J.: The Swarbrick Daylight Factor Theodolite. British Patent No. 356.822 (1929)
Swarbrick, J.: Easements of light. Vol. II. Batsford Ltd., London, Wykeham Press, Manchester (1933)
Tregenza, P.R. and Waters, I.M.: Daylight coefficients. Lighting Res. Technol., **15**, 2, 65–71 (1983)
Tregenza, P.R.: Modification of the split-flux formulae for mean daylight factor and internal reflected component with large external obstructions. Lighting Res. Technol., **21**, 3, 125–128 (1989a)
Tregenza, P.R.: Daylight measurement in models: New type of equipment. Lighting Res. Technol., **21**, 4, 193–194 (1989b)
Taylor, A.H.: A simple portable instrument for the absolute measurement of reflection and transmission factors. Scientific Papers of the Bureau of Standards, **16**, 391, 421–423 (1920) and **17,** 405, 1–6 (1922)
Taylor, A.H.: Errors in reflectometry. Journ. of the Opt. Soc. Amer., **25**, 51 (1935)
Turner, D.P.: Daylight Factor meter. Journ. Sci. Instr., **37**, 9, 316 and **38**, 1, 64 (1960, 1961)
Turner-Szymanowski, W.: A rapid method for predicting the distribution of daylight in buildings. Eng., Res., Bull., 17, Univ. of Michigan, Ann Arbor (1931)
Ulbricht, R.: Die Bestimmung der mittleren räumlichen Lichtintensität durch nur eine Messung. Elekrotechn. Zeitschrift, **29**, 595–597 (1900)
Ulbricht, R.: Das Kugelphotometer. R. Oldenbourg Verlag, Munich, Berlin (1920)
Vitruvius, M.P.: De architectura libri X. Rome (1487) English translation by F. Granger: Of Architecture, The ten books.. Heinemann, London (1970)
Ward, G., Rubinstein, F., Clear, R.: A ray tracing solution for diffuse interreflection. Computer Graphics, 22, 4, 85–92 (1988)
Yamauti, Z.: The light flux distribution of a system of interreflecting surfaces. J. Opt. Soc. Am., **13**, 561–561 (1926)
Yamauti, Z.: The amount of flux incident to rectangular floor through rectangular windows. Res. of the Electrotechn. Lab., Tokyo. No.250 (1929)

References to Appendix

Andersen, M., Ljubicic, D., Browne, C., Kleindienst, S., Culpepper, M.: Automated assessment of light redirecting properties of materials and sunlight penetration within scale models. Proc. ISES Solar World Congress (2005)

Andersen, M. and de Boer J.: Goniophotometry and assessment of bidirectional photometric properties of complex fenestration systems. Energy and Buildings, **38**, 836–848 (2006)

Courret, G., Paule, B., Scartezzini, J.L.: Anidolic zenithal openings: Daylighting and shading. Lighting Research and Technology, **28**, 1, 11–17 (1996)

Darula, S., Rybár, P., Mohelníková, J. Popeliš, M.: Measurement of tubular light guide efficiency under the artificial sky. Przeglad Elektrotechniczny, **86**, 10, 177–180 (2010)

Edmonds, I.R.: Performance of laser cut light deflecting panels in daylighting applications. Solar Energy Materials and Solar Cells, **29**, 1–26 (1993)

Gayeski, N. and Andersen, M.: New methods for assessing spectral, bi-directional transmission and reflection by complex fenestration systems. Proc. SOLAR 2007: Sustainable energy puts America to work, Cleveland, 2, 639–646 (2007)

Littlefair, P.J.: Innovative daylighting: Review of systems and evaluation methods. Lighting Res. Technol., **22**, 1, 1–17 (1990)

Müller, H.F.O.: Application of holographic elements in buildings for various purposes like daylighting, solar shading and photo-voltaic power generation. Renewable Energy, **5**, 2, 935–941 (1994)

Chapter 11
The Neurophysiology and Psychophysics of Visual Perception

11.1 Ancient Notions about Vision and Light Relations During Classical Antiquity and the Middle Ages

The Greco-Roman civilization, referred to as the period of classical antiquity, spanned from about 800 BC to AD 500. The Greek followed by the Roman philosophers began to develop a science of optics in their investigation of the natural sciences. However, the philosophers of that era were highly religious and that influenced the science developed, which was quite affected by spiritual and religious beliefs rather than by experiment, measurement, analysis, and then determination of only the facts. However, that is easier said than done for the philosophers were developing theories, initially from a void. In the period from about 440 BC to about 270 BC, Democritus followed by Epicurus concluded that for the eyes to see an object it had to come into physical contact with the eye by pressing the air between the eye and the object, so transmitting its color and shape to the eye. This is called the "intromission theory." However, in a variation of that intromission theory, Epicurus was of the opinion that it was not the compressed air between the object and eye that resulted in vision, but particles from the object traveling to the eyes. He hypothesized that produced vision by gradually shrinking objects through particles from the object filling the empty spaces in the object. Still, it was an intromission theory (Ackerman 1978).

Then Plato, followed by the great geometrician Euclid and much later, about AD 150, Ptolemy enhanced the intromission theory with a new "extramission theory," which states that the eye "sees" when light emanating from it hits an object. The eyes are thus light producers. Plato conjectured that when an object is seen it releases flame particles of different size which cause different colors (hues) and color depths (chroma and value).

Still in the same timeframe, Aristotle disagreed with Democritus, Epicurus, and Plato because he could not prove or disprove their theories on intromission or extramission by experiments; thus, he developed his own theories, rationally. He argued that the eyes could not possibly produce light or it would be possible to see

DOI 10.1007/978-1-4419-8816-4_11, © Springer Science+Business Media, LLC 2012

in the dark. And there was no evidence to support a theory that objects shrink to enter the eye. Instead, he observed that sunlight reflected from objects and that reflected light entered the eye and caused vision and that light is invisible (immaterial). Thus, Aristotle laid the foundation for modern optics!

Rufus of Ephesus (active late in 1st century BC until about mid 1st century AD), another of the great Greek physicians, was influenced by, but not always in agreement with, Hippocrates (ca. 460–377 BC), the acknowledged father of medicine. Likewise, Galen (ca. AD 133–200), whose influence spanned nearly two millennia and who was arguably the greatest of the medical researchers of antiquity, was also influenced by Hippocrates. Little is known about the actual work of Hippocrates, other than through the work of Galen, Rufus, and later others in the Middle Ages, particularly Alhazen (Lindberg 1967).

During the Middle Ages, the 900 dark years for European science, the Islamic scientists continued to explore and refine Greco-Roman thinking. The work of the ophthalmologist Hunayn ibn Ishaq was more akin to that of Galen, but some 700 years later. Although useful work and experience was gained, it did not make a material difference to that accomplished by the Greco-Romans; nevertheless, it kept the science alive. However, the great Muslim polymath Alhazen, born in Basra ca. AD 965, is credited with developing the "scientific method" – the systematic observation of phenomena and analysis of data in relation to theory. Through his research he discovered the principles of reflection, refraction, and transmission of light and applied these even to predict the depth of Earth's atmosphere. Alhazen's contributions to the science of optics were so significant that his work was the optics theory standard until Newton, some 600 years later, published *Philosophiae Naturalis Principia Mathematica* (Newton 1687).

11.2 The Renaissance Achievements in Explaining Visual Color Images

After the Muslim philosophers kept alive the work of the Greco-Roman philosophers, Europe reawakened from its near millennium of dark years' "science slumber." Probably the best known early works of the Renaissance were by Leonardo da Vinci, Johannes Kepler, René Descartes, and Isaac Newton.

Leonardo da Vinci's (1452–1519) knowledge of eye anatomy corrected and refined the much earlier work on eye anatomy of Rufus, Galen, Hunayn, and Alhazen. However, Leonardo was not satisfied with only the external look of eyes (Fig. 11.1); he used human external physical expression as a reflection of the inner workings of the mind. He tried to discover the inner path of images of what is seen by the eyes to understand perception in the brain. Leonardo also studied the eye's projection mechanism and how perspective distortions occur. The great Italian polymath tried to accomplish all of that by dissecting some 20 cadavers over 20 years of study, with his later drawings of the eye evolving with his

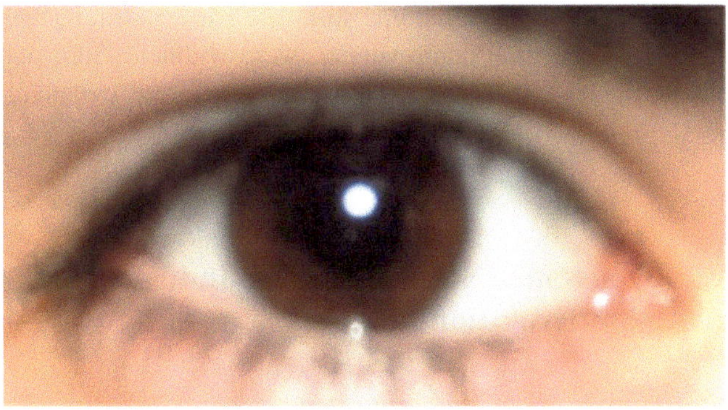

Fig. 11.1 Leonardo da Vinci believed that human eye and expression reflect the inner working of the mind

accumulating knowledge and all meticulously layered human external physical expressions as reflections of the inner workings of the mind. Leonardo knew much about the complex anatomical structure and function of the human eye (Ackerman 1978). Indeed, he knew much of present eye anatomy and vision and extended that knowledge to the geometrical proportions of the human body in his Vitruvian Man, for many still today “the” standard of architectural proportion.

Johannes Kepler’s (1571–1630) great primary work was on planetary motion; however, he also studied vision. He was the first person to suggest that the eye lens focuses images on the retina and, he produced eyeglass corrections for nearsighted and farsighted people. Kepler was the first to explain that depth perception required the use of both eyes, hence binocular vision, and he also explained the optics of the telescope.

René Descartes (1596–1650), the philosopher of “I think, therefore I am” fame, mathematician and dreamer, founded analytic geometry. On heated wax that changed its shape and destroyed the map of it made by the senses, Descartes remarked “And so something which I thought I was seeing with my eyes is in fact grasped solely by the faculty of judgment which is in my mind.” However, he also proved Kepler’s theory that the eye lens focuses images on the retina to be correct and discovered by experiment that the lens casts an inverted image on the retina. The views of Descartes on the senses and the pineal gland were far-reaching, complex, and profoundly affect modern studies involving human judgment and perception (Lokhorst and Kaitaro 2001).

Isaac Newton (1641–1727) produced the science that at last superseded that of Alhazen, which had been the seminal work in optics for 600 years. Newton investigated the refraction of light by a glass prism and developed increasingly elaborate experiments. He discovered measurable, mathematical patterns in the phenomenon of color and found white light to be a mixture of infinitely varied colored rays, with each ray definable by the angle through which it is refracted on entering or leaving a given transparent medium. He believed that light consisted of streams of minute particles. From his experiments he found that white light is selectively reflected off different colored surfaces. These findings, both theoretical and experimental, at the time controversial, were withheld from publication for a generation, until Newton's critics were dead, but were eventually published (Newton 1704).

11.3 Post-Renaissance Science and the Industrial Revolution/Evolution Progress

Naturally, there were many polymaths before this time and, there are still polymaths today. However, the universal nature of their deep understanding is more discipline-oriented in scientific discovery, and a new applied-science-oriented pragmatism developed technology during the Industrial Revolution (or evolution) at a pace never before experienced. By about 1750, science had at last advanced to a level where the old-style polymath could no longer preside over such a vast accumulation of knowledge. As a consequence of this new yardstick, scientists with a very wide curiosity spectrum tend only to publish in one of many specialized areas.

The time had come to subdivide the universe into science discipline areas and technology groupings, albeit with substantial overlaps facilitating cross-fertilization of ideas. This was also the beginning of the period of great large-scale technology inventions, with Britain being credited as the birthplace of the Industrial Revolution with such technology advances. Psychophysiophysics, psychophysics, and neurophysiology propelled knowledge and technology pertinent to daylight in humankind constructs. These will be addressed in three streams in subsections and then brought together later in the discussion of applications.

11.4 Psychophysiophysics

This term embraces the scope of research activities of the near-present-time polymaths of the old Renaissance type, with universal wide and deep knowledge and skills in the sciences of physics (natural philosophy), human physiology, and psychology. Purkinje and Helmholtz were two such latter-day scientists who made especially significant contributions to the science of visual perception involving skills in all three areas of psychophysiophysics.

Jan Evangelista Purkyně (in his native Czech), or alternatively Johannes Evangelist Purkinje (1787–1869), a physiologist and near-modern-day polymath, lent his name to many physiological discoveries. Among the most relevant is the long-famous Purkinje shift, which changes the perceived brightness of different colored surfaces, especially those rich in red or blue, as the eye adapts from black–gray nighttime to color daytime vision. That discovery led to much early work on the

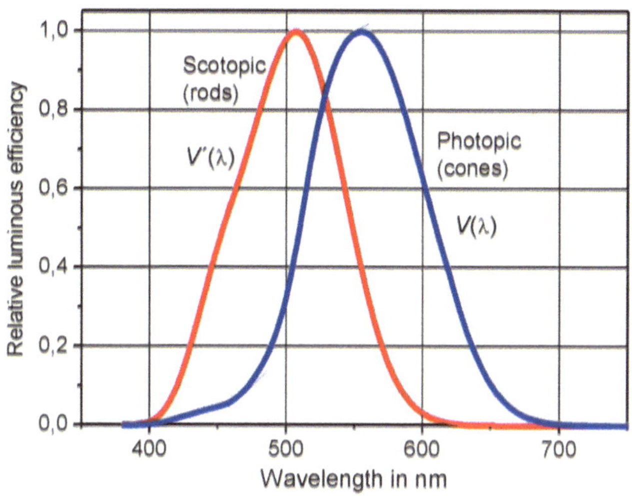

Fig. 11.2 Purkinje shift, refined by König, low luminance vision by rods to daytime vision via cones

relative luminosity function of the eye (Purkyně 1825; König and Dieterici 1886; König 1891) and to the discovery of, at first, two types of light sensor in the retina. Rods, the first sensor type, are bunched in groups and connected to the optic nerve via a ganglionic cell layer providing the high light sensitivity needed for visibility in low-light conditions, but relatively poor visual acuity. Cones, the second sensor type, are densely packed in the foveola, a small area in the center of the fovea where only cone cells exist. These have individual connections to the optic nerve via the ganglionic cell layer. Elsewhere across the retina the cones become increasingly less densely distributed, providing lower sensitivity but good visual acuity for high-light, daytime situations.

The tasks of standardizing the relative luminosity response of the receptors in the human retina became an urgent and continuing responsibility of the International Commission on Illumination (CIE). Obviously, if light were to be measured and compared, it was imperative that the same measuring standard prevailed. Likewise, the photometry required to measure light had to make spectral measurements using the same CIE relative luminosity function of the eye as the standard. The CIE again has this international responsibility and recommends both photometric and visual standards to the International Organization for Standardization (ISO). Using the response of the eye, based on visual investigations, extended from seminal work of Purkinje and refinement of König, the CIE standardized the daylight visibility function (photopic vision) in 1924 as $V(\lambda)$ and the eyes' sensitivity curve for rod vision (dark adaptation – scotopic vision) in 1954 as $V'(\lambda)$. These curves for scotopic vision at low luminance (rods) and photopic vision at high luminance (cones) are shown in Fig. 11.2.

Quantitative description of the transition from light-adapted vision to dark-adapted vision turned out to be a difficult task. This is partly because for light

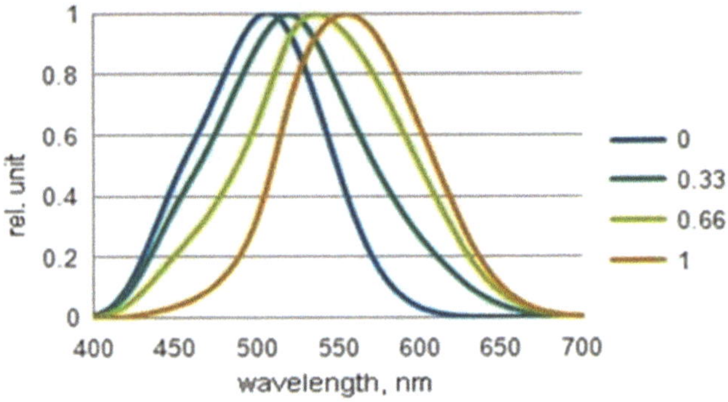

Fig. 11.3 Mesopic vision V_{mes} functions for different values of wavelength in relative terms

adaptation by $V(\lambda)$ was determined by flicker photometry, distinctness of the border, and similar visual investigations (thus it is responsible for task performance), whereas $V'(\lambda)$ is a sensitivity related to low luminance levels. Transition between the two is complicated, as low-brightness perception in the photopic range is a nonlinear, nonadditive phenomenon, whereas higher brightness and luminance are perceived as both linear and additive. However, many other complications exist. As brightness decreases, photopic vision descends into mesopic vision, with both cones and rods firing. As the decrease in brightness continues, the shape of the mesopic response changes as more rods and fewer cones fire until the scotopic range is reached, where only rods fire. That transition is shown in Fig. 11.3 and is explained further below.

Realizing the above, mesopic photometry developed recently in two directions: brightness description and photometry for task performance. CIE Technical Committee TC 1–37 "Supplementary system of photometry" is active in the field of brightness-related photometry that should encompass photopic, mesopic, and scotopic vision. In recent years, two groups performed investigations to determine a task-performance-related mesopic photometry, by trying to answer questions on how quickly an obstacle can be observed and how difficult it is to observe its details. An American group developed the so-called X-model; a European consortium developed the so-called MOVE model. Both models are very similar: both groups constructed the mesopic spectral visibility function from the $V(\lambda)$ and $V'(\lambda)$ functions. In a technical report (CIE 2010), a rationalized combination of the two systems was published. This CIE system defines as the lower limit of the mesopic visibility range with luminance 0.005 cd/m^2, and as the upper limit 5 cd/m^2.

The human sensitivity to light in the whole range is schematically shown in Fig. 11.4.

The relative luminosity function of the intrinsically photosensitive retinal ganglion cell is introduced in Fig. 11.5 (Gall and Biesk 2004). Current work on some of the myriad of nervous system functions of ganglia is likely to reawaken Macfarlane's work on "habituation," basic wiring, and the nature and speed of communication between the higher and lower central nervous system (CNS) and

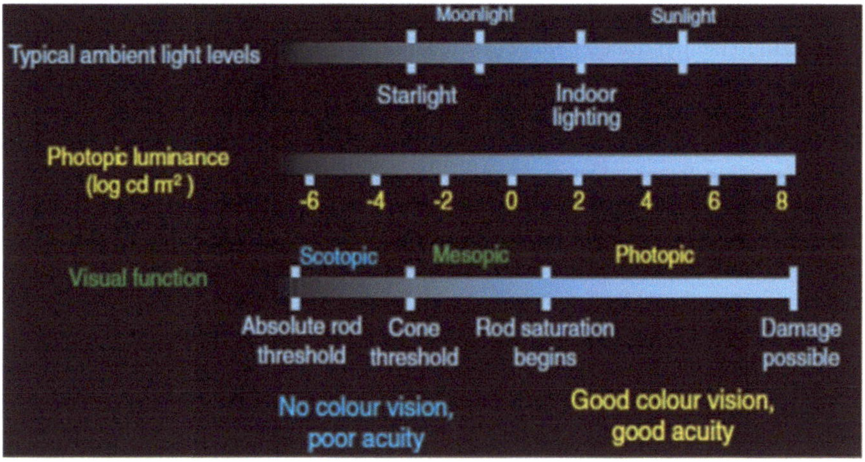

Fig. 11.4 Adaptation to different ambient or eye-incident light

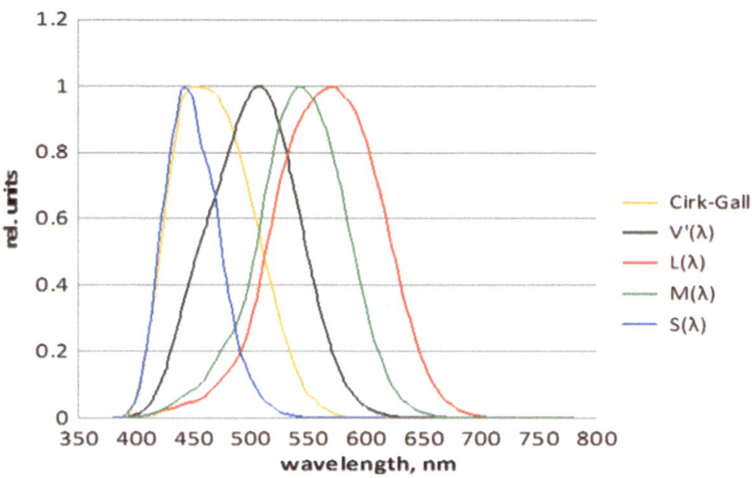

Fig. 11.5 Responsivity of light-sensitive cells in the retina

the role of the spinal cord in storing habituation memory of past environmental exposures, traumatic or climatic over long time periods, even eons.

11.5 Psychophysics of the Visual Environment

This discipline, which had yet to be established in the time of Purkinje and Helmholtz, although they were active in it, is concerned with the measurement of human psychological response to physical stimulus. Hermann Helmholtz

(1821–1894), a German physiologist and physicist, is the penultimate traditional polymath to be mentioned here because he inspired others to excel in fields beyond his own expertise. His famous achievements were in vision research and, as Fechner, on the relationship between perception and the laws of nature; again, as Fechner, with esthetics; in the conservation of energy, with which he is credited as its founding father; electrodynamics; chemical thermodynamics; and the invention of the ophthalmoscope, a huge advance which revolutionized ophthalmology. Helmholtz also inspired his student Wundt to ultimately become the acknowledged father of experimental psychology, and later in his line to influence Stevens via Titchener (1924) and Boring (1942, 1950).

Psychophysics is a specialized scientific area studying the quantitative links and relations between physical/objective stimuli and resulting subjective sensations that can be expressed by a subject. Its founder was Weber (1795–1878), a career physiologist and anatomist at the University of Leipzig. However, his primary area of study from the 1840s was perception of the "just noticeable difference" (jnd) for hearing, pain, sight, smell, taste, touch, and weight (Weber 1834). He found that his subjects did not notice absolute differences but that the jnd was a constant ratio of different stimulus magnitudes; different ratios for different stimuli and extremes at both ends produced no alteration of magnitude perception. He also found that such perceived change was greater if two or more senses were involved in the assessment, i.e., if a weight was lifted, involving both touch and muscle, the jnd was at a higher ratio than that for the same weight being placed on the hand, involving no lifting. Weber established a ratio link between physical stimulus magnitude and the just perceptible change of stimulus magnitude, which he called the "just noticeable difference" or $\Psi(\text{jnd})$. Weber's jnd was a constant ratio specific to stimulus type and magnitude; from these relationships he produced a set of fractions, later called the "Weber fraction." From the jnd data, Weber developed what Fechner later termed "Weber's law,"

$$\Psi(\text{jnd}) = \frac{\Delta\Phi}{\Phi + \Delta\Phi} \tag{11.1}$$

and Weber's continuum-dependent fraction is

$$k_{\text{W}} = \frac{\Delta\Phi}{\Phi + \Delta\Phi}, \tag{11.2}$$

where k_{W} is the Weber fraction for the stimulus type, Φ is the magnitude of the stimulus type (weight, heat, light, etc.), and $\Delta\Phi$ is the stimulus threshold below which no change could be detected for that stimulus type.

Typical Weber fractions are 1/60 for brightness, 1/40 for heaviness, and 1/20 for loudness. Such jnd fractions remained constant over a large intensity range over most continua but failed at extremes at both ends of the stimulus range.

Weber's slightly younger colleague Fechner (1801–1887), like Weber, studied medicine at Leipzig University. However, Fechner lost interest in medicine because of its poor science and turned to physics and mathematics and, by 1839 had developed a career as a physicist. When Fechner became ill then deeply religious, his views changed. He decided that all physical things were unreal and only mental images were real; he had entered a metaphysical world in which he attacked materialism. He argued that if mental states were correlated with neurological events, it should be possible to establish an equation which linked the neurological activity to the conscious event (which is now called neural coding). That was the beginning of the journey of this scientific genius into psychophysics and the universal translation of Weber's experiments on sensation, at least in part, into to a universal perception model to parallel the physical laws of the natural world. Weber's jnd approach had been criticized because in reality assessments were made not simply of the jnd, but over a wide range of physical differences, for instance, over the whole perceptual range of luminance, sound, and so forth found in nature. Fechner looked for universe-wide truths on perception which paralleled the physical laws of the natural universe, even beyond physics into esthetics, especially proportion (Fechner 1860, 1876, 1877, 1882). However, Weber's pioneering work provided the clues to enable Fechner to refine and extend that jnd fraction work. Fechner tested Weber's jnd and fractions over many continua and agreed with Weber's findings but found that although Weber's jnd and fraction experiments worked well for different continua for any specific magnitude in a continuum range, they did not reflect the perceived magnitude differences within the range itself. For instance, Fechner found that doubling of the "perceived" magnitude of change would yield a tenfold increase in physical stimulus magnitude, that is, Ψ was proportional to the logarithm of Φ, as $\log 10 = 1$, $\log 100 = 2$, $\log 1{,}000 = 3$, etc. Therefore, he extended what he had termed "Weber's law" beyond the jnd concept into a universe-wide "logarithmic" relationship between the stimulus ratio change and its perceived magnitude of change. Hence, Weber's jnd for different stimuli could be included as different k_{W} values and physical ratios Φ in a new Weber–Fechner law:

$$\Psi = k_{\mathrm{W}} \log_{10}\Phi, \tag{11.3}$$

where Ψ is the perceived (psychophysical) magnitude and k_{W} is the Weber fraction, i.e., constant depending on the continuum. However, the Weber–Fechner law has been more generally used with $k_{\mathrm{W}} = k$ and Φ as the ratio of stimulus (physical) magnitude to the predetermined or continuum threshold magnitude.

The Weber–Fechner universal (outer psychophysics) psychophysical law was generally accepted by 1860, even though both Weber and Fechner knew that although it was not universally true it was in the right direction. As a by-product it compressed to a more manageable level the huge range which exists in most physically measurable stimuli. Such scale adjustment by logarithmic compression and simple multiple expansions for adequate discrimination is used to construct appropriate, manageable-scale dimensions and must not be confused with the differently motivated and evolved Weber–Fechner law. That law was an attempt to order the perceived world in universal dimensions in a manner similar to the world's physical laws of nature. In so doing, Fechner is generally credited with founding the discipline of psychophysics, and with Weber its first "universal" law, reinforced some 70 years later by Thurstone (1931). That law remained virtually unchallenged for 100 years, perhaps only because the nature of the expression also found general scientific use as a scale compressor. Certainly, when one questions the continued use of the logarithmic scale in psychophysical expressions, the same "compression" excuse invariably follows.

Nevertheless, not all agreed that the Weber–Fechner law was the first such work. Some claimed (Masin et al. 2009) that an alternative derivation of the logarithmic relationship was performed by Bernoulli (1738), who "…derived the logarithmic law using principles other than Weber's law and that Fechner and Thurstone based their derivations on the principles originally employed by Bernoulli…" and "…concluded that awareness of researchers about Bernoulli's and Thurstone's derivations could expand the directions of research on the form of the psychophysical law." Unfortunately, science is full of "territorial" disputes, "lost" data, "overlooked" articles, "negligent" referencing, and "political" scientists. Perhaps such arguments could be placed in best perspective in the ever-diplomatic words of Stevens (1961a, b): "Whatever the subject matter addressed the common unifying theme of the Fechner–Thurstone development is the notion that units of a psychological scale can be fabricated from observations on variability, by means of a systematic analysis of one kind or another. Take away variability and there remains no measurement."

In America, James McKeen Cattell (1886), who collaborated with Galton (1883) (Sokal 1972), was appointed to the first Professorship in Psychophysics by the University of Pennsylvania, in 1889. Cattell, like Titchener (1924), Boring's mentor, was trained by Wilhelm Wundt at Leipzig University. Wundt was a student of both Hermann von Helmholtz and Johannes Peter Muller. Although Cattell was in fact the first Professor of Psychophysics, that appointment was short-lived, for Cattell moved to Columbia University, as Professor of Psychology, after being in post at the University of Pennsylvania for only 6 months.

Stevens (1906–1973) studied psychology in the Department of Philosophy at Harvard University under Boring, and physiology under Davis. Stevens's first venture into psychophysics was in psychoacoustics in the mid to late 1930s with Davis. Stevens's own academic neuropedigree is a gold standard; from Boring, Titchener, and via Wundt (pioneer of experimental psychology) to Helmholtz (pioneer of energy conservation). Stevens had three undeniable published strings and strengths to his bow: psychoacoustics (Stevens and Davis 1938); scales of measurement (Stevens 1946); and psychophysics (Stevens 1957; Stevens, JC and Stevens, SS 1963) with (Stevens 1970). Stevens is the undeniable founder and pioneer of modern psychophysics. Stevens's PhD research supervisor at Harvard, Boring (1942, 1950, 1961), had a practical no-nonsense engineering approach to science; he had a degree in electrical engineering before embarking on his route toward experimental psychology. That attitude undoubtedly influenced Stevens to broaden his skills base to combine practical needs with science. Therefore, in addition to his studies in experimental psychology and physiology at Harvard University, he studied physics and mathematics. Stevens's early work and

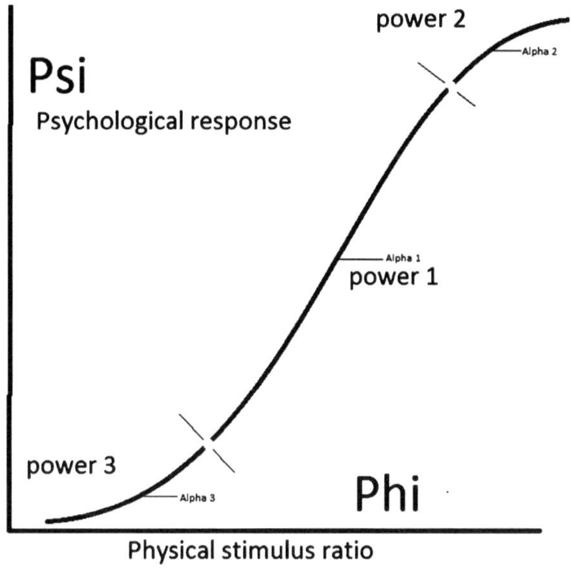

Fig. 11.6 Stevens's psychophysics power law scheme

consuming interest was in psychoacoustics and then he addressed the many measurement problems in psychology, before he finally turned his attention to psychophysics. His last work was published in 1975 in a book edited by his wife, Geraldine Stevens (1975), 2 years after the very premature death of this brilliant man, nearly, the world's first, but the established world's first professor of psychophysics.

Psychophysical experiments typically produce normal to elongated Gaussian S curves for both cumulative frequency of detecting a stimulus and perception of that stimulus from the onset to extreme ends of any continuum (Fig. 11.6). Cleverly, Stevens determined the best-fit equation for the middle elongated straighter part of the *S* for each continuum stimulus/response relationship, and repeated this experiment and analysis routine for each of the many continua he studied. Thus, he produced a family of adaptation-dependent continua stimulus/response relationships which he demonstrated obeyed a mathematical power law set, not a logarithmic law. However, the clever technique Stevens employed on the S curve meant that when the mid portion of the *S* began a pronounced deviation from one stimulus/response relationship it simply resulted in Stevens providing a new continuum name and power relationship or adaptation state for each end of the S curve, as Fig. 11.6 shows; the original (straighter part of the S curve) fit as power $\alpha 1$, then the upper deviation as power $\alpha 2$, and the lower deviation as power $\alpha 3$ and so forth over different continua. It may be that Stevens learned something from Fechner's adjustment of Weber's law to accommodate uncertainty (different continua) at extreme scale ends in his power law:

$$\Psi = k\Phi^{\alpha}, \tag{11.4}$$

where Ψ is the psychological response magnitude, k depends on the continuum, Φ is the physical magnitude ratio, and α is the power assigned to the continuum.

Then obviously such experiments would identify adjoining continua, one at each *S* tail deviation, for instance, as found below and above the thresholds of feeling or hearing onset and pain excess stimulation, or at the transition of low brightness through brightness to painful brightness. Stevens cleverly produced one function to accommodate human magnitude perception of all common physical stimuli. Thus, he produced his power law, which superseded the Weber–Fechner logarithmic law that had remained virtually unchallenged for a century. In general, ψ could be found from the α power of a low measured physical stimulus range minus the absolute threshold of perception for that continuum (Stevens, 1957), the basic Power Law; or from the α power resulting from a high luminous continuum stimulus range reference adaptation level (Hopkinson, 1957), also a power function; or from a high luminance specific continuum to a specific physical continuum reference stimulus ratio (MacGowan, 1984), still a power function. Thus, a selected datum rather than its threshold datum could equally apply to the Power Law. However, though the latter qualification is factual it still awaits peer recognition, probably because the mechanism which underpins the phenomenon remains unclear.

However, in determining brightness to luminance functions Stevens used Wright's technique, an unnatural *binocular* technique which employed a fixed adaptation right eye and an independently variably stimulated left eye (Wright 1939), criticised by (Hopkinson, 1939). At low adaptation levels, in dim conditions, the technique produced the power α=0.33. Independently, Hopkinson (Hopkinson, 1957), using his Luminosity Photometer, produced the same power α=0.33 power for low luminance to brightness magnitude conditions. However, Hopkinson's twin box brightness viewing device employed true binocular vision, and at high luminance and adaptation levels Hopkinson found the power to be α=1.0, which markedly differed from the power found via Wright's *binocular* routine, also employed by Stevens, J.C. and Stevens S.S. (1963), J.C., Stevens (Stevens J.C. and Marks L.E. (1999). The reason for such different α powers is still speculative and can't be addressed within the space and context of this book. However, the work of Wright, Nutting (Nutting 1920) and most others of the era predominantly addressed question of brightness perception at low luminances; rendering questions of perception at high luminances, counterbalance of dominant left or right eye and interaction between eyes at high luminances casualties of UK war-time priorities. However, Hopkinson used his Luminosity Photometer with more realistic, true binocular, viewing Hopkinson (1939) and (1957) to try to understand brightness magnitude assessment and later in the development and utility of his multi-criterion Discomfort Glare experiments (Hopkinson 1957), Hopkinson 1963, Part II, Section VII, p328–333), predominantly replicated in Discomfort Glare and evaluated for linearity using Stevens' magnitude estimation technique some two decades later by MacGowan (MacGowan et al. 1981 through 1983). Stevens determined a plethora of other α powers in which when any continuum type or character is markedly changed so α is changed. Stevens produced such continua α character tables as Table 11.1.

It should be understood that the JP to JI discomfort scale developed by Hopkinson (Hopkinson 1950) though referencing Abribat (Abribat 1935) who developed a scale of *contrast* based on Weber's *jnd*; the Weber-Fechner (Fechner

Table 11.1 Selected continua with α exponents from different techniques using Stevens' Magnitude Estimation

Continuum	Measured exponent α	Stimulus character
Brightness[a] daylight	1.0	$\geq$100 cd/m^2 (binocular)
Brightness mesopic	0.33	Low (dual monocular)
Discomfort (cold)	1.7	Whole body irradiation
Discomfort (warm)	0.7	Whole body irradiation
Electric shock	3.5	Current felt by fingers
Discomfort glare[b]	1.0	Luminance ($\geq$100 cd/m^2)

[a] Stevens, *via Wright's binocular (dual independent eye) technique*; Hopkinson (1963) *using his Luminosity Photometer*;
[b] MacGowan (1984, 1993) *unrestrained binocular in both the UW and SAS Discomfort Glare laboratories*

1860) decadic log scale of brightness perception, and Wright's *binocular* technique (Wright 1939), in fact RGH evaluated but employed none of these techniques in the development of his JP through JI Discomfort Glare scale. Instead, he used a practical scale in which three physical units of what he called Glare Constants could equal a quite noticeable difference in Discomfort Glare magnitude; not the magnitude of a *jnd* (as Weber), a *jnd* magnitude of equal contrast (as Abribat), a *decadic logarithmic* scale as (Fechner) nor, (Wright's) *binocular* technique; but a practical scale of Discomfort Glare spanning from the boundary of comfort to a boundary tolerable discomfort in 4 easily discernible steps reduced to 20 increments for engineering use. Nevertheless, the decadic log was used by Hopkinson to compress unwieldy large numbers, resulting from his empirical formulae, back into a manageable *perception* scale; as the deciBel is used in the acoustics dB and elsewhere, especially by Stevens, who was very fond of the dB, to convert unwieldy measurements into more *perceptible range* scales.

11.6 Neurophysiology and Problems of Neural Coding

The controversy was to result from Stevens's pronouncement of a new psychophysics law, for it was argued that the true psychophysics law was that which governed the neural activity relationship to perception of the outer environment which caused that neural activity, which, 100 years earlier, Fechner had called "inner psychophysics," which both Fechner and later Stevens thought might be linear but they had no means of determining that because the technology in neurophysiology did not exist.

In Galen's studies of eye anatomy, he categorized a phenomenon which he called "pneuma," as a spirit that connects the eye to the brain by traveling along the optic nerve to permit the soul to interact with the images on the eye. Was that the origin of studies of human "perception" and the more specific fields of ophthalmology, neurophysiology, psychology, psychometrics, psychophysics, and visual perception? Certainly, there are currently unresolved visual perception questions that involve a further understanding of the manner and order in which the CNS's and peripheral nervous system's mechanisms interact to create the mindful output we know as perceptual responses from sets of known physically measurable and presumed metaphysical stimuli. Determining the order and magnitude of these interactions at the neuronal level over the whole human body from "mind"-driven actions is necessary but is as beyond the science and technology of this day as was pneuma in Galen's time some 1,650 years before Fechner and 1,750 years before Stevens differentiated the perceived outer world from the inner world, albeit in Galen's more mystical manner. Was Fechner's "inner psychophysics," Galen's pneuma?

In years past, physical "outer world" phenomena sensed by human eyes, ears, nose, skin, and other sensors, translated and comprehended by the "inner world" CNS and mind into individual perceptions of reality, left huge knowledge gaps between the measureable and the imagined causes of such perceptions. Psychophysics was only 100 years old and neuroscience, in its infancy, lacked knowledge, technology, skills, and measuring equipment. Thus, the neuronal activity resulting from perceptions was not measurable in Fechner's time and was only crudely available near the time of Stevens's death and is hardly better in 2011. These great men could only speculate, as did Galen.

When Fechner separated the complete individual human response to any perceived environment into an "outer psychophysics" and an "inner psychophysics" he probably should have called the latter the human "neurophysiological" response. Significantly, both Fechner and Stevens suspected that the outer perception to the inner neurophysiological relationship was linear. That is, the inner neurophysiological response to any external stimulus would be linearly related to the psychophysical function that described the relationship between the perceived event and its physically measureable characteristics. However, to repeat, in Fechner's and most of Stevens's time, no technology was available to explore let alone determine the relationship between a complex measureable reality and the concomitant neurological activity provoked in any useful "environmental technology" manner. Nevertheless, following publication of the "To honor Fechner..." article (Stevens 1961), a storm was created. MacKay (1963) followed quickly by Werner and Mountcastle (1965) claimed that the psychophysics laws of both Fechner and Stevens were indistinguishable and equally incorrect for the true psychophysical law was a linear relationship between neural activity and the perception of an external event, which MacKay speculated and Mountcastle et al. to some extent demonstrated. These criticisms of Fechner and Stevens are explained in a nut shell by Johnson et al. (2002). The conclusions of MacKay, Mountcastle et al., and Johnson et al. seem not to recognize that Fechner and Stevens suspected, even expected, that the inner psychophysiological activity would be linearly related to the outer perception. However, it is that "outer" psychophysics model, the Stevens power law, that has been and remains so powerfully useful to environmental design, engineering, and their collateral domain professions.

Certainly, the motivators of perception need to be further understood, through a better understanding of the nature of all neurophysiological response mechanisms that drive or otherwise shape human perception of, especially, metaphysical

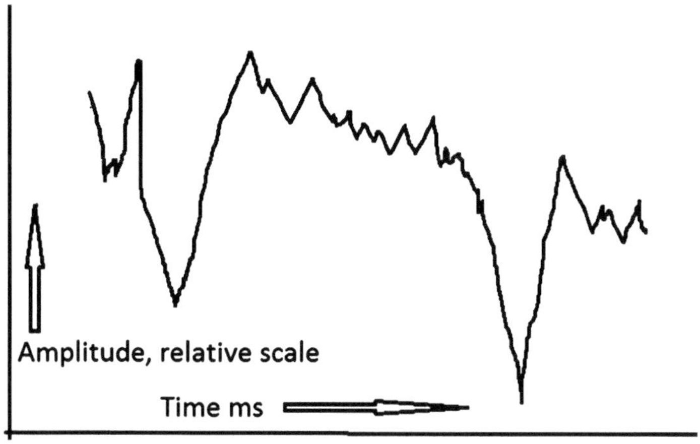

Fig. 11.7 Electroencephalography single channel response

environments. Neurons propagate signals rapidly over large distances in the body in relation to their size. They do this by generating electrical pulses called action potentials, spikes that travel down nerve fibers. These action potentials fire in bursts of spikes that relate to the manner of the stimulus input, such as light. The object of neural coding is to try to relate these bursts, trains of neuron profiles, to the object perceived to have caused such neuronal activity. This is done by presenting a brain with a very simple physical stimulus then encoding the associated neuronal activity across that brain by measuring the associated electrical activity generated and recording by electroencephalography the trains of responses similar to that in Fig. 11.7. Then, such resulting neural code is related to the physical object which provoked that neural activity. This encoding, mapping, of neuronal response to stimulus involves measuring the duration, amplitude, shape, and special frequency occurring within the trains of action potentials. The next step is to attempt to decode, reverse map, such encoded signals into the object which was thought to have caused the encoded neural activity. This is an attempt to find a more physiologically based model of human response to the environment beyond or complementary to the capability of psychophysical experimentation, which yields huge statistical variation among people.

Such efforts are worthwhile for they might ultimately reduce the variability involved in the results from human subject experiments. However, they are a long way from producing interpretable neural encoding to successful decoding/encoding on anything other than simplistic stimuli, such as roughness of texture or color of lights. Although encoding and decoding of neural activity can only deal with neuronal responses in the higher CNS and, to very simplistic physical stimuli, highly speculative extrapolations continue to be made about the power of such decoding, predominantly from the psychology community, who argued against the Stevens power law being the basic law of psychophysics; enough said.

Functional magnetic resonance imaging (fMRI), magnetoencephalography, and the much older and simpler electroencephalography are in common nonmedical use in attempts to relate their images to outer-world events, the former two at huge expense using primarily medical diagnostic equipment. Positron emission tomography cannot be used for nonmedical purposes because of the radiation danger inherent in repeated exposure to gamma rays. fMRI does show the aftereffects of stimulus input and that can be useful when interneuronal communication over the entire CNS is not required to answer the more fundamental questions on perception, such as information storage, mediation (as in habituation), and retrieval of previously stored information. Such imaging, so much needed for medical diagnostic purposes, is also regularly used by psychology groups in areas of signal-train rerouting, basic wiring alteration to restore physical function to normal after a stroke, and in some more frivolous cases to speculate about good and bad music and art, etc.

Magnetoencephalography is fast over the cortex, with good temporal resolution and is now commonly used in conjunction with fMRI to provide a better, more comprehensive medical diagnostic facility, but still the technology to trace the paths of action potentials over the whole CNS is beyond current science and technology. Nevertheless, any experimentation which leads to a better understanding of neural activity, and information storage and retrieval, in relation to perception must be engendered without its current value to technology being exaggerated, for it can complement, but not replace the practical applicability of conventional psychophysics.

Perhaps when negative refraction technology (Pendry 1996–2009) comes of age more answers will emerge; when work on the multitude of functions of ganglia becomes more enlightening (Hattar et al. 2002; Wong et al. 2005; Zaidi et al. 2007) more will become understood about upper (brain) to lower (spinal cord) communications in the CNS. In the meantime, another almost forgotten aspect of human physiology needs to be revisited to discover its implications for the perception. That neurophysiology resides under the heading of "habituation."

11.7 Habituation and Basic Human Wiring

Neurophysiology is a discipline within the now-established neurosciences, incorporating the older disciplines of ophthalmology, neurology, neurosurgery, and psychiatry. The discipline area is concerned with CNS neuronal activity in relation to external stimuli. Hence, it is central to discovering the actions of the biophysical drivers, storage, and routes of neuronal actions which manifest habituation, conditioning, learning, and arousal on actions of the mind; that is, answers to questions that are likely to remain a mystery for many decades into the future because of CNS action scanning technology limitations.

Walter Victor Macfarlane (1913–1982), born in Christchurch, New Zealand, had scholarly and curiosity interests in people, art, architecture, literature, history, theater, culture, ritual, animals, wind, weather, ecology, and indeed all aspects of science. Macfarlane was interested in and excelled in almost everything, a true twentieth century polymath. He graduated with an MA degree in zoology from Canterbury University College and received his MB and ChB degrees from the University of Otago. During his final year of medical studies he attended the seminars of the newly appointed John Carew Eccles (later knighted and a Nobel laureate) and joined him some years later as Research Professor at the Australian National University.

Macfarlane received his training in the delicate neurosurgical techniques required for intervention in brain, spinal cord, and nerve root surgery and, in electrical stimulation and recording of neuronal activity, as assistant to the neurophysiologist Murray Falconer (Macfarlane and Falconer 1947). Macfarlane received his MD degree from the University of Otago and later was elected Fellow of the Australian Academy of Science. These encounters with both Falconer and Eccles led to his lifelong interest in the mechanisms underlying nerve and brain function, and habituation, the field he pioneered with Glaser (Glaser 1966), a Slovak who qualified in medicine in Bratislava.

In the early 1970s to early 1980s, Macfarlane's interests in architecture, particularly in applying his neurophysiological experience to human habitat, found him teaching courses in architecture at the universities of Adelaide, Washington (Seattle), Manitoba, and California (Berkeley), where he always delighted and enlightened students with his unique manner of explaining functions of the CNS, particularly its role in habituation and its influence on perception.

The work of Macfarlane was pivotal in research on the neurophysiological action of habituation. By the late 1940s, the old subculture of neurophysiology was just entering its exponential phase and opened up a tremendously lively intellectual ground base for Macfarlane's earlier medical and surgical explorations, which were essential to his way of developing the field of habituation, elaborated by McIntyre (1985). Habituation is explained in simple layman terms in E.M. Glaser's book *The Physiological Basis of Habituation* (Glaser 1966), which readers are particularly encouraged to study; and in reviews by Harris (1943) and Thompson and Spencer (1966), shall be reintroduced here and in Chap. 12 with likely constant impact on the adjustment of human perception of external stimuli due to the inevitable changes of circumstance with the passage of time. The cortex is more complex than the lower centers of the CNS, such as the spinal cord, but all employ nerve circuits, differentiate stimuli, and use information storage and retrieval. However, the cerebral cortex is more involved with new information, new circuits, and interactive processes. Lower-level storage such as in the spinal cord depends to a greater extent on inherent pathways to store information about past events and act as a mediator on the higher-order response to environmental stimuli, such as meaning and associations with thermal, luminous, and/or sonic environments.

Habituation is one of three principal physiological storage responses to external stimuli. The other two are conditioning and learning. The three are often confused, but all three comprise stored information for use after the cessation of a stimulus. Storage of such information requires no response, whereas habituation, conditioning, and learning require a response. A persistent effect on the nervous system is not the storage, but can be a consequence of storage. For example, if a package is given to a courier with the instruction to take it by taxi to location *x* and the courier immediately travels by taxi to location *x*, then the journey to location *x* would be a persistent effect not requiring storage of the instruction. However, if the same instruction tells the recipient to travel on from location *x* to location *y* after arriving at location *x* and, if that instruction was remembered, then that would be a new response based on stored information, a latent response which persists outside the responding system after cessation of the stimulus, and would be based on stored information about the original instruction.

Pavlov and his team were the seminal workers in the area of conditioning, although not everyone agreed with all their findings (Pavlov 1927, p. 395), claiming discipline, training, education, etc. were simply conditioned reflexes (Vavilov 1952). Glaser (1966, pp. 16–17) explains that habits can be due to conditioned reflexes or learning. Conditioning is reinforcement or inhibition of an existing response by a new stimulus. This is also true when the origin of the response is acquired by learning or is inborn, in the basic wiring. Thus, conditioning changes the stimuli, not the response.

Acquired learning ability is understanding judged by response. In the former taxi example, when the courier arrived at location *x* and was not clear about the urgency of the delivery to location *y*, he called his instructor, because it was already beyond the regular delivery time, to ask if delivery should be delayed until the next day or if a surcharge should be levied for delivery that day. Such a response by the courier would be the result of previous learning and his further action would result in acquired ability, learning. If all that was remembered and stored, it would be apparent in the courier's response to the instructor. Thus, learning is a qualitative storage of response.

It is extremely important to realize that conditioning and learning are quite different from habituation. Habituation causes a gradual diminution of response to unwanted or irrelevant stimuli, never an amplification of response. It changes the response level, not the sensor signal level. For instance, in the perception of an uncomfortable but not dangerous thermal environment, stored habituation acts on the sensor information and reduces the signal between sensor and the brain through habituation's signal suppression action in the spinal cord. That is, the sensor signal never reaches the cortex before it is modified by habituation at a lower level of the CNS, in the spinal cord. That habituation response stored at spinal cord level is likely to affect (diminish) also uncomfortable brightness or luminance and annoying noise in much the same manner, but not qualitative signals, which are interactively processed at a higher CNS level without habituation suppression. Note that habituation is not a form of adaptation, conditioning, or learning for habituation does not affect sensor sensitivity.

Habituation storage is a necessary biological mechanism employed to suppress reaction to inconsequential stimuli to avoid unnecessary response; that is, to use stored information to suppress response to harmless stimuli that require no response, thus economizing on body energy resources (Porciatti et al. 2005), on light-sensitive retinal ganglia. However, in their article, Porciatti et al. "defined" habituation incorrectly. They "define" incorrectly the 2 min they quote a contrast amplitude took to decrease by 30% within a 4-min steady state test as habituation. Neither Kozak et al. (1962) with their surgical explorations, nor Glaser (1966) in his book of examples, nor Macfarlane (1974) in his update on habituation, all pioneers of habituation, refers to habituation in such a manner. Porciatti et al. seem to be referring to accommodation, which would be affected by the habituated state of the "basic wiring," but is not habituation, which is inherently a long, slow, imperceptible process of modification of information storage that nevertheless affects both perception and accommodation. Such activity requires no conscious action for it is an involuntary response and part of the basic wiring of the human system that has been generally accepted to take place during the first 6 years of life. Basic wiring can be changed by trauma and stressful or prolonged uncomfortable stimuli, and habituation is central to that change (Macfarlane 1964, 1974, 1978; Kozak et al. 1962; Ramachandran and Hubbard 2001). However, if concomitant neural activity associated with the Porciatti et al. type of 2-min delay to the accommodation experiment could be traced, then such work would indeed be truly enlightening.

Arousal is not the converse of habituation; however, it can inhibit habituation just as habituation can inhibit arousal (Welford 1968). Arousal differentiates stimuli which are relevant to the survival of the organism. Habituation inhibits responses that are irrelevant to the organism. For instance, an explosion causing noise (unwanted sound) would cause an arousal response to run and seek shelter, whereas the response to a perpetual loud noise (still unwanted sound) would be suppressed by habituation. The perpetual noise would be annoying; the explosion could kill the organism. That involuntary psychophysiological facility is part of an organism's intrinsic characteristic and is used to good effect for it alerts the organism to danger and suppresses dislikes if these present no danger to the organism, whether that organism is a rabbit or a human. Habituation desensitises, never sensitises perception – this can be used to good effect in environmental design or standards setting, but it takes time!

11.8 Partial Conclusions

Habituation causes a gradual change, diminution, of response to either a stressful event or a repeated or prolonged exposure to an uncomfortable event. It reduces the response to such events but does not change the strength of the signal at the sensor; it reduces the strength of the signal received by the brain. That appears to be accomplished at the spinal cord level (Kozak et al. 1962; Macfarlane 1964, 1972). Action in the spinal cord inhibits the signal strength, not action in the cerebrum, cerebellum, or brainstem, or by a change of receptor sensitivity. Where such habituation memory is stored for visual signals is only speculated. Nevertheless, its locations, by the nature of the signal, are of fundamental importance to researchers from many disciplines.

Unfortunately, when using currently available scanning equipment to try to understand the routes of action potentials or even the recent history of blood oxygen, images are in many cases pointless because of sluggish equipment, imaging speed, and lack of whole CNS scanning.

Habituation naturally has far-reaching consequences for human habitat design and operation. However, this helpful human facility has been neglected by recent scientists and technologists alike, who appear oblivious to this basic physiological facility and are bewildered to find people reacting differently to technology evolutions that impose different environmental stimulants or stresses on constantly habituating humans.

Chapter 12 demonstrates a consequence of disregarding habituation in current luminous environment design and the use of an evolved Stevens-based formula being developed to try and adapt to time- and circumstance-induced changes of human perception.

Appendix 11

Specific Research Area of Architectural Psychophysics

The general esteem and assessment of any architecture depends a great deal on its visual image, perception of its exterior, and impressions during the visit to its interior. The designed or already realized architectural concept of the overall architectural space and its functional and esthetic structure and form as well as the human (architectural) environment created are evaluated visually by investors, the general public, and each inhabitant or owner. As a consequence of such attention to "visual appraisal," several subjective appraisal studies of exterior or interior architectural spaces and their environmental influences were developed on the basis of psychological or psychophysical measuring methods.

In experimental psychology, factor analysis was applied, based on clearly identifiable evaluative polar scales expressed by adjectives with selective and high loading or meaning for the assessed object. Originally, such analysis was proposed by Thurstone (1948) and was later used by Osgood et al. (1957) with the semantic differential or judgment of subjectively felt meaning on a series of descriptive scales of a questionnaire-based experiment. Further studies have shown that people in over ten different language and cultural communities use quite similar semantic factors of the most salient type in meaningful judgment areas of evaluation, potency, and activity. A more sophisticated theoretical structure (Harman 1960) and application to architecture studies as by Vielhauer (1965) and Hershberger (1968) opened new possibilities of measurement of human responses to buildings and their interiors. A very important property of a building is its ability to provide an appropriate space and environment for its users, who are able to express their satisfaction or dissatisfaction (Canter 1968). Two experiments with architecture and nonarchitecture student groups by Canter (1969) indicated in the

Table A11.1 An example of a bipolar scale with dichotomization points for evaluating scores

Happy	1	2	3	4	5	6	7	Sad
Hot	1	2	3	4	5	6	7	Cold
Welcoming	1	2	3	4	5	6	7	Unwelcoming
Soft	1	2	3	4	5	6	7	Hard
Relaxed	1	2	3	4	5	6	7	Tense
Kind	1	2	3	4	5	6	7	Cruel
Sympathetic	1	2	3	4	5	6	7	Unsympathetic
Gentle	1	2	3	4	5	6	7	Rude
Friendly	1	2	3	4	5	6	7	Unfriendly
Interesting	1	2	3	4	5	6	7	Uninteresting

judgment of black and white slides of 20 building shown that from 45 adjective bipolar scales, the most relevant and meaningful considered were the factors of character and friendliness, coherence, and activity. In accordance with these findings, an ideal "friendliness" scale was adopted with the dichotomization points (Table A11.1) for subject sensitivity and satisfaction with different building types or designed alternatives (Canter and Wools 1969). This scale was used in a pilot study (Kittler and Wools 1969) in which an interior was shown to subject groups on slides in color, black and white, and with line drawings. The same interior was varied, presenting two seating arrangements, two window types, and a flat or sloping ceiling. Even more suitable ten bipolar semantic differential scales were applied to assess interior appearance (Loe et al. 1994, 2000). Unfortunately, no preference studies concerning windows, their size, and their placement under typical daylight conditions were made. Detail experiments structured to test if the provision of a view both in drawings and in models does affect the ratings of windows were included in psychophysical studies (Néeman and Hopkinson 1970; MacGowan et al. 1984).

In a different research area, on the experimental studies is the subjective rating of the dimensions of architectural spaces. Original experiments by Gilinsky (1951) have shown that the perceived size and distance of objects in the visual space are distorted. Exterior architectural spaces were compared subjectively as seen and drawn by architecture students with objectively measured dimensions (Koroyev and Fedorov 1954; Fedorov and Koroyev 1961). When these results were analyzed, the similarity with Stevens's power law was evident (Kittler 1968) and close to findings by Gilinsky. For instance, the subjectively perceived distance in nature d is related to the real distance D in meters:

$$d = 1.24D^{0.866}. \tag{A11.1}$$

The height of buildings (house front or spires) (h) is seen to be even more distorted in comparison with actual height H measured in meters as

$$h = 1.48H^{0.866}, \tag{A11.2}$$

if $H \geq 30$ m.

So, the proportion h/d seems to be

$$\frac{h}{d} = 1.2\left(\frac{H}{D}\right)^{0.866}. \tag{A11.3}$$

The third group of psychophysical scaling uses an arbitrary numbered scale for subjectively felt just noticeable differences (jnd) related to the physically measured ratio. In this area is also the multicriterion scale of glare which causes visual discomfort without necessarily impairing the vision of objects (Hopkinson 1950, 1957, 1963 and Hopkinson et al. 1971). For the discomfort glare index (GI), changes were determined by RGH fixed reference values representing Hopkinson's multicriterion scale points:

- Just perceptible with GI = 8
- Just acceptable with GI = 15
- Just uncomfortable with GI = 21
- Just intolerable with GI = 28

These first critical points roughly correspond to the straight line on Stevens's S sensation curve for average observers (Hopkinson 1957).

Further details and progress in glare determination are given in Chap. 12.

References

Ackerman, J.S.: Leonardo's Eye. Journal of the Warburg and Courtauld Institutes. 41 (1978)

Bernoulli, D.: Specimen theoriae novae de mensura sortis. Commentarii Academiae Scientiarum Imperialis Petropolitanae 5, 175–92, St. Petersburg (1738.) English translation by Sommer, L.: Exposition of a New Theory on the Measurement of Risk. Econometrica, **22**: 23–36 (1954)

Boring, E.G.: Sensation and perception in the history of psychology. Appleton-Century-Crofts, New York (1942)

Boring, E.G.: A history of experimental psychology. Appleton-Century-Crofts, New York (1950)

Boring, E.G.: History, Psychology, and Science Selected Papers. Wiley, New York (1961)

Cattell, J. McK.: Psychometrische Untersuchungen. Philosophische Studien, **3**, 305–335, 452–492 (1886)

CIE: Principales décisions V(λ) function, CIE Proc. 6th Session 1924 p. 67–70 Cambridge University Press, (1926)

CIE: Recommended system for mesopic photometry based on visual performance. CIE Publ. 191: (2010)

Fechner, G.T.: Elemente der psychophysik II, Breitkopf & Hartel, Leipzig (1860)

Fechner G.T.: Vorschule der Aesthetik.I.II. Breitkopf & Hartel, Leipzig (1876)

Fechner, G.T.: In Sachen der Psychophysik. Breitkopf & Hartel, Leipzig (1877)

Fechner, G.T.: Revision der Hauptpuncte der Psychophysik. Breitkopf & Hartel, Leipzig (1882)

Gall, D. and Biesk, K.: Definition and measurement of circadian radiometric qualities. Light and Health: non-visual effects. CIE x027 (2004)

Galton, F.: Inquiries into the human faculty and its development. Macmillan, London (1883)

Glaser, E.M.: The Physiological Basis of Habituation. Oxford Medical Publications, Oxford University Press, London, 1–102 (1966)

Hattar, S., Liao, H.W., Takao, M., Berson, D.M., Yau, K.W.: Melanopsin-containing retinal ganglion cells: architecture, projections, and intrinsic photosensitivity. Science, **295**, 5557, 1065–70 (2002)

Hopkinson, R.G.: Architectural Physics: Lighting, London: HMSO, Part II, Section VII, 329–336 (1963)

Johnson, K.O., Hsiao, S.S., Yoshioka, T.: Neural Coding and the Basic Law of Psychophysics. Neuroscientist, **8,** 111–121 (2002)

Keep, P.V., James, J., Inman, M.: Windows in the intensive therapy unit. Anaesthesia, **35**, 257–262 (1980)

König, A. and Dieterici, C.: Die Grundempfindungen und ihre Intensitäts-Vertheilung im Spectrum. Sitz. Akad.Wissensch. Berlin, 805–829 (1886)

König, A.: Über den Hellikeitswert der Spektralfarben bei verschiedener absoluter Intensität. Beiträge zur Psychologie und Physiologie der Sinnesorgane, (1891)

Kozak, W., Macfarlane, W.V., Westerman, R..: Long-lasting reversible changes in the reflex responses of chronic spinal cats to touch, heat and cold. Nature, **193**, 171 (1962)

Lindberg, D. C.: Alhazen's theory of vision and its reception in the West. Isis. **58**, 3 (1967)

Lokhorst, G.J.C. and Kaitaro, T.T.: The originality of Descartes' theory about the pineal gland. Journ. for the History of the Neurosciences, **10**, 6–18 (2001)

Macfarlane, W.V., and Falconer M.A.: Diverticum of the lateral ventricle extending into the posterior cranial fossa: report of a case successfully relieved by operation. J. Neurol. Neurosurg Psychiatry, **10**, 3, 100–106 (1947)

Macfarlane, W.V.: Habituation to heat and cold at the spinal cord level. Environmental Physiology and Psychology in Arid Conditions. UNESCO. Paris, 351 (1964)

Macfarlane, W.V.: Personal communication to David MacGowan on Habituation. University of Adelaide (1972)

Macfarlane, W.V.: Habituation. Progress in Biometeorology, **1**, 462 (1974)

Macfarlane, W.V.: Thermal comfort studies since 1958. Arch. Sci. Rev. **21**, 86 (1978)

MacGowan, D., Thompson, H.E., Eaton, M., Harris, A.: Window Architectonics, Design Guidelines, University of Manitoba report, to Public Works Canada, Contract No.330-993, Ottawa, Canada, 4, 50–67 and 1–83 (1984)

MacGowan, D., Pulpitlova, J., Šubova, A..: Time and Circumstance, and the perception of discomfort glare. Lux Europa Proc., **1**, 193–211 (1993)

McIntyre, A.K.: Walter Victor Macfarlane 1913–1982. Biographical Memoirs of Deceased Fellows. Australian Academy of Science Project (1985)

MacKay, D.M.: Psychophysics of perceived intensity, a theoretical basis for Fechner's and Stevens' laws. Science, **139**, 1213–16 (1963)

Masin, S.C., Zudini, V., Antonelli, M.: Early alternative derivations of Fechner's law. J Hist. Behav. Sci., **45**, 1, 56–65 (2009)

Mountcastle, V.B., Poggio, G.F., Werner, G.: The relation of thalamic cell response to peripheral stimuli varied over an intensive continuum. J Neurophysiol. **26**, 807–34 (1963)

Newton, I.: Philosophiae Naturalis Principia mathematica. J. Streater for the Royal Society, London (1687) English translation (1729) and American edition (1846)

Newton, I.: Optics or a treatise of the reflexions, refractions, inflexions and colours of light. Print S. Smith and B. Walford for the Royal Society, London (1704) Russian translation by Vavilov, S. I.: Optika. Gosizdat, Moscow (1927)

Pavlov, L.P.: Conditioned Reflexes. An Investigation of the Physiological Activity of the Cerebral Cortex. English translation by. Anrep, G.V., London (1927)

Pendry, J.B., Schurig, D., Smith, D.R.:: Controlling electrical fields. Science, **312:** 1780–1782 (2006)

Porciatti, V., Sorokac, N., Buchseer, W.: Habituation of Retinal Ganglion Cell Activity in Response to Steady State Pattern Visual Stimuli in Normal Subjects. Invest Ophthalmol Vis., **46**, 4, 1296–1302 (2005)

Purkyně, J.E.: Zur Physiologie der Sinne. Prague (1825)

Ramachandran, V.S. and Hubbard, E.M.: Synaesthesia—A Window Into Perception, Thought and Language. Journal of Consciousness Studies, **8**, 12, 3–34 (2001)
Sokal, M.: Psychology at Victorian Cambridge – the unofficial laboratory of 1887–1888. Proceedings of the American Philosophical Society, **116**, 145–147 (1972)
Stevens, S.S. and Davis H.: Hearing: its Psychology and Physiology. New York. Wiley (1938)
Stevens S.S.: On the theory of scales of measurement. Science. **103**, 677–80 (1946)
Stevens, S.S.: On the psychophysical law. Psychol Rev., **64**, 1, 53–81 (1957)
Stevens, S.S.: Toward a Resolution of the Fechner-Thurstone legacy. Psychometrica, **26**,1 (1961)
Stevens, S. S: The psychophysics of sensory function. Sensory communication. Cambridge, MA: MIT Press, 1–34 (1961)
Stevens, S.S.: To honor Fechner and repeal his law. Science, **1**, 33, 80–6 (1961)
Stevens, J.C. and Stevens, S.S.: Brightness function: Effects of adaptation. Journ. Opt. Soc. Amer., **53**, 3, 374–380 (1963)
Stevens, S.S.: Neural events and the psychophysical law. Science. **170**, 1043–50 (1970)
Stevens, S.S.: Psychophysics: Introduction to its Perceptual, Neural, and Social Prospects. Wiley, New York (1975)
Thurstone, L.L.: "The Indifference Function."Jour. Soc. Psychol., **2**, 139–167 (1931)
Thurstone, L.L.: Psychological method, 5: Methods of Psychology. Wiley, New York (1948)
Titchener, E.B.: A text-book of psychology. Macmillan, New York (1924)
Vavilov, S.I.: Deysviye sveta. (In Russian: Action of light). Gosizdat, Moscow (1952)
Weber, Ernst Heinrich (1795–1878): Leipzig physiologist. **199**, 4, 272–3 (1967)
Weber, E.H.: De tactu. Annotationes anatomicae et physiologicae. Leipzig, (1834)
Welford, A.T.: Fundamentals of Skill. Methuen, London (1968)
Werner, G. and Mountcastle, V.B.: Neural activity in mechanoreceptive cutaneous afferents: stimulus-response relations, Weber functions, and information transmission. J Neurophysiol, **28**, 359–97 (1965)
Wong, K.Y., Dunn, F.A.; Berson, D.M.: Photoreceptor Adaptation in Intrinsically Photosensitive Retinal Ganglion Cells. Neuron, **48**, 6, 1001–1010 (2005)
Zaidi, F.H., Hull, J.T., Peirson, S.N., Wulff, K., Aeschbach, D., Gooley, J.J., Brainard, G.C., Gregory-Evans, K., Rizzo, J.F., Czeisler, C.A., Foster, R.G., Moseley, M.J., Lockley, W.: Short-wavelength light sensitivity of circadian, pupillary, and visual awareness in humans lacking an outer retina. Curr. Biol., 17, 24, 21, 22, 8 (2007)

References to Appendix

Canter, D.: The study of meaning in architecture. Res. Rep. Build. Perform. Unit, University of Strathclyde (1968)
Canter, D.: An intergroup comparison of connotative dimensions in architecture. Environment and Behaviour, 1, 1,37-48 (1969)
Canter, D. and Wools, R.: A technique for the subjective appraisal of buildings. Res. Rep. Build. Perform. Unit, University of Strathclyde (1969)
Fedorov, F.V. and Koroyev, Y.I.: Obyemno-prostranstvennaya kompoziciya v proyekte i v nature (In Russian: volume-spatial composition in design and in nature). Gosstroyizdat, Moscow (1961)
Gilinsky, A.S.: Perceived size and distance in visual space. Psychol. Rev., **58**, 4, 460 (1951)
Harman, H.: Modern factor analysis. University of Chicago Press, Chicago (1960)
Hershberger, R.G.: A study of meaning in architecture. Man and his environment, **1**, 6, 6–7 (1968)
Hopkinson, R.G.: The multiple criterion technique of subjective appraisals. Quart. J. Exp. Psychol., **2**, 124 (1950)
Hopkinson, R.G.: The evaluation of glare. Illum. Eng., **52**, 305 (1957)
Hopkinson, R.G.: Architectural physics: Lighting. HMSO, London (1963)

Hopkinson, R.G., Longmore, J., Woods, P.C. Craddock, J.: Glare discomfort from Windows. Paper 71.37, Proc. CIE Gen. .Session, Barcelona (1971)
Kittler, R.: Vizuálne pôsobenie hlavných rozmerov v architektonickom priestore a rôznych formách jeho zobrazenia. (In Slovak: The visual effect of the pricipal dimensions in architectural space, and in the various forms if its representation). Architectúra a urbanizmus, **2**, 4, 33–45 (1968)
Kittler, R. and Wools, R.M.: A cross cultural pilot study using a Guttmann scale to asses representations of the physical environment. Res. Rep., Strathclyde University, (1969)
Koroyev, Y.I. and Fedorov, F.V.: Architektura i osobennosti zritelnogo vospriyatiya. (In Russian: Architecture and specialty of visual perception). Gos. izdat. po stroyit. i archit., Moscow (1954)
Loe, D.L., Mansfield, K.P., Rowlands, E.: Appearance of lit environment and its relevance in lighting design: experimental study. Lighting Res. Technol., **26**, 3, 119–133 (1994)
Loe, D.L., Mansfield, K.P., Rowlands, E.: A step in quantifying the appearance of lit scene. Lighting Res. Technol., **32**, 4, 213–222 (2000)
Neéman, E. and Hopkinson R.G.: Critical minimum acceptable window size: a study of window design and provision of a view. Lighting Res. Technol., **2**, 1, 17–29 (1970)
Osgood, C.E., Suci, G.J., Tannenbaum, P.H.: The measurement of meaning. University of Illinois Press, Urbana (1957)
Thurstone, L.L.: Psychophysical methods. Methods of psychology, 124–157. T.G. Andrews Press, New York (1948)
Vielhauer, J.A.: The development of a semantic differential scale for the description of the physical environment. Ph.D. Dissertation, Louisiana State University (1965)

Chapter 12
Discomfort and Disability Glare in the Visual Environment

12.1 Recent History

It is important to understand recent lighting and fenestration history and its development. Before the era of on-tap electrical energy supply for lighting, when artificial lighting sources were inefficient, fenestration was used as a means of providing working illuminance in addition to its many other well-documented functions. Daylighting technology developed firstly from a need to ease natural light into buildings in order to illuminate tasks and interior spaces. High ceiling heights with tall windows were favored and roof lighting, to provide working illuminance, became extensively used in industrial premises. With the advent of the fluorescent lamp, and its rapid rise to prominence during the 1950s, buildings were freed from many configuration constraints and the use of fenestration, as a means of providing working illuminance, diminished. Within a further decade, artificial lighting levels soared, especially in the USA. Levels of 2,000 lux were not uncommon as the philosophy that more light is better propagated rapidly. Controversy abounded concurrently with these developments. One area of controversy was associated with the trend toward seemingly unnecessary ever higher interior illuminances. The lighting industry was blamed for producing a body of opinion which favored higher illuminance and, hence, higher financial returns to the industry. Then the 1970s energy crisis materialized. It was already known that the then popular deep office building, illuminated almost totally by artificial lighting, was a prime offender in the excess energy consumption league. The high lighting levels used in these buildings came under immediate attack, resulting in large-scale delamping, which resulted in widespread but scarcely acknowledged complaints about difficulties due to observed fenestration brightness.

In the same time period, the use of reduced-transmittance window glass became popular, although its application was clumsy and insensitive toward external view because of the unnecessary but regularly used excessive light transmittance reduction, commonly down to only 8%. A 75% reduction in the brightness is hardly noticeable without simultaneous contrast against clear glass and would have

R. Kittler et al., *Daylight Science and Daylighting Technology*,

DOI 10.1007/978-1-4419-8816-4_12,

reduced the window brightness by a factor of at least 3, while preserving the exterior view (MacGowan et al. 1984, 1993). Nevertheless, the crude but architecturally fashionable technique answered, to some extent, the complaints about fenestration systems under those lower levels of illuminance from manufactured sources. This fashion was to be discarded later as daylighting gained new popularity when it was "rediscovered" that fenestration could be used as part of a building's heating system, via solar means, and seemingly with the intended added bonus of the provision of free, comfortable, working illuminance.

Long before all of these events transpired, Hopkinson (1950) published his concept of a solution to that visual discomfort malady, in the form of a discomfort glare scale based on a multiple criterion technique of subjective appraisal of discomfort glare. The multiple criterion steps were just imperceptible (discomfort glare) acronym JI; just acceptable, acronym JA; just uncomfortable, acronym JU; and just intolerable, acronym JI. These criteria spanned the whole range of discomfort glare from its initial perception to the limit of its tolerability. Because the acronym for "just imperceptible" was the same as that for "just intolerable" it was changed to just perceptible, acronym JP. These discomfort glare rating scale points were then translated to luminous environment physics in controlled laboratory luminous environments and later comprehensively validated in field studies. Laboratory and extensive field validation work in the UK produced and verified in use what became a set of empirical equations with numerical relationship values called discomfort glare constants (GC) for single sources, which were summed and brought to conveniently sensible numbers on a glare index (GI) scale ranging from 8 to 28 as $GI = 10\log_{10} \sum GC$. The numbers become significant only when their meaning can be translated to their associated multiple criterion scale points of $JP = 8$, $JA = 15$, $JU = 21$, and $JI = 28$. Then it was desirable for consistency to cast these numbers and their perceptual associations in stone, as the upper, intermediate, and lower limits of discomfort glare perception. That GI psychophysical system evolved into an internationally accepted specification system of permissible discomfort glare indices related to the nature and location of the task.

From about 1960 through to early 1980, experimental work to advance the state of the art of discomfort glare work was sparse – although much speculative comment was made about the discomfort glare index. In addition, by the middle of 1970 it was thought that discomfort glare could have a substantial impact on the energy-trade-off daylighting design of, presumed, energy-efficient office buildings. By 1980 it was apparent from the latter and the growth of contradictory publications, authored by both players and actors in the field, that the phenomenon of discomfort glare required reinvestigation, especially in light of the energy/green imperative, evolving experience and new cognitive settings. What was known in 1980 that previous workers in the field did not know or make clear was that predictions from the original Hopkinson (1950) formula for small area sources,

$$GI = 10\log_{10} \sum \frac{k {L_s}^{1.6}\, \omega^{0.8}}{L_b}, \tag{12.1}$$

where k is a scale-unit-correcting constant, L_s is the luminance of the glaring source, L_b is background luminance, and ω is the solid angle of the glaring source, overestimated discomfort glare perception for windows as shown by Hopkinson and Bradley (1960). But at the same time that was thought to be a source size and window contribution to interior light problem, which resulted in the development of the large-area-source formula

$$GI = 10\log_{10} \frac{0.478\, L_s^{1.6} \Omega^{0.8}}{L_b + 0.07 L_s \omega^{0.5}}, \tag{12.2}$$

where Ω is the solid source angle corrected for p, the position of the source element (Luckiesh and Guth 1949).

However, even though this formula was said to overestimate discomfort glare perception from windows, about another 10 years later Hopkinson (1970, 1971) revised and adopted a new glare scaling formula:

$$\mathrm{GI_n} = 1.5\mathrm{GI_a} - 14 \leqslant 28, \tag{12.3}$$

where $\mathrm{GI_n} = \mathrm{GI}$ for daylight situation, and $\mathrm{GI_a} = G$ for artificial lighting.

By that time three other developments had taken place:

1. Chauvel et al. (1980) pronounced that the area of window-size glare sources did not matter, only the source luminance mattered. Naturally, if that were true, it would simplify the glare formulae.
2. Stevens (1957, 1961) produced his psychophysics power law after (11.4).
3. Stevens and Hopkinson (1963, Sect. VII, pp. 329–336) found in the late 1950s that luminance and brightness at typical task levels and beyond were linearly related, that is, $a = 1.0$ The attraction was then to try to express the GI relationship in terms, not of an entirely empirical formula, but in terms of an empirical formula that had become a psychophysics law.

Hopkinson wanted to comprehensively pursue the linearity revelation but was about to move from the Building Research Station (BRS) to University College, London, as a professor. Therefore, he could not undertake such a lengthy new discomfort glare study because of administrative duties and because his former colleagues at the depleted BRS, about to become the Building Research Establishment, had no enthusiasm for such a large undertaking after just finishing the protracted Cornell/BRS large glare source work. MacGowan, without even knowing the latter history at the time, but spurred on by Hopkinson's publication (Hopkinson 1963) and Stevens's work, particularly (Stevens 1946–1972), realized the possibilities for a new formula based on the power law. Then, in 1972, MacGowan learning at first hand of Macfarlane's habituation work (Macfarlane 1964, 1972; Kozak et al. 1962), it became clear that the most likely cause of overestimations of discomfort glare which were becoming increasingly apparent from the application of the Hopkinson formulae, and were a mystery at the time, may not have been due to basic shortcomings or errors in the Hopkinson the formulae, as criticized, but were caused by a failure to provide an unanticipated facility to accommodate the daunting task of impact of Time, Circumstance and Habituation driven evolutionary changes on human perception within the technology! That realization led to the development of a research program at the University of Washington (UW) in 1974 and postponed to 1982 on energy savings through daylighting in commercial buildings. Fortunately, the program was able to encourage Hopkinson and Macfarlane to be the senior consultant in their fields on the proposed new work, which received funding initially at UW from the US National Science Foundation. However, the work spanned from 1974 and continues, slowly, to this date. Very sadly, Macfarlane died very prematurely in 1982, not long after he had helped shape what he wanted to have in the UW research program, then equally tragically Hopkinson died in June 1994, after providing invaluable advice to the program for 13 years.

12.2 Further Progress in Discomfort Glare Research

Hopkinson's large-area formula in (12.2) was the starting point at UW. A fixed glare source size of $\omega = 0.21$ sr was used during the investigation (MacGowan 1980; MacGowan et al. 1984). Respecting the foregoing, it was decided to assign $\mathrm{GI} = \Psi$; to replace the log function by the exponential, so the decadic log was discarded as irrelevant; retain k; discard $\Omega^{0.8}$ as irrelevant, but test the influence of both glare source size and position later. The glare source luminance was retained and divided by the luminance of the task, thus the physical (luminance) component of Adaptation was to be accommodated not as Stevens' Φ, but as (Ls/Lt). The glare source luminance L_s was retained and divided by the luminance of the task L_t. The position index was also disregarded as irrelevant, although later it was to be tested. Thus, the GI expression could be rewritten (MacGowan 1980; MacGowan et al. 1993) and tested as

$$\mathrm{GI} = k\left(\frac{L_\mathrm{s}}{L_\mathrm{t}}\right)^{\alpha}. \tag{12.4}$$

Therefore, the relationship to be tested in the laboratory in Seattle was simply the question whether (12.4) is true and valid.

New experiments were designed on the above reasoning. The experimental work began by predetermining comfort luminances for the laboratory, where controlling task luminance and allowing subjects to adjust the ambient laboratory lighting to a comfortable level resulted in the following task to ambient luminances:

- Task set at 40 cd/m^2, preferred ambient luminance 75 cd/m^2
- Task set at 60 cd/m^2, preferred ambient luminance 60 cd/m^2
- Task set at 120 cd/m^2, preferred ambient luminance 50 cd/m^2

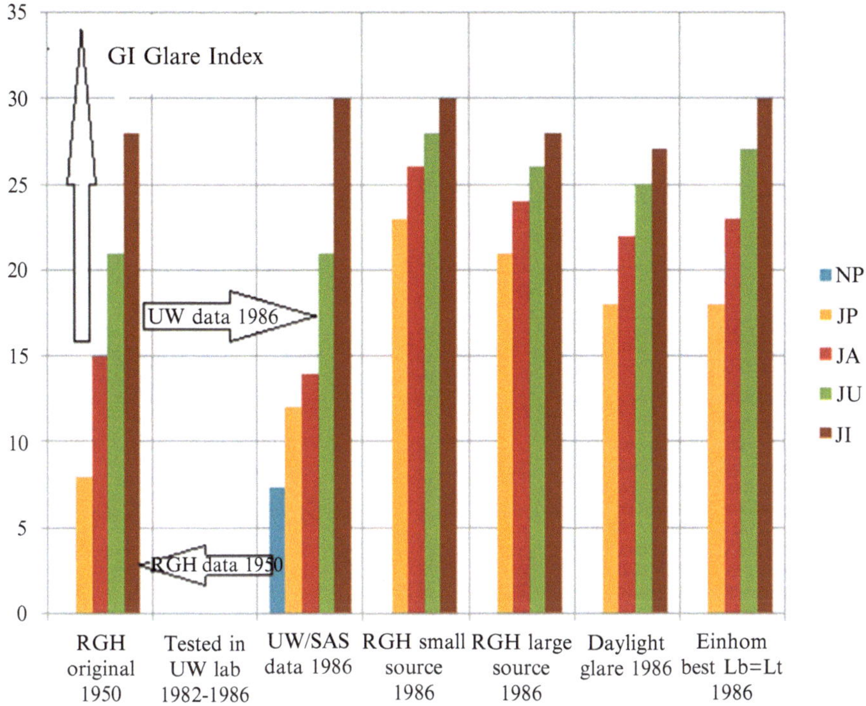

Fig. 12.1 Discomfort glare index confusing value changes as named formulae change in time, against RGH fixed reference values of glare indices of 8, 15, 21, and 28, representing Hopkinson's multiple criterion scale points of just perceptible (JP), just acceptable (JA), just uncomfortable (JU), and just intolerable (JI)

These results have implications for the use of task luminance to reduce ambient lighting requirements. There were no significant differences, at $p < 0.05$, when 32 subjects were employed, in the experiment that exposed subjects to daylighting or artificial lighting prior to the adaptation time allowed for the experiments (MacGowan et al. 1984, 1986). The main work to address the basic thesis began in earnest at UW, Seattle, in 1982 through late 1986, then continued in early 1988, through a Royal Society and Czechoslovak Academy of Sciences peer-reviewed study agreement. The study extended the UW work at the Institute of Construction and Architecture, Slovak Academy of Sciences (SAS), Bratislava. The initial 8-year Royal Society/SAS program agreed was to address glare source position and source size effects in light of past and present experience of the UW work. Results of that UW work are given in the tabulations of computations based on the 1982–1986 UW experiment data shown in Fig. 12.1. These demonstrate quite clearly that when applying UW laboratory data in both Hopkinson's, formulae (12.1) and (12.2) by Hopkinson (in 1950–1972), with comparison to the formula of

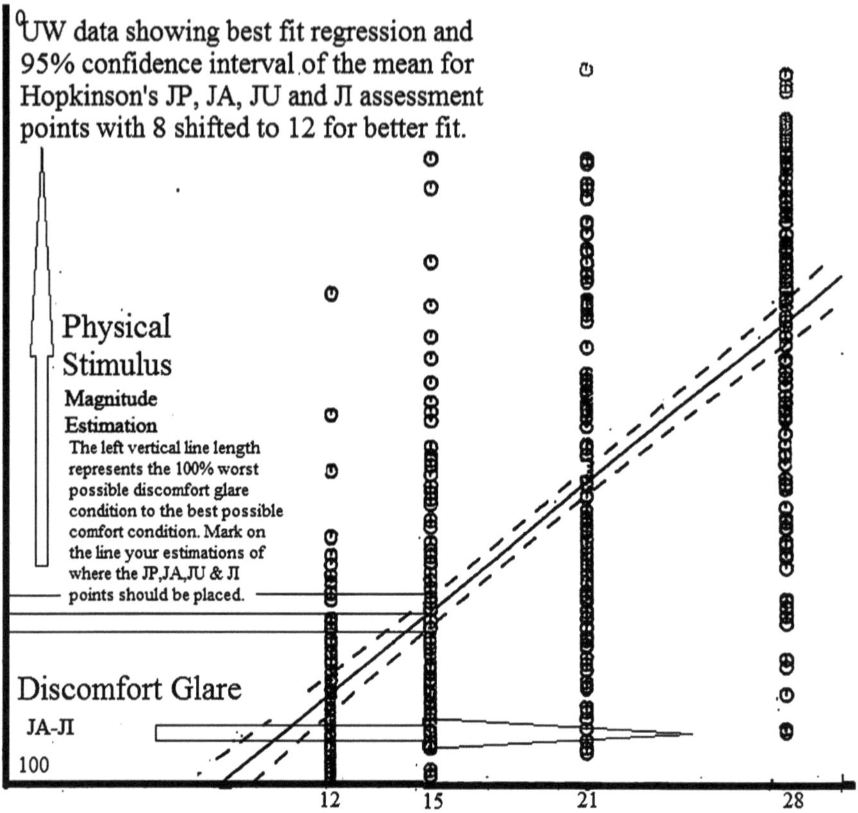

Fig. 12.2 The 95% confidence interval of the means at JP, JA, JU, and JI points

Einhorn (1969a, b, 1979) and formula facilitations why there was discontent with the application of these formulae some 30 years after Hopkinson's initial work produced appropriate discomfort glare predictions. Almost certainly the perceptions of people had changed (MacGowan et al. 1993; Nelson et al. 2002; Kittler et al. 2010) and that was most likely due to "habituated" exposure to ever-evolving lighting technology.

Figure 12.2 shows an example from the visual discomfort studies conducted by the latter authors using Hopkinson's multiple criterion JP, JA, JU, and JI elaborations and protocol after Hopkinson (1950) for Stevens-type physical stimulus function, visual comfort index, and magnitude estimations.

The elongated S curve of the "psychophysics power law" influence at the lower end of the continuum can be observed (see Chap. 11). These data show the nature of the typical spread of ψ responses in psychophysics experimentation, and they show the greater uncertainty in the subject responses at the lower and higher ends of the visual discomfort scale.

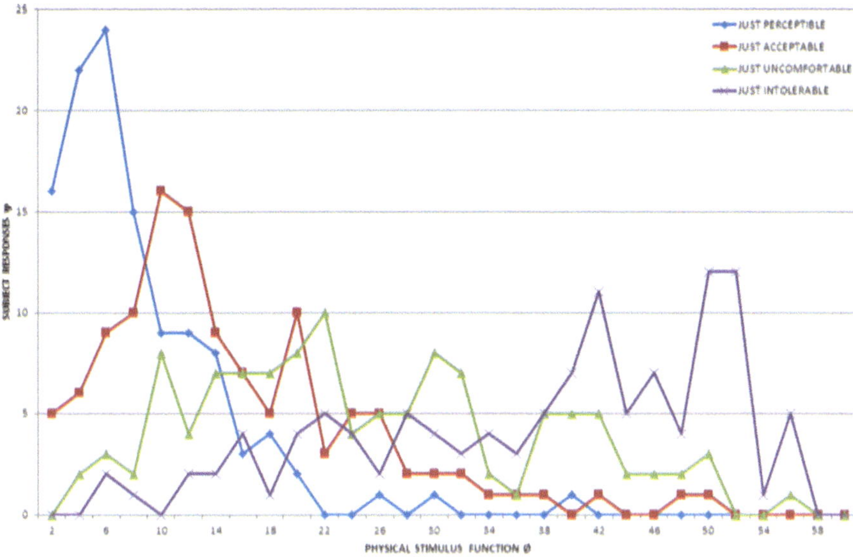

Fig. 12.3 Frequency polygons of the discomfort glare multiple criterion point subject evaluations using Hopkinson's JP, JA, JU, and JI levels of discomfort

Statistical power analysis Cohen (1977) revealed 78% at JP, 95.5% at JP, 92.5% at JU, and 57% at JI. The results in the center are excellent; the result at JP is good but perhaps it is influenced by an "elongated-*S*" tail. The result at JI is less certain, again perhaps owing to an "elongated-*S*" tail.

Figure 12.2 shows that the 95% confidence interval of the mean regression is particularly good in the mid-range section, although overall confidence in the population mean regression is excellent, especially in view of the typically wide range of responses from naive subjects. It should also be noted that all that can be achieved to date and likely in the foreseeable future from well-conducted visual discomfort psychophysical experimentation are application tools which satisfy the requirements of 50% of the population, because to satisfy even one standard deviation beyond the mean would be grossly uneconomic – just look at the evidence in the raw data frequency in Fig. 12.3, which is also evidence against the validity of any claimed 95% visual comfort probability.

Figure 12.3 also shows the very substantial variation in the responses from subjects, selected to reduce bias, and the need for power analysis to determine the minimum acceptable number of subjects required to provide reliable data from such experiments.

Figure 12.4 shows a pronounced effect of bias; the upper regression is from unbiased subjects, whereas the lower regression shows the results of "conditioned" subjects. It is further proof, if that were needed, that accidents can sometimes produce results that lead to better understanding. The bias, so clearly seen between two groups of subjects, was due to "conditioning." The data on the lower regression were the result of an inadvertent break in the UW subject recruitment protocol,

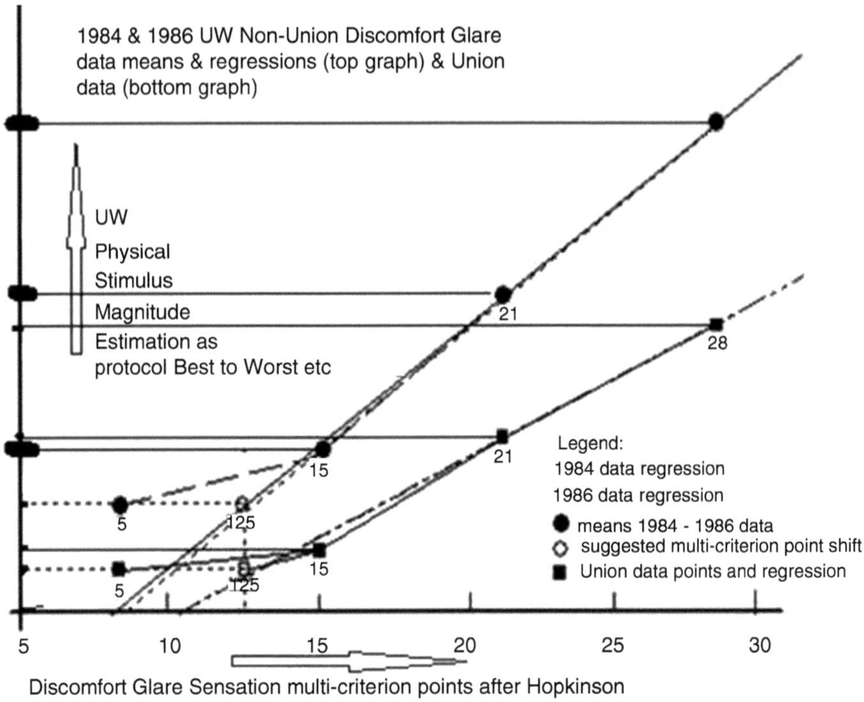

Fig. 12.4 University of Washington effect of subject bias (MacGowan et al. 1993)

which allowed subjects who had been exposed to trade union publicity on the adverse effects of working with VDUs to be misconstrued and, in the experimenter's opinion, led to less tolerance of bright screens, hence the UW discomfort glare source. However, the only reason for the difference in results that is important was a break in the very specific subject recruitment protocol that introduced bias.

12.3 Position Index Experiments

The protocol for the experiment of Luckiesh and Guth (1949) was based on "shock" exposure of subjects to a luminous source. They initially employed 50 subjects, and obtained minimum luminance of 1,079 cd/m^2 and maximum luminance of 5,482 cd/m^2 with a geometric mean of 2,844 cd/m^2 and an arithmetic mean of 3,053 cd/m^2. The geometric means were used in their quest to determine the borderline between comfort and discomfort due to glare source position. Condensed and reformatted results from Luckiesh and Guth (1949) borderline between comfort and discomfort (BCD) single criterion glare protocol experimentation, which led to their position index, are $\omega = 0.0011$ sr. and $L_b = 35$ cd/m^2. For vertical and horizontal displacement from the line of vision, the luminances of the glare source are as follows:

$$0°\text{V: } L_s = 2844 \text{ cd/m}^2,\ 0°\text{H: } L_s = 2840 \text{ cd/m}^2,$$
$$30°\text{V: } L_s = 9470 \text{ cd/m}^2,\ 30°\text{H: } L_s = 4297 \text{ cd/m}^2,$$
$$40°\text{V: } L_s = 15254 \text{ cd/m}^2,\ 50°\text{H: } L_s = 26091 \text{ cd/m}^2,$$
$$60°\text{V*: } L_s = 47968 \text{ cd/m}^2,\ 60°\text{H: } L_s = 8792 \text{ cd/m}^2,$$
$$90°\text{H*: } L_s = 51394 \text{ cd/m}^2 (^*\text{limit of field of view for some subjects}).$$

These results show staggering differences obtained from the SAS position experiment data, primarily due to rejecting the Luckiesh and Guth protocol in favor of the modified Hopkinson protocol. Rejection of flash-on flash-off glare source presentations might be appropriate under different, direction-critical, intermittent high contrast, luminance regimes, such as nighttime driving but i not appropriate, or even relevant for daytime noncritical attention tasks such as office, school, and indeed most paper and video tasks. The main glare sources are windows and the sky luminance in their solid angles, which usually do not change abruptly. It was considered unrealistic to expect the attention of people performing relatively safe visual tasks to be encumbered by a restriction of their direction of view, otherwise why should there be a proven need for window contact with the exterior (Marcus 1965; Keep 1977; Taylor 1979; Keep et al. 1980)? Therefore, dropping the requirement for the subject to fixate on a point during the experiments and replacing that protocol with an instruction to the subjects to imagine that they would be required to perform such a task in a specific direction of view for much or all of every working day in the presence of such a light source and so allow the subject to consider the task in relation to the overall work environment resulted in the markedly different data obtained (see Figs. 12.5 and 12.6) (Šubova et al. 1991; MacGowan et al. 1993; Nelson et al. 2002).

By comparing the Luckiesh and Guth data above with the SAS data below, one can only plot the 0° in vertical displacement on the same altitudinal graph, whereas on the azimuthal SAS graph, one can only plot one point, at 0°.

Further, the BCD single criterion adopted by Luckiesh and Guth does not provide any means to widen a discomfort glare specification to accommodate the ease or difficulty of the task where acceptance levels of discomfort glare would very likely differ. Therefore, in contrast to Luckiesh and Guth's experiment, the glare source position experiment at the SAS employed a modified RGH multiple criterion (fourth step) protocol and used the Bratislava artificial sky with 8-m diameter as the "position" laboratory (Kittler 1974). For the observers sitting in the sky center with their visual work, the uniform sky luminance presented a rather boring background scene; thus, the position of the glare source in different horizontal positions was quite attractive. The glare source position experiment used the same $\omega = 0.0011$ sr and $L_b = 35$ cd/m^2 used originally by Luckiesh and Guth (1949). However, data

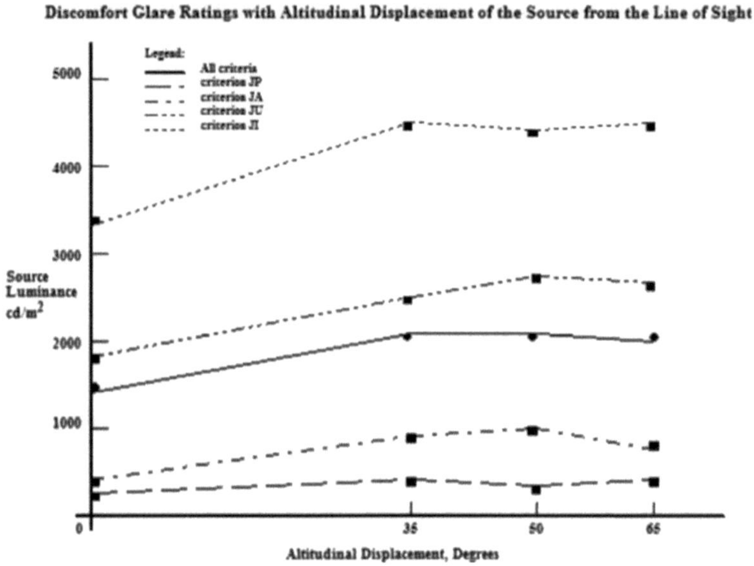

Fig. 12.5 Discomfort glare changes with azimuthal displacement of the glare source from the direction of work view

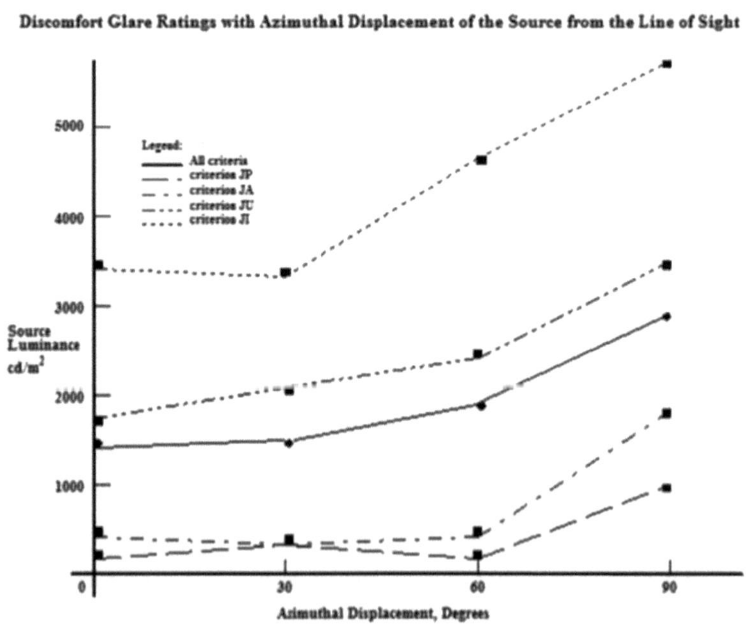

Fig. 12.6 Discomfort glare changes with altitudinal displacement of the glare source from the direction of work view

from these psychophysical experiments using unbiased human subjects can only be assumed to be useful if they are subjected to assessment by Cohen's statistical power analysis. Such analysis was not refined by Cohen (1977), so it was unavailable to Luckiesh and Guth, whose results by 1980 would likely fail rigorous scrutiny. Cohen's analysis permits the determination of a required subject population size from the results of a pilot experiment using similar methodology and the same subject pool and the traditional statistics obtained. To have sufficient numbers of subjects to achieve the degree of certainty required of discomfort glare experimentation using unbiased subjects usually means at the outset employing 32 subjects, although more or fewer may be required or suffice if the experiment's variables can be reduced. Luckiesh and Guth at first used 50 subjects, but because the hardware technology made their experiments very difficult and time-consuming, that number had to be reduced to make the work practicable. Hence, they concluded that because the subject arithmetic mean and median and geometric mean were close, they had a "normal" distribution and could reduce their subject size to ten using "selected representatives" of the data from the 50 subjects. It is easy to see how ten subjects closest to these means could be selected from the results from the 50. However, they did not do that; they selected ten subjects with the same characteristic distribution of data, so they tried to be impartial. However, they would not likely represent the population data spread from these 50 subjects at all times. Nevertheless, because of the primitive technology used, it would have been virtually impossible to conduct their experiments with even 32 subjects. There is no evidence to suggest that the ten subjects were conditioned or that they were selected because of their consistency, equivalent of Hopkinson's "human meters" (Hopkinson 1955), a technique we rejected in the Seattle and Bratislava experiments as potentially biased, in favor of the required subject population size computed in accord with Cohen's statistical power analysis.

However, such errors arising from reduced numbers have to be placed in perspective for many current psychophysical experimenters reported using only one, two, or three subjects, a grossly inadequate number, well below the threshold for Cohen's statistical power analysis and, often also using biased "conditioned" subjects. Such "quickie" jobs producing meaningless data are at best only useful in some cases as a dry run prior to the actual experiment, to help the experimenter.

12.4 Glare Source Size and Task Orientation Experiments

The SAS protocols used were Slovak translations of those used at UW (supplied by Hopkinson), and in the SAS discomfort chamber laboratory (Fig. 12.7). An artificial window 2 m$\times$1.2 m closely packed by vertically placed Osram 36-W, 5,000 K fluorescent tubes behind a diffuse glazing formed a uniform large source, the luminance of which could be operated via a Lutron dimmer and solid-state ballasts by the observer/controller of the discomfort glare source. Simultaneously, the experimenter measured the glare source luminance and observer's task luminance

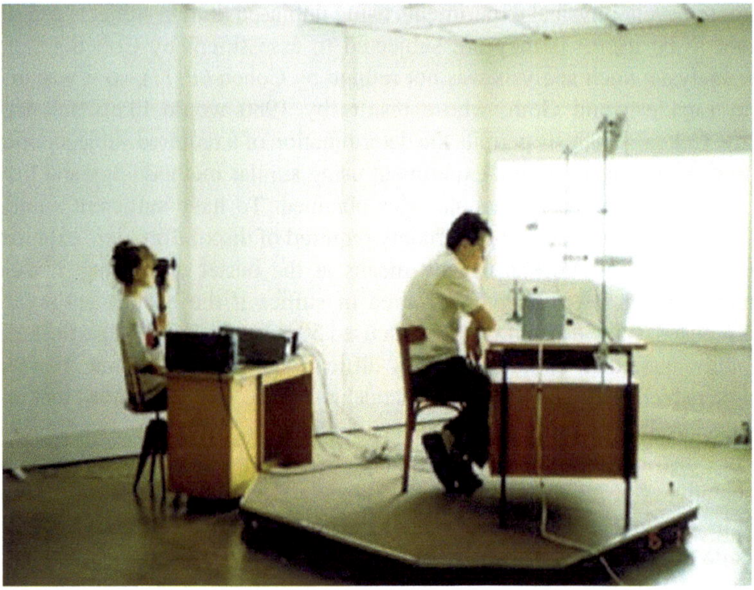

Fig. 12.7 The Slovak Academy of Sciences large-area-source discomfort glare laboratory

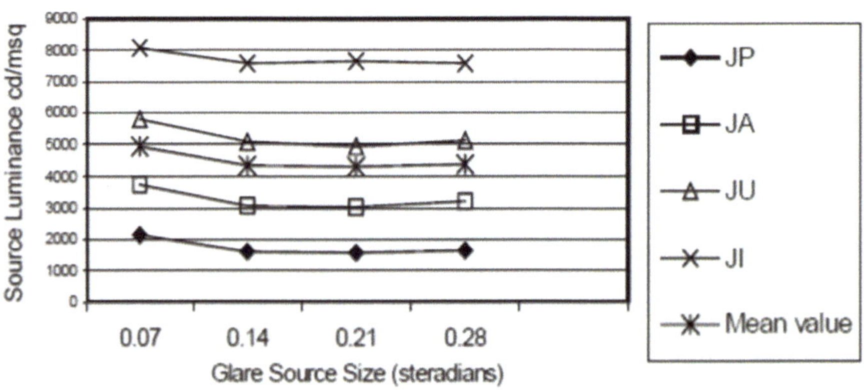

Fig. 12.8 Change of perceived glare discomfort with source size

and illuminance. Glare source shielding and segmental switching of the artificial window enabled the solid angle of the glaring source to be changed. The statistically evaluated experiments' data on discomfort glare are shown in Figs. 12.8 and 12.9. The tests with window size (Fig. 12.8) demonstrate that there is no change of discomfort glare rating owing to window size for all "acceptable" window sizes, thus validating the conclusions of Chauvel et al. (1980). However, there is a slight

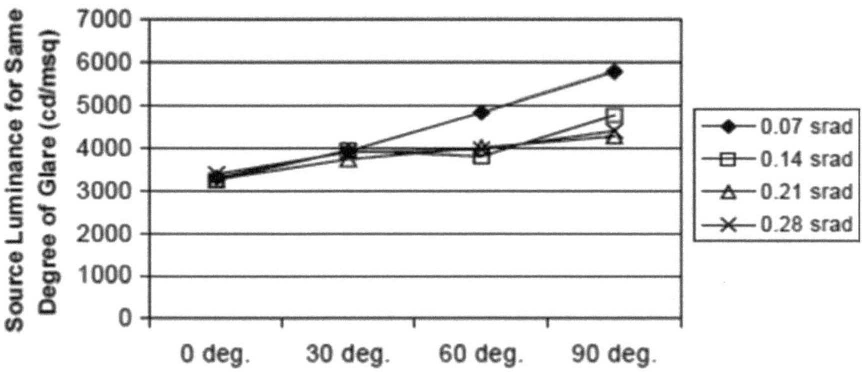

Fig. 12.9 Influence of vertical source size and azimuthal displacement from work view on discomfort glare perception

size effect below 0.14 sr. However, owing to practical, mainly financial, constraints the experiment could not reach below 0.07 sr (Nelson et al. 2002). Windows with good views require even larger solid angles above 0.21 sr, as 1984 ref. above, but there is no source size effect on discomfort glare perception beyond $\omega = 0.14$ sr, where the UW/SAS GI formula predicted values can be used with confidence. Guidance on how to accommodate window sizes that are below the minimum acceptable window size is beyond the scope of this publication but can be found in MacGowan et al. (1984), available from the lead author or perhaps from Public Works or Health and Welfare Canada, the original sponsors and owners of that minimum window size work, which remained substantially unpublished.

As evident from Fig. 12.9, use of the Luckiesh and Guth position index should be discontinued for all tasks which do not involve safety of life or limb in which sight must be constantly directed toward the task. The SAS data from the glare source position experiments provide sufficient reason for that recommendation. However, the natural involuntary retinal reflex response of the eye (Šubova 1991) is an additional supportive reason to discontinue use of such long-standing flawed technology. In addition, no direct task lighting should be applied from above the line of sight to the task; down-lighters should be replaced by above the line of sight "up-lighters" and appropriately placed below the line of sight task lighting. The source of any luminaire should not be visible to the work position eye; this general rule would apply to all office, school, and hospital facilities, indeed all occupancy situations which would involve the specification of a GI below JI (MacGowan et al. 1984; 1986, 1993; Nelson et al. 2002). Tasks which are only casual would normally not require a GI to be specified; therefore, they would be excluded from the recommendation.

Some years after work on the Stevens form of discomfort glare expression began, a continuation of the old Fechner form of expression attempted to make use of the Hopkinson formulae easier and to reconcile computed values with apparently changed perceptions of discomfort glare magnitude. The new, old format, formula is called the unified glare rating (UGR) adopted by the International Commission on Illumination (CIE 1995):

$$\mathrm{UGR} = 8\log\left[\frac{0.25}{L_\mathrm{b}}\sum\frac{L_\mathrm{s}{}^2\omega}{p^2}\right], \tag{12.5}$$

where L_b is the background luminance in candelas per square meter, L_s is the luminance of the glare source in the direction of the observer's eyes in candelas per square meter, ω is the solid angle of the glare source at the observer's eye in steradians, and p is the Guth position index for glare source displacement from the line of sight (Guth 1946; Luckiesh 1949).

Firstly, because the UGR was spawned after the UW and its forerunner work had been published (MacGowan et al. 1980, 1981, 1984, 1986, 1993; Emery and MacGowan 1984) in Europe and North America, then further developed by the SAS and published in Europe and Japan (Šubova et al. 1990, 1991; Pulpitlova et al. 1992), there was no reason to introduce the UGR formula (12.5) at least for window glare evaluations.

Secondly, Sörensen (1987), Akashi et al. (1996), Einhorn (1998), and the CIE promoted a regression of science, apparently in ignorance or in denial of Stevens's power law, back to Fechnerite thinking despite the available evidence against their promotions. As aptly stated by Ekman and Sjőberg (1965): "After a hundred years of almost general acceptance and practically no experimentation, Fechner's logarithmic law was replaced by the power law. The amount of experimental work performed in the 1950's on this problem by Stevens and other research workers was enormous....The power law was verified again and again, in literally hundreds of experiments. As an experimental fact, the power law is established beyond any reasonable doubt, possibly more firmly established than anything else in psychology." Yet Einhorn promotes the Fechner form of the UGR as a "bonus" even though it fails to address the nature of the scale magnitude increments, the ω relationship, or the position of source questions.

Figure 12.1 clearly shows how the perception of discomfort glare had changed consistently with time and almost certainly with circumstance (MacGowan et al. 1993). Therefore, it matters not that $10\log L^{1.6}\omega^{0.8}$ may be more simply stated as $8\log L^2\omega$, for the nature of such expressions was found to be wrong, and the continued use of both ω and Guth's p was questioned in 1984 as components of GI formulae. Einhorn and his CIE committee knew that in 1993 when they received the raw data from the Seattle and Bratislava studies and Einhorn et al. witnessed the presentation of these facts in Edinburgh (MacGowan et al. 1993) and thereafter by correspondence.

It is unfortunate that experiments of Akashi et al. (1996) revealed that the UGR vastly overpredicted discomfort glare but they blamed this on their naive subjects. Even worse, because of that they conducted further experiments using only "expert" subjects. In fact, they initially used 56 naive subjects and five "expert" subjects were just enough to pollute their results and the understanding of their meaning. Had they paid closer attention to articles by Šubova et al. (1991) and MacGowan et al. (1993), they would have used only naive subjects totally free of lighting design knowledge to avoid the pollution that destroyed the value of their work. However, that work was at least an attempt to discover how human subjects responded, in that time and circumstance, to discomfort glare in their controlled luminous environment.

To achieve accurate comparison of data from different experiments with the same goal, the experiment protocols must be rigorously replicated. Scales cannot be modified, extended, or truncated without expressing these changes in relation to the original scale. In an attempt to assist such replication, Hopkinson's precise discomfort glare protocol was published by MacGowan (2010). It has to be realized that when computed GI numbers no longer represent the "number associated" perceived discomfort glare level, then it is the formula that must change, not the meaning of the numbers.

12.5 Partial Conclusions and Future Research Needs

Orientation toward bright luminance sources is a basic wiring retinal reflex response. It matters not whether the discomfort is due to overhead, in front of the head, to the side of the head, or behind the head glare. It is the same discomfort glare and one formula, correct in the same technology timescale, is all that is needed.

However, discomfort glare from considerably below the line of sight is likely to be different because the habituated basic wiring of humans evolved in an environment in which bright light was and still is in the upper hemisphere (on or above the line of sight), except for light which is specularly reflected. Therefore, it is reasonable to suspect that discomfort glare arising from the lower hemisphere could be different from and probably would be more severe than that which produces discomfort from the upper hemisphere.

Stevens did not immerse himself in the perception of visual comfort; he left such visual studies to Hopkinson, with whom he fruitfully communicated in the mid-1950s to the mid-1960s on brightness perception and Hopkinson's visual discomfort scales. They agreed through corroborating research on the luminance to brightness relationship that it was not as the Weber–Fechner formula predicted but was, in an important visual work range, linear (Stevens's $\alpha = 1.0$) (Hopkinson 1963; Stevens JC and Stevens SS 1963). These communications, coupled with Macfarlane's input to the research program on habituation as the likely cause of visual comfort perception shift provoke further experimentation to ultimately discover, not without further questions that we strive to answer, whether discomfort glare in the important daytime work brightness range is linearly proportional to luminance and also whether this is true for discomfort glare from daylight.

Discomfort glare experimentation needs to continue, to answer further basic questions and extend the work to embrace multiple small source lighting systems.

Two sets of new experiments are needed to understand the flexibility of a basic equation for time and circumstance shift, as proposed earlier (MacGowan et al. 1993).

There is no irrefutable evidence that habituation to new environmental lighting technology and workforce preferences has caused the discomfort glare formulae of the 1950s, and their evolutions, to predict ever-more-conservative answers as has time rolled into the second decade of the third millennium, some 60 years on. Nevertheless, habituation occurs involuntarily and imperceptibly. Perhaps it is time to abandon the GI system of numbering while retaining the Hopkinson multi-criterion JA to JI routine as the Discomfort Glare sensation anchor descriptors, and revert to the PSF, a simple ratio scale of L_s to L_t. For instance, a UW/SAS GI = 16 would simply be a DGR (Discomfort Glare Ratio) of 20:1 (glare source to task luminance). Thus making determination of a DGR a very simple elementary-school level mental arithmetic task which would lend itself to the accommodation of Time, Circumstance and Habituation changes required of the DGRs in accord with current international differences and those that are bound to arise in the future due to such involuntary unavoidably changing influences?

Finally, it should be kept in mind that all plausible empirical discomfort glare formulae have, or should have been developed large subject number experiments, with resulting highly variable. Thus, it is not economically feasible to apply even a standard deviation from the mean of such data. Therefore, one could raise the GI requirement by dropping the GI by one or two levels, but also bear in mind that habituation will desensitise the glare discomfort in time if it is not harmful to the organism. However, to compound the situation, anyone super-sensitive, or who thinks they are super-sensitive to *glare* will involuntarily enact arousal, which will overpower habituation and make the situation worse.

Appendix 12

Comparison of Changing Glare Situations Under Various Daylight Conditions

In an artificially illuminated interior, e.g., during evenings, the overall luminance distribution is almost stable and constant in time. So, the glare situation can be unchanged, with probably only interference with temporary changes caused by the TV monitor excitation. However, during daytime, owing to outdoor daylight condition changes, i.e., gradual or abrupt increasing or moving cloudiness patterns and the sun position movements, the window luminance and consequently also glare situations can vary. The window luminance pattern is influenced by sky luminance in its solid angle and the window orientation. So, in a side-lit room the GI can reach different values indicating glare-free situations under overcast skies or extreme discomfort or disability conditions when the sun appears in the window view. Such varieties can be documented in an example of a room with one window earlier used in the appendix in Chap. 8 applying the MAM modeller computer possibilities (Roy et al. 2007) to check window luminance under typical sky standards.

The glare source luminance, the maximum luminance in the window solid angle $L_{W\,max}$ seen from the three elements of the working plane as task luminance L_t that can cause to rise the GI in the simple formula (12.5) to rise will be evaluated:

$$\mathrm{GI} = 0.8\frac{L_{\mathrm{Wmax}}}{L_{\mathrm{t}}}. \tag{A12.1}$$

The momentary maximum luminance within the window solid angle was estimated from the upper corner of the window image in MAM while the task luminance of common reading print with reflectance $\rho_t = 0.7$, illuminated by window luminance with horizontal sky illuminance E_{vS}, was increased by interreflection r. Thus, the task luminance for the printed text on a horizontal table desk is

$$L_{\mathrm{t,text}} = \frac{E_{\mathrm{vS}}\,\rho_{\mathrm{t}}\,r}{\pi}\ (\mathrm{cd/m^2}), \tag{A12.2}$$

where approximate r values rise from the elements close to the window to the element deep inside.

To enable the comparison with the example in a previous article (Kittler et al. 2010), the same interior and window size as well as the elements on the working plane were chosen:

- An administrative room 3 m wide and 4.5 m deep with one unobstructed window 1.5 m in width and 1.8 m high was chosen as representative for office work.
- In this room three reference points on the working plane 0.85 m above the floor were placed along the window axis noted as A at 1-m distance from the glazing, B at 2-m distance, and C at 3-m distance, respectively.
- Sky luminance seen through the glazing is reduced by normal glass transmittance of 80%.
- Paper work and text reading are still very common visual tasks. The reflectance of printed paper of 70% was considered.
- The room surface reflectance for the ceiling (80%), for walls (50%) and for the floor (20%) represent most common cases, and exterior terrain reflectance was taken as 20%.
- As MAMmodeller calculates only the horizontal sky illuminance E_{vS} in elements A, B, and C these, were multiplied by factor r, i.e.,
  - At A: $r = 1.15$, ; thus, $L_{\mathrm{t,text}} = E_{\mathrm{vS}}\,\rho_{\mathrm{t}}\,r/\pi = 0.256\,E_{\mathrm{vS}}$.
  - At B: $r = 1.35$, i.e., $L_{\mathrm{t,text}} = 0.3\,E_{\mathrm{vS}}$.
  - At C: $r = 1.75$ and $L_{\mathrm{t,text}} = 0.39\,E_{\mathrm{vS}}$.

Regular office, school, and other cerebral work requires a mixture of both paper and video display reading tasks to be accomplished without inducing undue visual comfort. In addition to satisfactory task visibility requirements, a range of task-related GIs are employed to ensure freedom from excessive visual discomfort. Therefore, daylighting levels must be determined in partnership with the GI levels they induce. There are two obvious ways to reduce an excessive GI: reduce the luminance of the source or increase the luminance of the task. The first should be a last resort if daylighting is prioritized and building energy use parametric simulations, for instance, by eQUEST (http://doe2.com/equest/index.html), are used to provide an agreeable energy cost. The second choice, also with energy cost implications, is to increase the task luminance by either supplementary task lighting, for paper tasks, or by increasing video monitor display luminance. The following tables provide a snapshot of the visual alternatives to be contemplated against incentives such as building energy use, maintenance, visual quality, amenity, and workplace productivity. Bear in mind, with the latter, the very high financial cots of people in the workplace, in relation to the cost of materializing and operating a large-scale work environment. The following tables only sketch the possibilities for intervention to correct a wayward design, but hopefully illustrate a virtually intuitive approach to dealing with the potential pitfalls of daylighting design of cerebral work environments.

The screen luminances have been selected after sampling literature on readily available computer video display luminance and contrast capabilities of readily available video monitors. Also, bear in mind that LED backlit monitors are proliferating, are energy-efficient, and will produce luminances far beyond the luminances selected for this demonstration: 100, 200, and 800 cd/m^2 were adopted as L_{tC} in candelas per square meter.

Owing space restrictions in this appendix only one window orientation could be considered in order to demonstrate that any room orientation can be chosen, in this example a random option is the window azimuth $A_W = 140°$ from north. Furthermore, to demonstrate the dependence of GI on the geographical latitude of the locality, the 45° latitude was chosen, along which lie North American cities such as Minneapolis/St. Paul, Ottawa, and Montreal and European centers such as Bordeaux, Turin, and Belgrade. As in MAMmodeller, a certain location has to be logged. In the following example was assumed a locality east of Bordeaux with geographical latitude 45° and longitude 0°. Of course, because of different climate and weather influences besides the relevant sun-path similarities, the glare problem can be also different in the momentary, daily, or seasonal occurrence of sky patterns, cloudiness, and sunshine duration expectancies. Experienced designers know that the roughly southeastern window has to have some protection against excessive heat gains and glare during sunny conditions. In Design, appropriate external shading and building configuration alternatives should have been explored and deployed. If the window is causing discomfort glare use venetian blinds, reduced transmittance glass, retrofit film or other shading to alleviate the malady while maintaining the external view (Köster 2004). If the GI rating is too high, try raising the video luminance and interior lighting will also reduce the window and

Table A12.1 Daytime changes of glare situations on a winter day under sky type 1

Time	γ_s	Position	$D_{v\,ext}$ (lx)	$E_{v\,int}$ (lx)	$L_{W\,max}$ (cd/m^2)	L_{tP} (cd/m^2)	L_{tC} (cd/m^2)	GI_P	GI_C
8:00	3.3°	A	775	86	235	22	100	9	2
		B		24	199	7		23	2
		C		8	166	3		44[a]	1
9:00	11°	A	2,544	281	771	72	100	9	6
		B		79	654	24		22	5
		C		27	545	11		40[a]	4
10:00	17°	A	3,885	429	1,178	110	100	9	9
		B		121	997	36		22	8
		C		42	832	16		42[a]	7
11:00	20.6°	A	4,708	520	1,427	133	100	9	11
		B		146	1,210	44		22	10
		C		51	1,009	20		40[a]	8
12:00	21.7°	A	4,954	547	1,502	140	100	9	12
		B		154	1,273	46		22	10
		C		53	1,061	21		40[a]	8

[a]Glare index values over 28

luminaire effective impact on the rating on paper tasks. Do not reduce the window transmittance below about 25% or you will depreciate the view (MacGowan 1984). A favourable energy balance between the efficiency and the utility of windows usually requires a computer simulation, then the most favourable luminance for a work environment can be achieved. The tables provided in this appendix are only simplistic outlines to be played with in order to engender understanding by providing an intuitive feel for the intervention strategies.

For this example only, several outdoor conditions and ISO/CIE sky types have been chosen to demonstrate and test glare conditions calculating GI values either for paperwork (GI_P) or for visual work on a computer screen (GI_C):

Under the overcast sky ISO type 1 (ISO 2004), if $D_v/E_v = 0.1$, $\tau_g = 0.8$, on December 15 in Table A12.1.

In this first test the lowest winter daylight conditions were taken into account as well as the lowest computer screen luminance $L_{tC} = 100$ cd/m^2 . In this example, interior illuminance levels are so low that certainly artificial illumination will be used, which will better the overall satisfaction. Therefore, also better daylighting under overcast sky type 4 was tested assuming the same day in December but a higher illuminance level in the following example:

Under the overcast sky ISO type 4 (ISO 2004), if $D_v/E_v = 0.22$, $\tau_g = 0.8$, on December 15 in Table A12.2.

From Tables A12.1 and A12.2 it is evident that point C, with low interior illuminance, causes the low task luminance and in comparison wit window luminance also intolerable glare (indicated by “a”). Some brighter sky patches under sky type 4 and $D_v/E_v = 0.22$ are in the window solid angle at 10:00–12:00, the computer screen luminance becomes too low, and GI$\geqslant$28.

Table A12.2 Daytime changes of glare situations on a winter day under sky type 4

Time	γ_s	Position	$D_{v\,ext}$ (lx)	$E_{v\,int}$ (lx)	$L_{W\,max}$ (cd/m^2)	L_{tP} (cd/m^2)	L_{tC} (cd/m^2)	GI_P	GI_C
8:00	3.3°	A	1,705	283	833	72	100	9	7
		B		102	862	31		22	7
		C		48	861	17		41[a]	7
9:00	11°	A	5,596	982	2,744	251	100	9	22
		B		362	2,751	109		20	22
		C		156	2,752	61		36[a]	22
10:00	17°	A	8,547	1,511	4,263	387	100	9	34[a]
		B		544	4,280	163		21	34[a]
		C		225	4,261	88		39[a]	34[a]
11:00	20.6°	A	10,357	1,738	5,270	445	100	9	42[a]
		B		578	4,892	173		23	39[a]
		C		225	4,527	88		41[a]	36[a]
12:00	21.7°	A	10,898	1,637	5,260	419	100	10	42[a]
		B		513	4,363	154		23	35[a]
		C		196	3,942	76		41[a]	32[a]

[a]Glare index values over 28

Table A12.3 Daytime changes of glare situations on a spring day under sky type 13

Time	γ_s	Position	$D_{v\,ext}$ (lx)	$P_{v\,ext}$ (lx)	$E_{v\,int}$ (lx)	$L_{W\,max}$ (cd/m^2)	L_{tP} (cd/m^2)	GI_P
8:00	17.5°	A	16,104	6,603	4,996	53,012[a]	1,279	33[b]
		B			2,074	41,307	622	53[b]
		C			981	33,210	383	69[b]
9:00	26.8°	A	22,350	17,896	8,120	51,073[a]	2,079	20
		B			3,993	51,428[a]	1,198	34[b]
		C			1,874	48,241	562	69[b]
10:00	34.7°	A	25,970	28,946	9,657	47,383[a]	2,472	15
		B			4,708	47,507[a]	1,412	27
		C			2,031	40,200	792	41[b]
11:00	40.3°	A	27,515	36,846	9,125	43,303[a]	2,336	15
		B			3,584	41,611	1,075	31[b]
		C			1,391	30,959	543	46[b]
12:00	42.6°	A	27,945	40,150	6,713	41,215[a]	1,719	19
		B			2,110	23,843	633	30[b]
		C			840	18,585	328	45[b]

[a]The solar disc is present in the window solid angle
[b]Unacceptable glare conditions when $GI_P > 28$

Under the clear sky ISO type 13 (ISO 2004), if $T_v = 5.5$, $\tau_g = 0.8$, on March 1. The window orientation to 140° exposes it to direct sunshine, which is evident in Table A12.3.

In spite of the gradual decreasing effect of sunbeams on the window façade in afternoon hours, the unacceptable glare conditions are only slightly lowered.

Table A12.4 The glare situation on March 15 at 8 a.m. under sky type 13 when the window is shaded and the display luminance is 800 cd/m²

Time	γ_s	Position	$D_{v\,ext}$ (lx)	$P_{v\,ext}$ (lx)	$E_{v\,int}$ (lx)	L_{Wmax} (cd/m²)	L_{tP} (cd/m²)	GI_P $L_{tC} = 800$
8:00	17.5°	A	16,104	6,603	2,186	23,193	556	23
		B			907	18,072	272	18
		C			429	14,529	168	15

Under sunshine and clear sky conditions, the glare problem can be solved by several alternative possibilities, e.g., applying window transparent shading ($\tau_g = 0.35$) while still retaining an acceptable view (MacGowan et al. 1984). At the same time, if computer screens are manually or automatically controlled to respond to higher window luminances, by display luminance adjustment, e.g., to 800 cd/m², then the reduction from GI_P to GI_C would be significant as shown in Table A12.4.

Shading alone will not result in the solution of disability glare because the interior illuminance also drops because of the window's shading. Thus, additional artificial illuminance of the visual task is one simple solution. For instance, only turning on night lighting with an additional 200-lx level to complement the shaded window's contribution would result in a quite high temporarily satisfactory GI = 24. Hence, the example above shows how higher interior artificial illuminance and video display luminance could be used to lower the GI under sunny conditions and shading while preserving a key function of the window view.

If a video display screen has a "controlled" working luminance of up to 800 cd/m², then an insolated window's interior luminance could reach 12,000 cd/m², GI = 12, and be unlikely to cause any perceptible discomfort for most people who work in such a "daylit" luminous environment.

As soon as the sun path goes over the window head in summer, the glare situation is less severe owing to the absence of the solar disc in the window solid angle. For critical point A, the window head on the centerline is $\varepsilon = \text{arc cotan } 1/1.8 = 60.97°$, whereas in the relevant upper window corner it is $\varepsilon = \text{arc cotan}\left(\sqrt{1 + 0.9^2}\right)/1.8 = 53.2°$.

Such a situation can be documented, e.g., by the following example.

Under the clear sky ISO type 12 (ISO 2004), if $T_v = 4$, $\tau_g = 0.8$, on June15. The window orientation to 140° exposes it to direct sunshine, but under relatively high solar altitudes. The glare situation is evident from inspecting Table A12.5.

When the sun position is not passing the window solid angle, the glare situation better resembles the conditions under cloudy sky type 4 in December. However, owing to relatively higher illuminance levels outdoors, the maximum window luminance is also high and window shading with additional artificial lighting is recommended especially in position C.

Every window orientation is exposed to a variety of daylight conditions and sun-path changes, as documented by interior glare situations in the example above.

Table A12.5 The glare situation on June 15 under sky type 12 when the window at orientation 140° is unshaded

Time	γ_s	Position	$D_{v\,ext}$ (lx)	$P_{v\,ext}$ (lx)	$E_{v\,int}$ (lx)	L_{Wmax} (cd/m²)	L_{tP} (cd/m²)	GI_P
8:00	37.1°	A	20,254	41,601	918	5,200	235	18
		B			541	4,623	162	23
		C			303	4,920	118	33[a]
9:00	47.6°	A	22,602	57,355	996	4,818	255	15
		B			562	4,446	169	21
		C			311	4,774	121	32[a]
10:00	57.3°	A	23,629	69,868	1,023	4,806	262	15
		B			559	4,109	168	20
		C			307	4,561	120	30[a]
11:00	65.0°	A	23,866	77,875	1,022	4,512	262	14
		B			548	3,865	164	19
		C			301	4,465	117	31[a]
12:00	68.3°	A	23,833	80,649	1,017	4,107	260	13
		B			543	3,145	163	15
		C			297	4,327	116	30[a]

[a]Glare index values over 28

References

Akashi, Y., Muramatsu, R., Kanays, S.: Unified Glare Rating (UGR) and subjective appraisal of discomfort glare. Lighting Res. Technol., 28, 4, 199–206 (1996)

Chauvel, P., Collins, J.B., Dogniaux, R., Longmore, J.: Glare from windows: current views of the problem. Proc. Symposium on Daylight: Physical, Psychological, and Architectural Aspects, Institut für Lichttechnik der TU Berlin, 294–302 (1980)

CIE – Commission Internationale de l'É clairage: Discomfort Glare in Interior Lighting. CIE Publ. 117 CB CIE Vienna (1995).

Cohen, J.: Statistical Power Analysis for the Behavioural Sciences. Academic Press, New York (1977)

Einhorn, H.D.: An analytical expression for the glare position index. Illuminating Engineering, 64, 228–229 (1969a)

Einhorn, H.D.: A new method for the assessment of discomfort glare. Lighting Res. Technol., 1, 4, 235–247 (1969b)

Einhorn, H.D.: Discomfort glare: a formula to bridge differences. Lighting Res. Technol., 2, 2, 90–94 (1979)

Einhorn, H.D.: Unified glare rating (UGR): Merits and application to multiple sources. Lighting Res. Technol. 30 (2) 89–93 (1998)

Ekman, G., and .Sjőberg L.: Scaling. Annual Review of Psychology 16 451–474 (1965)

Emery, A.F., and MacGowan, D.: Windows in Internal Thermal Load Dominated Buildings. Proc. Windows in Building Design and Maintenance, CIB-84 Gothenburg, Sweden (1984)

Guth, S.K.: Discomfort Glare and Angular Distance of Glare-source. Illuminating Engineering, 41, 6, 485 (1946)

Hopkinson, R.G.: The multiple criterion technique of subjective appraisal. Quarterly Journal of Experimental Psychology, 2, 124 (1950)

Hopkinson, R.G.: Subjective judgements – some experiments involving experienced and inexperienced observers. British Journal of Psychology, 46, 262 (1955)
Hopkinson, R.G. and Bradley, R.C.: Glare from very large sources. Illuminating Engineering, 55, 288–297 (1960)
Hopkinson, R.G.: Architectural Physics: Lighting, London: HMSO, Part II, Section VII, 329–336 (1963)
Hopkinson, R.G.: Glare from windows. Construction Research and Development Journal (CONRAD), 2, 3, 98–105; 2, 4, 169–175; 3, 1, 23–28 (1970–1971)
Keep, P.V.: Stimulus deprivation in windowless rooms. Anaesthesia, 32, 598–602 (1977)
Keep, P.V., James, J., Inman, M.: Windows in the intensive therapy unit. Anaesthesia, 35, 257–262 (1980)
Kittler, R.: New artificial 'overcast and clear' sky with an artificial sun for daylighting research. Lighting Research and Technology, 6, 4, 227–229 (1974)
Kittler, R., Darula, S., MacGowan, D.: The Critical Window Luminance Causing Glare in Interiors. Selected papers of the Light and Lighting Conference with Special Emphasis on LEDs and Solid State Lighting, 2009, Budapest, CIE publ. x034: 63–73 (2010)
Kozak, W., Macfarlane, W. V., Westerman, R.: Long-lasting reversible changes in the reflex responses of chronic spinal cats to touch, heat and cold. Nature, 193, 171 (1962)
Luckiesh, M. and Guth, S.K.: Illuminating Engineering, 43, 10 (1949a)
Luckiesh, M. and Guth, S.K.: Brightness in visual field at Borderline between Comfort and Discomfort (BCD). Illum. Engineering, 44, 11, 650–670 (1949b)
Macfarlane, W.V.: Habituation to heat and cold at the spinal cord level. Environmental Physiology and Psychology in Arid Conditions. UNESCO. Paris, 351 (1964)
Macfarlane, W.V.: Personal communication to David MacGowan on Habituation. University of Adelaide (1972)
MacGowan, D.: An Hypothesis - A new discomfort glare formula arising from a revision of the BRS/Cornell discomfort glare formulae in the light of past and present experience. Report Lawrence Berkeley Laboratory, University of California, Berkeley (1980)
MacGowan, D.: Determination of modified external components of daylight factor and glare index using LIAM diagrams. Report Lawrence Berkeley Laboratory, University of California, Berkeley (1981)
MacGowan, D., Thompson, H.E., Eaton, M., Harris, A.: Window Architectonics, Design Guidelines. University of Manitoba report to Public Works Canada, Contract No.330-993, Ottawa, Canada, 4, 50–67 and 1–83 (1984)
MacGowan, D., Kraft, C.L., Emery, A.F.: The Discomfort Glare Study at the University of Washington. Report to the USA National Science Foundation. Department of Mechanical Engineering University of Washington, 1–68 (1986)
MacGowan, D., Pulpitlova, J., Subova, A.: Visual Work Environment and VDUs. Abstract Book, Work with Display Units, WWDU '92. A47. Technische Universitat Berlin, (1992)
MacGowan, D., Pulpitlova, J., Šubova, A.: Time and Circumstance, and the perception of discomfort glare. Lux Europa Proc., 1, 193–211 (1993)
MacGowan, D.: Discomfort glare criteria. Lighting Res. Technol., 42, 1, 121–122 (2010)
Marcus, T., A.: The significance of sunshine and view for office workers. Sunlight in Buildings. 59–93. Bouwcentrum International Rotterdam, (1965)
Nelson, P., MacGowan D., Pulpitlova J., Šubova A., Kittler R.: Unpublished data from the discomfort glare, joint Slovak Academy of Sciences and Royal Society countersigned 1990 research programme. Proc. CIE/Arup Visual Environment Symposium, London. CIE Central Bureau, Vienna. Publ. CIE x024 : 88–94 (2002)
Pulpitlova, J., Subova, A., MacGowan, D.: Psychologische Blendung in Innenraumen. Proc., Licht 92, Saarbrucken, Germany (1992)
Sörensen, K.: A modern glare index method. Proc. CIE Venice Vols. I, II, 108–111 (1987)
Stevens S.S.: On the theory of scales of measurement. Science, 103, 677–680 (1946)
Stevens, S.S. (ed.): Handbook of Experimental Psychology. Wiley, New York (1951)

Stevens, S.S.: On the psychophysical law. Psychol Rev., 64, 153–181 (1957)
Stevens, S. S: The psychophysics of sensory function. Sensory communication. MIT Press, Cambridge, pp, 1–34 (1961)
Stevens, S.S.: To honor Fechner and repeal his law. Science, 133, 80–86 (1961)
Stevens, J.C. and Stevens, S.S.: Brightness function: Effects of adaptation. Journ. Opt. Soc. Amer., 53, 3, 374–380 (1963)
Šubova, A. and MacGowan, D.: Comfort driven controls for 'Intelligent Buildings'. Proc. Vnútorná klíma budov, Tatranska Lomnica, 116–123 (1990)
Šubova A., Kittler R., MacGowan D.: Results of on-going experiments on subjective response to discomfort glare. Proc. IES Japan Conf., Tokyo, 129–130 (1991)
Taylor, L.: The natural history of windows: a cautionary tale. British Medical Journal, 1, 870–875 (1979)
Wilson, L.M.: Intensive care delirium. The effect of outside deprivation in a windowless unit. Archives of Internal Medicine, 130, 225–226 (1972)

References to Appendix

ISO – International Standardisation Organisation: Spatial distribution of daylight – CIE Standard General Sky. ISO Standard 15409:2004 (2004)
Kittler, R., Darula, S., MacGowan, D.: The Critical Window Luminance Causing Glare in Interiors. Selected papers of the Light and Lighting Conference with Special Emphasis on LEDs and Solid State Lighting, 2009, Budapest, CIE publ. x034:2010, 63–73 (2010)
Köster, H.: Dynamic daylighting archtiecture - basics, systems, pojects. Birkhäuser Verlag, Basel (2004)
MacGowan, D., Thompson, H.E., Eaton, M., Harris, A.: Window Architectonics, Design Guidelines. University of Manitoba report to Public Works Canada, Contract No. 330-993, Ottawa, Canada, 4, 50–67 and 1–83 (1984).
Roy, G.G., Kittler, R., Darula, S.: An implementation of the Method of Aperture Meridians for the ISO/CIE Standard General Sky. Light. Res. Technol., 39, 3, 253–264 (2007)

Index

DOI 10.1007/978-1-4419-8816-4, © Springer Science+Business Media, LLC 2011

M

N

O

P

MIX
Papier aus verantwortungsvollen Quellen
Paper from responsible sources
FSC® C105338

If you have any concerns about our products,
you can contact us on
ProductSafety@springernature.com

In case Publisher is established outside the EU,
the EU authorized representative is:
Springer Nature Customer Service Center GmbH
Europaplatz 3, 69115 Heidelberg, Germany

Printed by Libri Plureos GmbH
in Hamburg, Germany